Recovery and Recycling of Valuable Metals

Recovery and Recycling of Valuable Metals

Editor

Dariush Azizi

MDPI • Basel • Beijing • Wuhan • Barcelona • Belgrade • Manchester • Tokyo • Cluj • Tianjin

Editor
Dariush Azizi
SGS Canada Inc.
Canada

Editorial Office
MDPI
St. Alban-Anlage 66
4052 Basel, Switzerland

This is a reprint of articles from the Special Issue published online in the open access journal *Metals* (ISSN 2075-4701) (available at: https://www.mdpi.com/journal/metals/special_issues/recovery_recycling_valuable_metals).

For citation purposes, cite each article independently as indicated on the article page online and as indicated below:

LastName, A.A.; LastName, B.B.; LastName, C.C. Article Title. *Journal Name* **Year**, *Volume Number*, Page Range.

ISBN 978-3-0365-3034-5 (Hbk)
ISBN 978-3-0365-3035-2 (PDF)

Contents

About the Editor

Dariush Azizi is a metallurgist working in the department of extractive metallurgy at SGS Natural Resources, Lakefield, Canada. He received his PhD in chemical engineering from Laval University, Quebec, Canada, in 2018. His research areas mostly focus on the integration of the concepts of sustainability in mining and the metal sector, specifically, in mineral processing and extractive metallurgy process development to produce critical metals from secondary and primary resources. Since 2013, he has worked in different research institutes and private companies in Canada, and during this period, he has been involved in a variety of R&D and industrial projects with outstanding outputs.

Editorial

Recovery and Recycling of Valuable Metals

Dariush Azizi

Metallurgy & Mineral Processing, IGS Impact Global Solutions, Greater Montréal Metropolitan Area, Montreal, QC J5B 1V4, Canada; Dariush.Azizi@inrs.ca; Tel.: +1-41-8265-1079

Citation: Azizi, D. Recovery and Recycling of Valuable Metals. *Metals* **2022**, *12*, 91. https://doi.org/10.3390/met12010091

Received: 20 December 2021
Accepted: 22 December 2021
Published: 4 January 2022

Publisher's Note: MDPI stays neutral with regard to jurisdictional claims in published maps and institutional affiliations.

1. Introduction and Scope

Metals have always played a significant role in human life; contemporary global growth and prosperity are directly dependent on these materials. With the rapidly growing global demand for metals, their extraction from natural minerals (as their primary sources) has been enhanced. This has caused significant reductions in the grade and quality of the ores in ore deposits. In this context, huge amounts of waste can be generated through mining and metallurgical activities which need management. Although these huge quantities of waste generated through all steps of metal production are known to be a source of environmental pollution, their valorization can create value via recycling metals or even through use in the production of other valuable materials. Such waste valorization is also in line with the United Nations' Sustainable Development Goals (SDGs), as well as implementation of the Paris Agreement. On this matter, the recycling of end-user products in order to reproduce valuable metals can also create significant values and reduce mining activities, and accordingly, their harmful consequences all around the world. Therefore, research and development into state-of-the-art technologies for the recovery and recycling of metals are absolutely necessary [1–4]. In this regard, more novel ideas focusing on the development of advanced metal recovery technologies from primary and secondary sources can lead the mining and metal sectors towards increased sustainability in the future.

In this Special Issue, the endeavor was to collect a range of articles covering different aspects of valuable metal recovery and recycling from primary and secondary sources. The objective was to decipher all new methods, processes, and knowledge in the production of valuable metals from various sources. We hope that this open access Special Issue will represent a great opportunity to demonstrate the prestigious work of active researchers in the field, contributing towards advancements in the mining and metal sectors.

2. Contributions

Seventeen articles have been published in this Special Issue of *Metals*, encompassing the fields of mineral processing and extractive metallurgy [5–21]. These articles cover a wide range of topics in the field and provide some ideas for active researchers who are working on the production of valuable metals from primary and secondary resources. Similarly, researchers from different disciplines can contribute to the enrichment of this Special Issue, through sharing multidisciplinary views on the outputs of each article.

3. Conclusions and Outlook

Different interesting research subjects focusing on metal production process development and identification are published in the present Special Issue of *Metals*, with endeavors of demonstrating new insights towards future advancement of the field. Although most articles focused on hydrometallurgical operational units and processes, mineral processing procedures were also taken into account by some of them. The articles had specific views regarding waste recycling and valorization, indicating the importance of such topics between decision makers in the mining and metal sectors from economic and environmental point of views.

As Guest Editor of this Special Issue, I am so proud of the outcome, and I have high confidence that such effort can help to make future advancement of the field. I deeply appreciate all the authors for their great efforts and contributions to this Special Issue, especially during the difficult situation caused by the COVID-19 pandemic. Their commitment and fantastic results have resulted in the publication of this Special Issue with great success. Likewise, I would like to take this opportunity to warmly thank the many anonymous reviewers who assisted me in the reviewing process. Without their help, this Special Issue could not have been finalized. Sincere thanks also goes to the Editors of *Metals* for their continuous help, and to the *Metals* Editorial Assistants for their valuable support during the preparation of this Special Issue. In particular, my sincere thanks goes to Mr. Toliver Guo for his great help and support from the beginning until the end of the project.

Funding: This research received no external funding.

Conflicts of Interest: The author declares no conflict of interest.

References

1. Green Mining Innovation. Available online: https://www.nrcan.gc.ca/our-natural-resources/minerals-mining/mining-resources/green-mining-innovation/8178 (accessed on 20 December 2021).
2. Evaluation Report: Green Mining Initiative. Available online: https://www.nrcan.gc.ca/maps-tools-publications/publications/minerals-mining-publications/evaluation-report-green-mining-initiative/17190 (accessed on 20 December 2021).
3. Ilyas, S.; Kim, H.; Srivastava, R.R. *Sustainable Urban Mining of Precious Metals*, 1st ed.; CRC Press: Boca Raton, FL, USA, 2021.
4. Cossu, R.; Williams, I.D. Urban mining: Concepts, terminology, challenges. *Waste Manage.* **2015**, *45*, 1–3. [CrossRef] [PubMed]
5. Chen, W.; Lee, C.; Chung, Y.; Tien, K.; Chen, Y.; Chen, Y. Recovery of Rubidium and Cesium Resources from Brine of Desalination through t-BAMBP Extraction. *Metals* **2020**, *10*, 607. [CrossRef]
6. Zhao, Q.; Li, M.; Zhou, L.; Zheng, M.; Zhang, T. Removal of Metallic Iron from Reduced Ilmenite by Aeration Leaching. *Metals* **2020**, *10*, 1020. [CrossRef]
7. Tran, L.; Tanong, K.; Jabir, A.; Mercier, G.; Blais, J. Hydrometallurgical Process and Economic Evaluation for Recovery of Zinc and Manganese from Spent Alkaline Batteries. *Metals* **2020**, *10*, 1175. [CrossRef]
8. Entezari-Zarandi, A.; Azizi, D.; Nikolaychuk, P.; Larachi, F.; Pasquier, L. Selective Recovery of Molybdenum over Rhenium from Molybdenite Flue Dust Leaching Solution Using PC88A Extractant. *Metals* **2020**, *10*, 1423. [CrossRef]
9. García, A.; Latifi, M.; Amini, A.; Chaouki, J. Separation of Radioactive Elements from Rare Earth Element-Bearing Minerals. *Metals* **2020**, *10*, 1524. [CrossRef]
10. Terrones-Saeta, J.; Suárez-Macías, J.; Iglesias-Godino, F.; Corpas-Iglesias, F. Evaluation of the Use of Electric Arc Furnace Slag and Ladle Furnace Slag in Stone Mastic Asphalt Mixes with Discarded Cellulose Fibers from the Papermaking Industry. *Metals* **2020**, *10*, 1548. [CrossRef]
11. Boisvert, L.; Turgeon, K.; Boulanger, J.; Bazin, C.; Houlachi, G. Recovery of Cobalt from the Residues of an Industrial Zinc Refinery. *Metals* **2020**, *10*, 1553. [CrossRef]
12. Choi, S.; Jeon, S.; Park, I.; Ito, M.; Hiroyoshi, N. Enhanced Cementation of Co^{2+} and Ni^{2+} from Sulfate and Chloride Solutions Using Aluminum as an Electron Donor and Conductive Particles as an Electron Pathway. *Metals* **2021**, *11*, 248. [CrossRef]
13. Álvarez, M.; Fidalgo, J.; Gascó, G.; Méndez, A. Hydrometallurgical Recovery of Cu and Zn from a Complex Sulfide Mineral by Fe^{3+}/H_2SO_4 Leaching in the Presence of Carbon-Based Materials. *Metals* **2021**, *11*, 286. [CrossRef]
14. Caplan, M.; Trouba, J.; Anderson, C.; Wang, S. Hydrometallurgical Leaching of Copper Flash Furnace Electrostatic Precipitator Dust for the Separation of Copper from Bismuth and Arsenic. *Metals* **2021**, *11*, 371. [CrossRef]
15. Liu, W.; Zhang, J.; Xu, Z.; Liang, J.; Zhu, Z. Study on the Extraction and Separation of Zinc, Cobalt, and Nickel Using Ionquest 801, Cyanex 272, and Their Mixtures. *Metals* **2021**, *11*, 401. [CrossRef]
16. Zhou, Z.; Tang, P. Optimization on Temperature Strategy of BOF Vanadium Extraction to Enhance Vanadium Yield with Minimum Carbon Loss. *Metals* **2021**, *11*, 906. [CrossRef]
17. Lv, N.; Du, C.; Kong, H.; Yu, Y. Leaching of Phosphorus from Quenched Steelmaking Slags with Different Composition. *Metals* **2021**, *11*, 1026. [CrossRef]
18. Bonin, M.; Fontaine, F.; Larivière, D. Comparative Studies of Digestion Techniques for the Dissolution of Neodymium-Based Magnets. *Metals* **2021**, *11*, 1149. [CrossRef]
19. Kim, Y.; Min, D. Viscosity and Structural Investigation of High-Concentration Al_2O_3 and MgO Slag System for FeO Reduction in Electric Arc Furnace Processing. *Metals* **2021**, *11*, 1169. [CrossRef]
20. Moldoveanu, G.; Papangelakis, V. Chelation-Assisted Ion-Exchange Leaching of Rare Earths from Clay Minerals. *Metals* **2021**, *11*, 1265. [CrossRef]
21. Nam, S.; Park, S.; Rasheed, M.; Song, M.; Kim, D.; Kim, T. Influence of Dysprosium Compounds on the Extraction Behavior of Dy from Nd-Dy-Fe-B Magnet Using Liquid Magnesium. *Metals* **2021**, *11*, 1345. [CrossRef]

Article

Recovery of Rubidium and Cesium Resources from Brine of Desalination through t-BAMBP Extraction

Wei-Sheng Chen, Cheng-Han Lee *, Yi-Fan Chung, Ko-Wei Tien, Yen-Jung Chen and Yu-An Chen

Department of Resources Engineering, National Cheng Kung University, No.1, DaxueRoad, Tainan City 70101, Taiwan; kenchen@mail.ncku.edu.tw (W.-S.C.); eddie21039@gmail.com (Y.-F.C.); n46071172@gs.ncku.edu.tw (K.-W.T.); qaz5932201@gmail.com (Y.-J.C.); evan860102@gmail.com (Y.-A.C.)
* Correspondence: happy980074@gmail.com; Tel.: +886-6-2757575 (ext. 62828)

Received: 23 April 2020; Accepted: 6 May 2020; Published: 8 May 2020

Abstract: 50 billion cubic meters of brine every year creates ecological hazards to the environment. In order to reuse brine efficiently, rubidium and cesium were recovered in this experiment. On the other hand, the main impurities which were needed to be eliminated in brine were lithium, sodium, potassium, calcium, and magnesium. In the procedure, seawater was distilled and evaporated first to turn into simulated brine. Perchloric acid was then added into simulated brine to precipitate potassium perchlorate which could reduce the influence of potassium in the extraction procedure. After that, t-BAMBP and ammonia were separately used as extractant and stripping agent in the extraction and stripping procedures to get rubidium hydroxide solutions and cesium hydroxide solutions. Subsequently, they reacted with ammonium carbonate to get rubidium carbonate and cesium carbonate. In a nutshell, this study shows the optimal parameters of pH value to precipitate potassium perchlorate. Besides, pH value in the system, the concentration of t-BAMBP and ammonia, organic phase/aqueous phase ratio (O/A ratio), reaction time, and reaction temperature in solvent extraction step were investigated to get high purities of rubidium carbonate and cesium carbonate.

Keywords: solvent extraction; t-BAMBP; rubidium; cesium; brine; chemical precipitation; recovery

---

1. Introduction

According to the investigation of the United Nations, more than 1 billion people in the world currently live in the areas with scarce water resources. Moreover, this number will reach 1.8 billion by 2025 [1]. In response to the shortage of freshwater resources, seawater desalination technology has developed rapidly since the beginning of the 20th century [2]. However, as seawater desalination is common now, waste brine also causes considerable harm to the environment. For example, waste brine will change the composition of seawater and affect the ecosystem. In order to reuse waste brine, some elements such as lithium, magnesium, calcium, and chlorine are recycled from brine recently [3]. In this study, rubidium and cesium were extracted from brine through the hydrometallurgy method. It is expected to achieve the goal of recovering valuable metals, reducing waste, and protecting the environment.

Based on the report of the U.S. Geological Survey (USGS), the rubidium and cesium resources are mainly from primary minerals such as carnallite, garnet, and lepidolite [4–8]. The main reservoirs are Namibia, Zimbabwe, Afghanistan, and some other countries. Rubidium and cesium respectively have only 90,000 tons of reserve, and the difficulty of mining makes them rarer. Although rubidium and cesium are not common, they are valuable and useful in many areas. Therefore, it is important to recover rubidium and cesium or its compounds from waste brine which can reduce the reliance of primary mineral and create the economic value of brine.

Rubidium and cesium have a wide range of using [9–12]. For industry activities, rubidium and cesium metals are the material of television, radar, infrared filter, radiant energy receiver, and reconnaissance telescope [13]. On the other hand, rubidium carbonate (Rb_2CO_3) and cesium carbonate (Cs_2CO_3) can be raw materials of glasses and increase stability and durability. Additionally, rubidium carbonate and cesium carbonate are easy to turn into other compounds such as rubidium chloride (RbCl) and cesium nitrate ($CsNO_3$). Rubidium chloride can be used to produce sleeping pills, sedatives, and treatment of bipolar disorder [14–17]. Cesium nitrate can be used as a light refraction regulator in the optical fiber industry and glass industry [18]. Due to the unlimited development of rubidium and cesium compounds, many countries try various methods to get rubidium and cesium and apply them in industries.

Nowadays, rubidium and cesium are mainly recovered from saline lake and ores through a solvent extraction method with t-BAMBP extractant [19–22]. On the other hand, because impurities could be removed efficiently through chemical precipitation and make higher purification of compounds, it was chosen to be the procedure before solvent extraction in this experiment. During the chemical precipitation process, perchloric acid ($HClO_4$) was added into the brine to selectively precipitate potassium perchlorate ($KClO_4$). Due to the reduction of potassium, it could avoid the adverse effects which potassium create in the follow-up processes. Moreover, t-BAMBP and ammonia were used in the solvent extraction procedure to separate rubidium, cesium, and other impurities such as lithium, sodium, potassium, calcium, and magnesium efficiently. To sum up, the chemical precipitation method and solvent extraction method were used in this experiment to get high purities of rubidium and cesium resources and made them be able to reuse in the industries.

2. Material and Methods

2.1. Materials

In this experiment, the simulated brine was got from seawater through distillation and evaporation. The metal compositions of simulated brine are measured by inductively coupled plasma optical emission spectrometry (ICP-OES) and the concentrations are shown in Table 1.

Table 1. Metal compositions of simulated brine

Element	Li	Na	K	Ca	Mg	Rb	Cs
Concentration (mg/L)	182	501,30	5914	702	139,80	7.1	43.6

In the whole process, perchloric acid ($HClO_4$) was purchased from Sigma-Aldrich (St. Louis, MO, USA) (70%) to selectively precipitate potassium perchlorate. Sodium hydroxide (NaOH) and sulfuric acid (H_2SO_4) were separately acquired from Applichem Panreac (Barcelona, Spain) (≥98%) and Sigma-Aldrich (St. Louis, MO, USA) (≥98%). They were used in the extraction process to adjust the pH value. Kerosene was purchased from CPC Corporation (Kaohsiung, Taiwan) to dilute the extractant. t-BAMBP was purchased from Realkan Corporation (Beijing, China) (≥90%) for the extraction process, and ammonia (NH_4OH) was purchased from Sigma-Aldrich (St. Louis, MO, USA) (30–33%) for the stripping process. In the final process, ammonium carbonate (($NH_4)_2CO_3$) was acquired from Sigma-Aldrich (St. Louis, MO, USA) (≥90%) to produce rubidium carbonate and cesium carbonate. During the analysis procedure, ICP rubidium standard solution, ICP cesium standard solution, and ICP multi-element standard solution were acquired from High-Purity Standards, Inc. (North Charleston, SC, USA). The nitric acid (HNO_3) was purchased from Sigma-Aldrich (St. Louis, MO, USA) ($HNO_3 \geq 65\%$) and diluted to 1% to be the background value and thinner for ICP analysis.

2.2. Equipment

The materials and products were analyzed by X-ray fluorescence spectrometer (XRF; ZSX100s, SPECTRO Analytical Instruments Inc., Kleve, Germany) and inductively coupled plasma optical emission spectrometry (ICP-OES; Varian, Vista-MPX, PerkinElmer, Waltham, MA, USA). In the separated process, 445 mm × 730 mm × 505 mm of funnel shaker (FS-12; Shin Kwang Precision

Industry Ltd., New Taipei City, Taiwan) was used to shake 500 mL of separating funnels at 3000 rpm. On the other hand, the thermostatic bath (XMtd-204; BaltaLab, Vidzemes priekšpilsēta, Rīga, Latvia) was used to maintain the temperature during the extraction process and stripping process. In the procedure of producing compounds, rotary evaporators (BUCHI R-300; BÜCHI Labortechnik AG, Flawil, Switzerland) was used to evaporate solutions under low pressure and high purity of rubidium carbonate and cesium carbonate could be precipitated.

2.3. Experimental Procedures

2.3.1. Chemical Precipitation

Chemical precipitation procedure was used to remove potassium in this experiment. In the process, the potassium which was in simulated brine would be selectively precipitated as potassium perchlorate by adding perchloric acid. Removal of potassium could reduce the co-extracted effect in the solvent extraction process. The parameter of chemical precipitation such as effect of pH value was investigated. Precipitation rate was calculated according to Equation (1) and the flow diagram is shown in Figure 1.

$$P(\%) = \frac{[M]_0 - [M]}{[M]_0} \times 100 \tag{1}$$

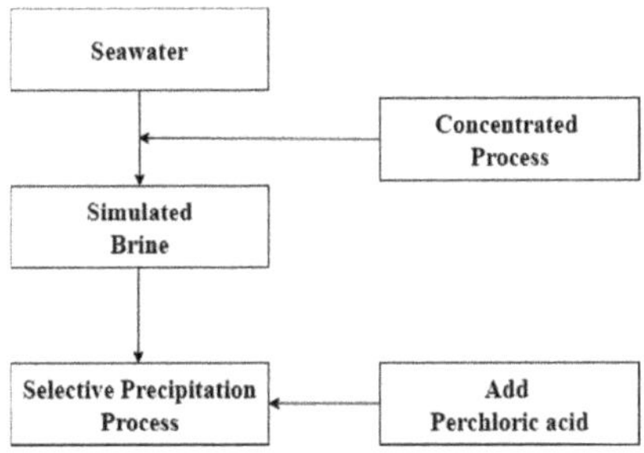

Figure 1. Flow diagram of chemical precipitation.

P is Precipitation rate, $[M]_0$ is metal concentration of leach liquor, $[M]$ is metal concentration of leach liquor after precipitation.

2.3.2. Solvent Extraction–Extraction Process

In this study, t-BAMBP was diluted into kerosene and used as the extractant to separate rubidium and cesium from other impurities such as lithium, sodium, potassium, calcium, and magnesium. The extraction process was divided into two stages. Cesium was extracted in the first stage, and Li, Na, K, Ca, Mg, and Rb were still in the aqueous phase. Rubidium was then extracted in the second stage, and other impurities were in the aqueous phase as well.

To calculate the efficiency of extraction, distribution ratio and extraction percentage were used in this process. Distribution ratio, D, was the concentration ratio of the metal in the organic phase to the metal in the aqueous phase at equilibrium. The distribution ratio can be written as Equation (2).

$$D = \frac{[M]_{org}}{[M]_{aq}} = \frac{C_i - C_f}{C_f} \times \frac{V_{aq}}{V_{org}} \tag{2}$$

C_i is the initial concentration of metal ions in aqueous phase, C_f is the equilibrium concentration of metal ions in aqueous phase. V_{aq} and V_{org} are separately the volume of aqueous phase and organic phase.

Based on the distribution ratio, the extraction percentage can be written as Equation (3).

$$E_{ex} (\%) = \frac{D}{D + V_{aq}/V_{org}} \cdot 100 \tag{3}$$

D is the distribution ratio. V_{aq} and V_{org} are separately the volume of aqueous phase and organic phase.

In the extraction process, the parameters such as the pH value of aqueous phase, the concentration of t-BAMBP, the O/A ratio, the reaction time, and the reaction temperature were all presented in the form of extraction percentage.

2.3.3. Solvent Extraction–Stripping Process

Ammonia was the stripping agent in this experiment which could strip the rubidium and cesium from the organic phase. The stripping efficiency is written as Equation (4).

$$E_{st} (\%) = \frac{\Sigma M_{aq}}{\Sigma M_{org}} \times 100 \tag{4}$$

ΣM_{aq} is the concentration of metal ion in aqueous phase after stripping, and ΣM_{org} is the concentration of metal ion in organic phase before stripping.

The flow diagram of whole solvent extraction process is shown in Figure 2. After the first and the second stages of extraction, the concentration of ammonia, the O/A ratio, and the reaction temperature were investigated in the form of stripping percentage in the stripping process.

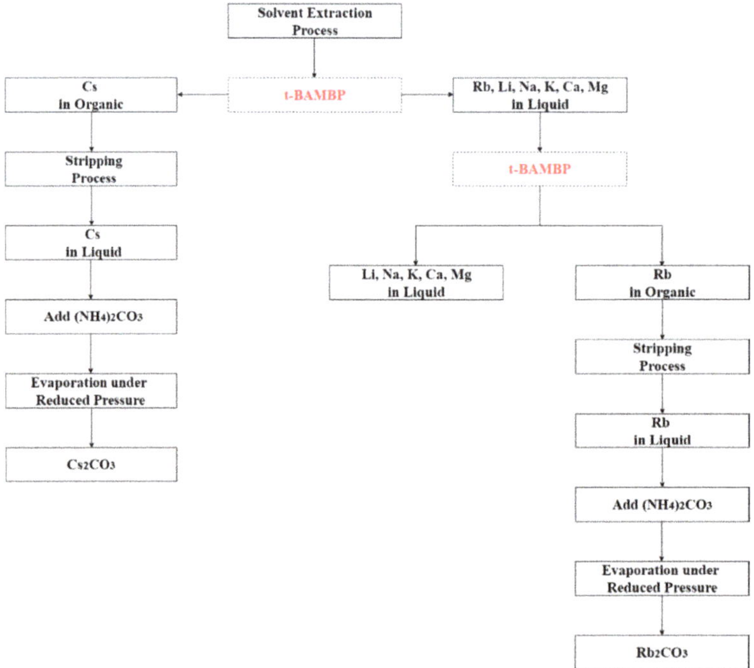

Figure 2. Flow diagram of whole solvent extraction process.

3. Results and Discussion

3.1. Removal of Potassium

Because potassium would be co-extracted with rubidium and cesium in the extraction process, the perchloric acid (HClO$_4$) was added into simulated brine to selectively precipitate potassium perchlorate (KClO$_4$) at −5 °C and 10 min. The precipitation percentages in different pH values are shown in Figure 3 and the final compositions of simulated brine are shown in Table 2. In this procedure, the precipitation percentage of KClO$_4$ was 98.5% at −5 °C and pH 2.

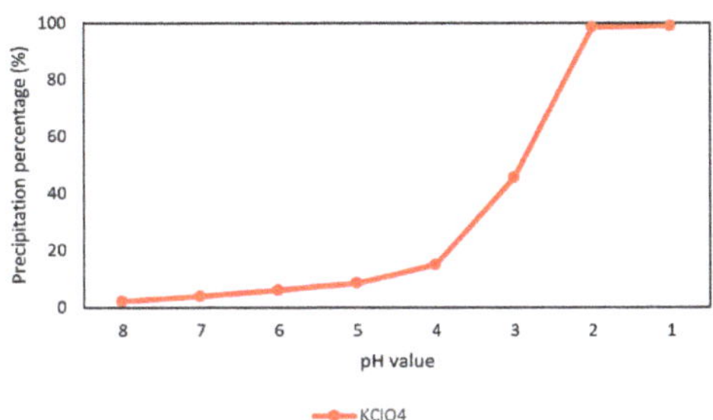

Figure 3. Precipitation percentage of KClO$_4$ in different pH value at 10 min.

Table 2. Final composition of simulated brine

Elements	Li	Na	K	Ca	Mg	Rb	Cs
Concentration (mg/L)	167	491,80	91	622	135,70	6.94	42.14

3.2. First Stage of Solvent Extraction for Cs

In the first stage of solvent extraction, t-BAMBP was used to separate Cs and other metal ions. Due to the property of t-BAMBP, only K, Rb, and Cs were able to be extracted efficiently. In this study, the values of them in the first stage of solvent extraction were analyzed by ICP-OES and turned into the extraction percentage.

3.2.1. Effect of pH Value of the Aqueous Phase

Because t-BAMBP is an extractant which is suitable for alkalic condition, the pH values were adjusted to 8–14 with 0.1 M t-BAMBP and O/A ratio 1:1 at reaction time 15 min and 25 °C in this experiment. Figure 4 shows that the extraction percentage of K$^+$ and Rb$^+$ were observed almost 0% to 5% at any pH value and the percentage of Cs$^+$ was observed above 99% from pH 8 to pH 12. In the condition of pH 13 and pH 14, emulsification happened and reduced the extraction percentage. Due to these situations, the optimal pH value of the aqueous phase was set to pH 8 which has a high extraction percentage of Cs$^+$ and could reduce the usage of sodium hydroxide.

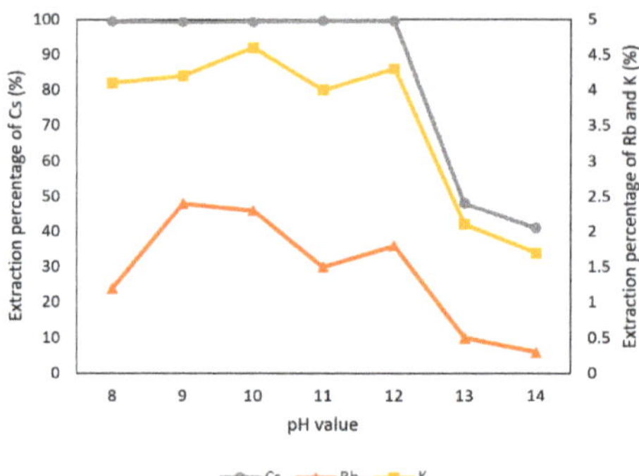

Figure 4. Extraction percentage of metals at different pH values in the first stage.

3.2.2. Effect of t-BAMBP Concentration

The conditions of t-BAMBP concentration were set up from 0.001 M to 1 M at pH 8 and O/A ratio 1:1 at reaction time 15 min and 25 °C in this step. Figure 5 shows that the extraction percentage of Cs^+ from 0.001 M to 0.1 M increased gradually and became stable. The reason was that a higher concentration of the extractant enabled more Cs^+ to be caught. However, the extraction percentage of K^+ and Rb^+ started to increase with a higher concentration of extractant. This was because K^+ and Rb^+ were extracted by excess extractant and made an adverse effect on this system. Due to this condition, the optimal parameter of t-BAMBP concentration was chosen as 0.1 M.

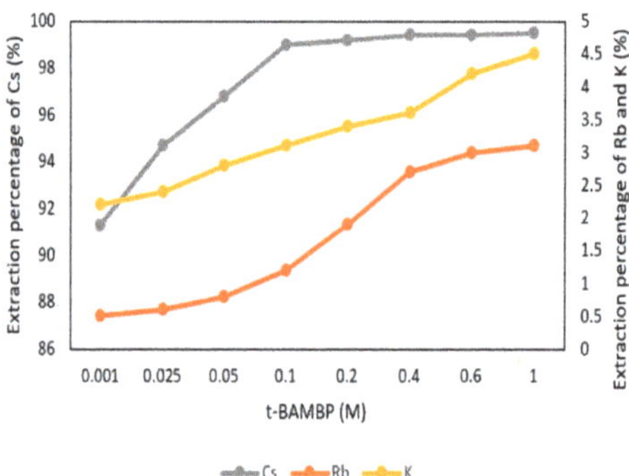

Figure 5. Extraction percentage of concentration of t-BAMBP in the first stage.

3.2.3. Effect of O/A Ratio

Figure 6 shows the O/A ratios were set from 0.1 to 2 with 0.1 M t-BAMBP at pH 8 and at reaction time 15 min and 25 °C. The result shows that the extraction percentages of Cs^+ maintain above 99% with different O/A ratios, which means that Cs^+ were almost extracted. However, when the O/A ratio was greater than 0.5, the extraction percentage of K^+ and Rb^+ increased gradually above 2%. This was because K^+ and Rb^+ were extracted by excess extractant as well. Hence, in order to get a high concentration of Cs^+ and avoid the impurities, the O/A ratio of 0.1 was an optimal parameter in this step.

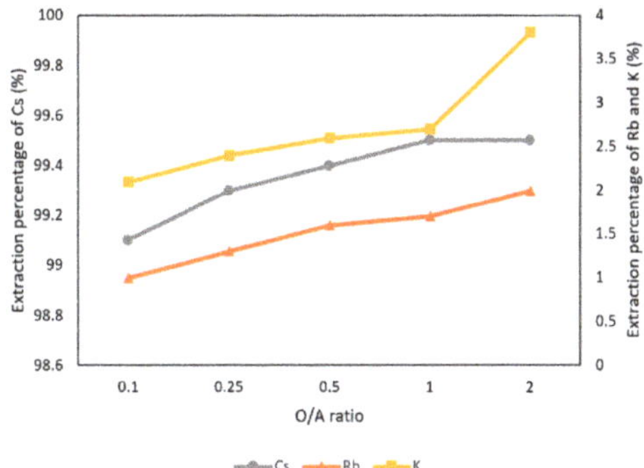

Figure 6. Extraction percentage of O/A ratio in the first stage.

3.2.4. Effect of Reaction Time

The effect of reaction time was set from 3 min to 60 min with 0.1 M t-BAMBP at pH 8 and O/A ratio 1:1 at 25 °C. In Figure 7, the extraction percentage of Cs^+ was very stable from 3 min to 60 min. K^+ and Rb^+ were at equilibrium and low extraction percentage as well. It shows that the reaction of t-BAMBP was fast and reaction time was not a significant influence in the extraction process. On account of this condition, 3 min was chosen.

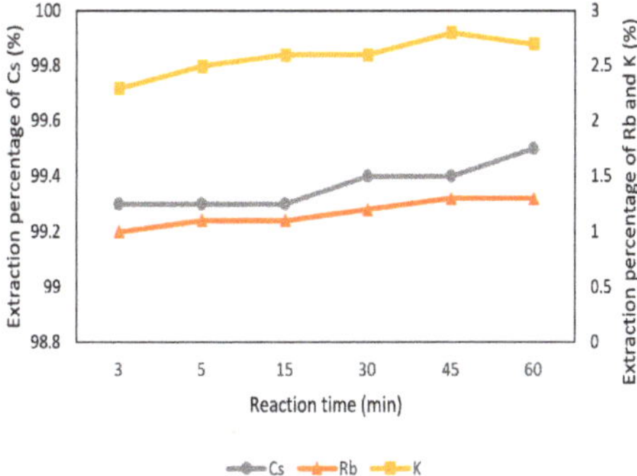

Figure 7. Extraction percentage of reaction time in the first stage.

3.2.5. Effect of Reaction Temperature

Figure 8 shows the reaction temperature was a significant parameter in the extraction process. The effect of reaction temperature was set from 5 °C to 65 °C with 0.1 M t-BAMBP at pH 8 and O/A ratio 1:1 at 3 min. The percentage of extraction decreased drastically from 35 °C to 45 °C. This is because t-BAMBP extracted metal ions with exothermic reaction and the high temperature caused the decrease of distribution ratio. Finally, the extraction percentages of Cs^+, K^+, and Rb^+ were respectively 99.8%, 5.2%, and 1% with the condition of pH 8 of the aqueous phase, 0.1 M t-BAMBP, O/A ratio of 0.1, 3 min reaction time, and 35 °C reaction temperature. Compared to other studies, this study shows the high extraction percentage of Cs^+ with lower pH value and O/A ratio at the low temperature and Cs^+ could be separated from Rb, K, and other impurities such as lithium, sodium, potassium, calcium, and magnesium.

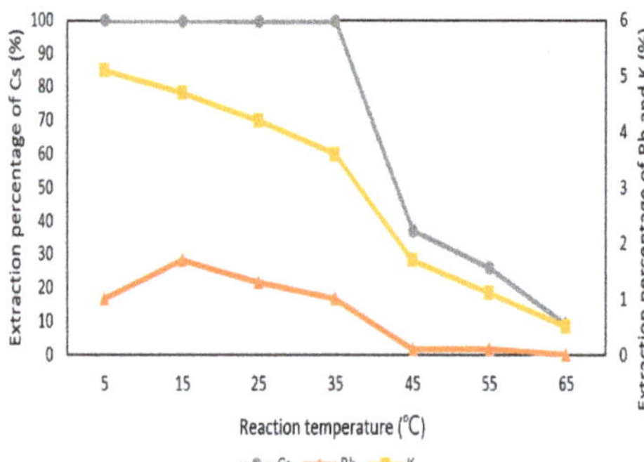

Figure 8. Extraction percentage of reaction temperature in the first stage.

3.2.6. Stripping of Cs from the Organic Phase through Ammonia

After the extraction process, 420 mg/L of cesium was in the organic phase and needed to be stripped. In this process, NH_4OH was chosen as a stripping agent and the effect of NH_4OH was presented in Figures 9–11. Because the stripping efficiencies of rubidium and potassium were low in this procedure, only stripping efficiency of cesium was investigated. Figure 9 shows that when the concentration of NH_4OH increased, the stripping efficiency increased as well and become equilibrium in the situation of 1 M, 2 M, and 5 M. Therefore, the optimal concentration of NH_4OH was 1 M in the procedure. Figure 10 shows that the stripping efficiency decreased drastically when O/A ratio was 4. It means that insufficient NH_4OH was unable to strip the Cs^+. Due to this condition, O/A ratio 2 was the optimal condition to get a high concentration of Cs^+. Figure 11 presents the stripping efficiency with reaction temperature. Because t-BAMBP has great extracting ability at lower temperatures, it made NH_4OH unable to strip Cs^+. However, if the temperature was above 35 °C, NH_4OH decomposed to NH_3 first which was unable to strip Cs^+. Hence, the stripping temperature was chosen at a room temperature of 25 °C in this step. Finally, the stripping efficiency of Cs^+ was almost 99.9% in the first stripping process.

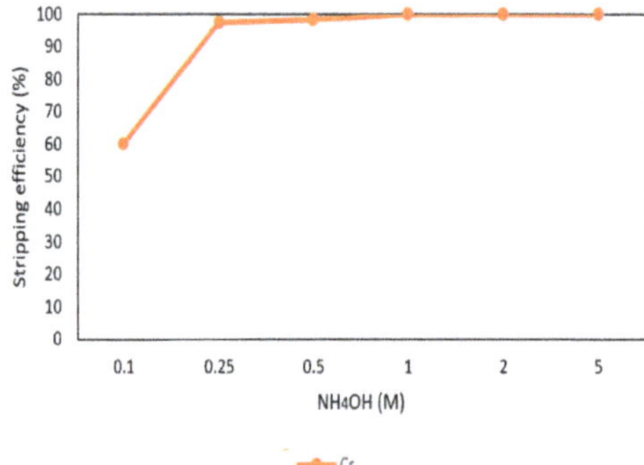

Figure 9. Stripping efficiency of concentration of NH_4OH in the first stage.

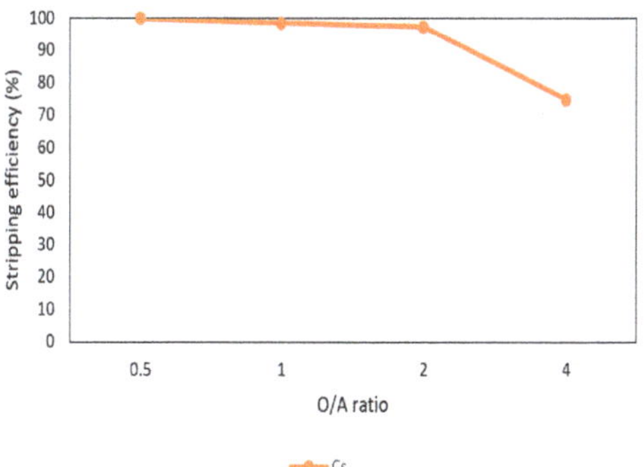

Figure 10. Stripping efficiency of O/A ratio in the first stage.

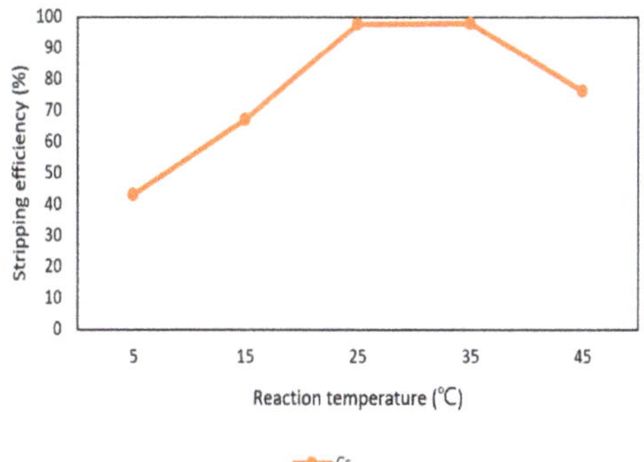

Figure 11. Stripping efficiency of reaction temperature in the first stage.

3.3. Second Stage of Solvent Extraction for Rb

After the first stage of solvent extraction, metals were separated into two systems. One system was cesium, and another system was rubidium with other impurities such as lithium, sodium, potassium, calcium, and magnesium. In the second stage of solvent extraction, t-BAMBP was also chosen as extractant to separate rubidium with other impurities. Among the rest of the impurities, only K could be extracted by t-BAMBP due to the property of extractant. In this case, the values of K and Rb were presented to analyze the optimal condition.

3.3.1. Effect of pH Value of the Aqueous Phase

The effect of pH value of the aqueous phase in the extraction and separation of Rb^+ and K^+ was shown in Figure 12. The pH values were adjusted to 8 to 14 with 0.1 M t-BAMBP and O/A ratio 1:1 at reaction time 15 min and 25 °C. The extraction percentage of Rb^+ and K^+ increased when the pH value raised up and declined at pH 13 and pH 14 due to the emulsification. In order to extract more Rb^+, pH 12 value was chosen in this procedure.

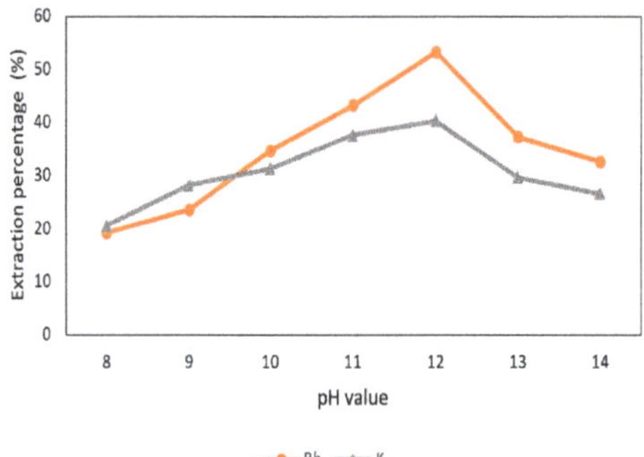

Figure 12. Extraction percentage of metals at different pH value in the second stage.

3.3.2. Effect of t-BAMBP Concentration

The conditions of t-BAMBP concentration from 0.01 M to 1 M at pH 12 and O/A ratio 1:1 at reaction time 15 min and 25 °C were set in this step. Figure 13 shows that the extraction percentages of Rb$^+$ from 0.01 M to 1 M were about 50% and the percentages of K$^+$ were about 40%. In order to get more rubidium, 0.5 M t-BAMBP was chosen in this procedure.

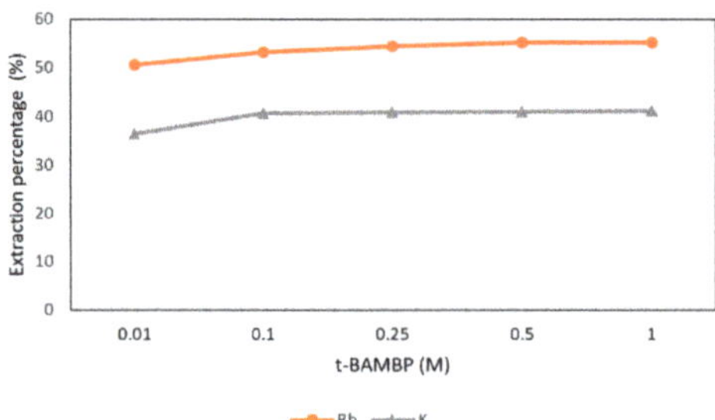

Figure 13. Extraction percentage of concentration of t-BAMBP in the second stage.

3.3.3. Effect of O/A Ratio

Figure 14 shows the O/A ratio was set from 0.1 to 2 with 0.5 M t-BAMBP at pH 12 and at reaction time 15 min and 25 °C. The result shows that the extraction percentages of Rb$^+$ maintain above 50% with different O/A ratios. However, when the O/A ratio was greater than 0.1, the extraction percentage of K$^+$ increased gradually. Hence, in order to get a high concentration of Rb$^+$ and avoid the impurities, the O/A ratio 0.1 was an optimal parameter in this step.

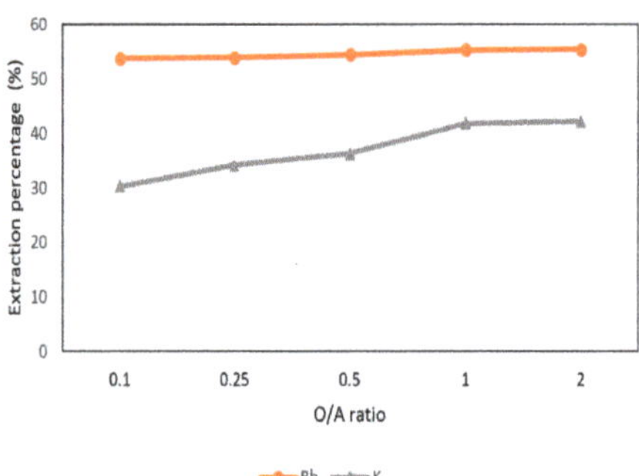

Figure 14. Extraction percentage of O/A ratio in the second stage.

3.3.4. Effect of Reaction Time

The effect of reaction time was set from 3 min to 60 min with 0.5 M t-BAMBP at pH 12 and O/A ratio 0.1:1 at 25 °C. In Figure 15, the extraction percentage of Rb$^+$ increased gradually from 3 min to 15 min and became stable. On the other hand, the extraction percentage of K$^+$ was almost 30% from 3 min to 60 min. Hence, 15 min of reaction time was chosen in this process to extract rubidium.

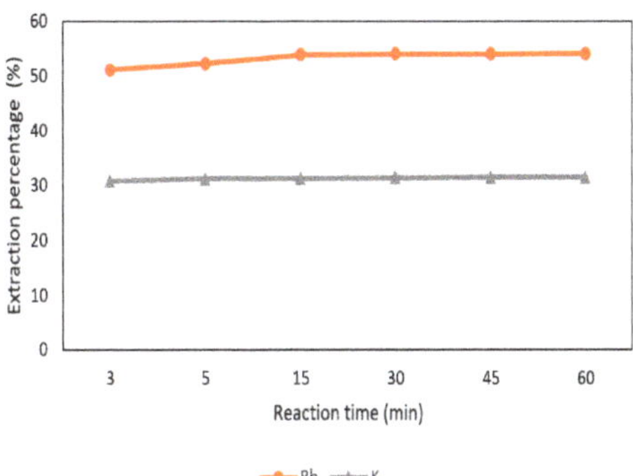

Figure 15. Extraction percentage of reaction time in the second stage.

3.3.5. Effect of Reaction Temperature

In Figure 16, it shows that Rb$^+$ was influenced by the reaction temperature. The effect of reaction time was set from 5 °C to 65 °C with 0.5 M t-BAMBP at pH 12 and O/A ratio 0.1:1 at 15 min. The extraction percentage of Rb$^+$ was about 98% at 5 °C and decreased drastically after 5 °C. The percentage of 35 °C was 56.8% and only about 10% at 45 °C. In order to extract more Rb$^+$, 5 °C was chosen as the optimal parameter. Finally, the extraction percentage of Rb$^+$ and K$^+$ and were respectively 98.3% and 41.3% with the condition of pH 12 of the aqueous phase, 0.5 M t-BAMBP, O/A ratio of 0.1, 15 min reaction time, and 5 °C reaction temperature. Compared to other studies, this study shows the almost same extraction percentage of Rb$^+$ with lower pH value and O/A ratio at low temperature.

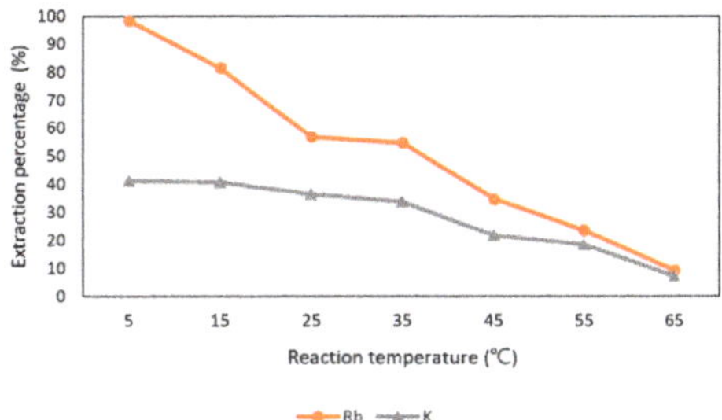

Figure 16. Extraction percentage of reaction temperature in the second stage.

3.3.6. Stripping of Rb from the Organic Phase through Ammonia

After the second stage of the extraction process, 70 mg/L of rubidium was in the organic phase and needed to be stripped. In this process, NH$_4$OH was chosen as a stripping agent as well and the effect of NH$_4$OH concentration was presented in Figures 17–19. Because the stripping efficiencies of potassium was low in this procedure, only stripping efficiency of rubidium was investigated. Figure 17 shows that when the concentration of NH$_4$OH increased from 0.1 M to 0.5 M, the stripping efficiency of Rb$^+$ increased as well. However, efficiency became equilibrium in the situation of 0.5 M, 1 M, 2 M, and 5 M. Therefore, the optimal concentration of NH$_4$OH was 0.5 M in this step. Figure 18 shows that the stripping efficiency of Rb$^+$ was above 90% and decreased at O/A ratio 4. Due to this condition, O/A ratio 2 was the optimal condition. Figure 19 presents the stripping efficiency with reaction temperature. Like the situation of Cs$^+$, the stripping efficiency of Rb$^+$ was lower at the low temperature and gradually rose with increase in temperature. Hence, the stripping temperature was chosen as 35 °C in this procedure. Finally, the stripping efficiency of Rb$^+$ was almost 95% in the second stripping process. After the extraction and stripping of Cs$^+$ and Rb$^+$, the optimal parameters of solvent extraction are shown in Tables 3 and 4, and the components of stripping solutions are shown in Tables 5 and 6.

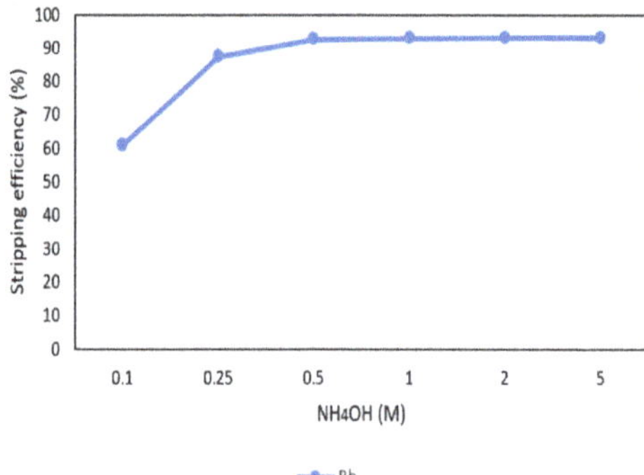

Figure 17. Stripping efficiency of concentration of NH$_4$OH in the second stage.

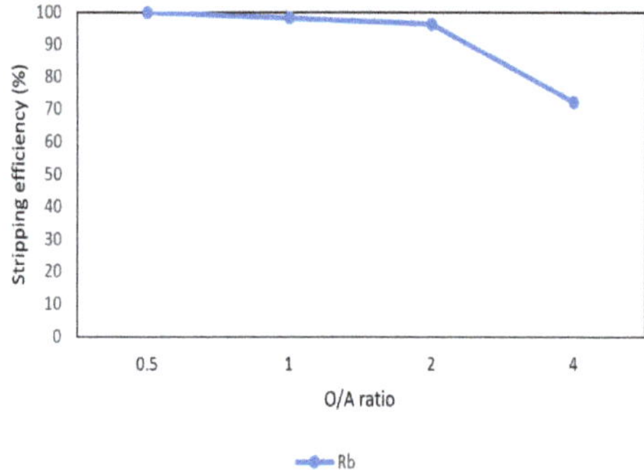

Figure 18. Stripping efficiency of O/A ratio in the second stage.

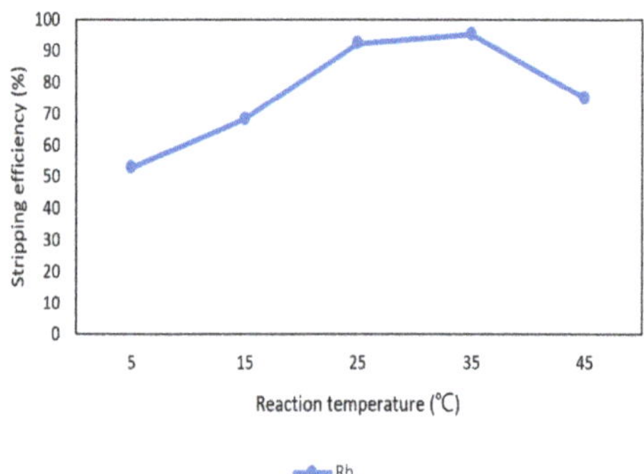

Figure 19. Stripping efficiency of reaction temperature in the second stage.

Table 3. Optimal parameters of solvent extraction of cesium

Extractant and Stripping Agent	pH Value of Aqueous Phase	Concentration (M)	O/A Ratio	Reaction Time (min)	Reaction Temperature (°C)
t-BAMBP	8.0	0.1	0.1	3	35
Ammonia	-	1	2	-	25

Table 4. Optimal parameters of solvent extraction of rubidium

Extractant and Stripping Agent	pH Value of Aqueous Phase	Concentration (M)	O/A Ratio	Reaction Time (min)	Reaction Temperature (°C)
t-BAMBP	12.0	0.5	0.1	15	5
Ammonia	-	0.5	2	-	35

Table 5. Metal compositions of the first stage of stripping (cesium hydroxide solutions)

Element	Li	Na	K	Ca	Mg	Rb	Cs
Concentration (mg/L)	N.D.	8.13	10.33	N.D.	N.D.	1.39	1635

Table 6. Metal compositions of the second stage of stripping (rubidium hydroxide solutions)

Element	Li	Na	K	Ca	Mg	Rb	Cs
Concentration (mg/L)	N.D.	0.02	6.02	N.D.	N.D.	293.68	N.D.

3.4. Production of Rubidium Carbonate and Cesium Carbonate

In order to get stable compounds of rubidium and cesium, ammonium carbonate was added into rubidium hydroxide solutions and cesium hydroxide solutions to produce rubidium carbonate solutions and cesium carbonate solutions. The reaction situations were shown in Equations (5) and (6).

$$(NH4)_2CO_{3(s)} + RbOH_{(aq)} \rightarrow Rb_2CO_{3(aq)} + NH_{3(aq)} + H_2O_{(l)} \tag{5}$$

$$(NH4)_2CO_{3(s)} + CsOH_{(aq)} \rightarrow Cs_2CO_{3(aq)} + NH_{3(aq)} + H_2O_{(l)} \tag{6}$$

The rubidium carbonate solutions and cesium carbonate solutions were then put into rotary evaporators and evaporated under low pressure and temperature and precipitate the purity of 98.0% of rubidium carbonate and 98.9% of cesium carbonate. After chemical precipitation and solvent extraction, the ICP-OES analyses are shown in Table 7.

Table 7. ICP-OES analyses of rubidium carbonate and cesium carbonate

Compounds	Li	Na	K	Ca	Mg	Rb	Cs
Rb_2CO_3	N.D.	0.1%	1.9%	N.D.	N.D.	98.0%	N.D.
Cs_2CO_3	N.D.	0.4%	0.5%	N.D.	N.D.	0.2%	98.9%

N.D.: Not-detected.

4. Conclusions

Hydrometallurgy method was used to separate rubidium and cesium effectively from the desalination brine in this study. The recommended recovery process is shown in Figure 20. The brine was treated by chemical precipitation and solvent extraction to recover rubidium and cesium. Perchloric acid was added into brine to pH 2 at −5 °C to precipitate potassium perchlorate which could reduce the influence of potassium in the extraction procedure. After that, t-BAMBP and ammonia were used as extractant and a stripping agents. The results show that 0.1 M t-BAMBP was used as extractant to separate cesium and impurities under the optimal condition of pH 8, O/A ratio 0.1, 3 min reaction time, and 35 °C reaction temperature. Then, cesium was stripped by using 1 M ammonia under O/A ratio 2 and 25 °C reaction temperature. On the other hand, the results show that 0.5 M t-BAMBP was used as extractant to separate rubidium and potassium under the optimal condition of pH 12, O/A ratio 0.1, 15 min reaction time, and 5 °C reaction temperature. In the stripping process, 0.5 M ammonia under O/A ratio 2 and 35 °C reaction temperature are the optimal parameters. Finally, the rubidium carbonate and cesium carbonate were produced by adding ammonium carbonate. By these processes, the purity of rubidium carbonate and cesium carbonate were 98.0% and 98.9%.

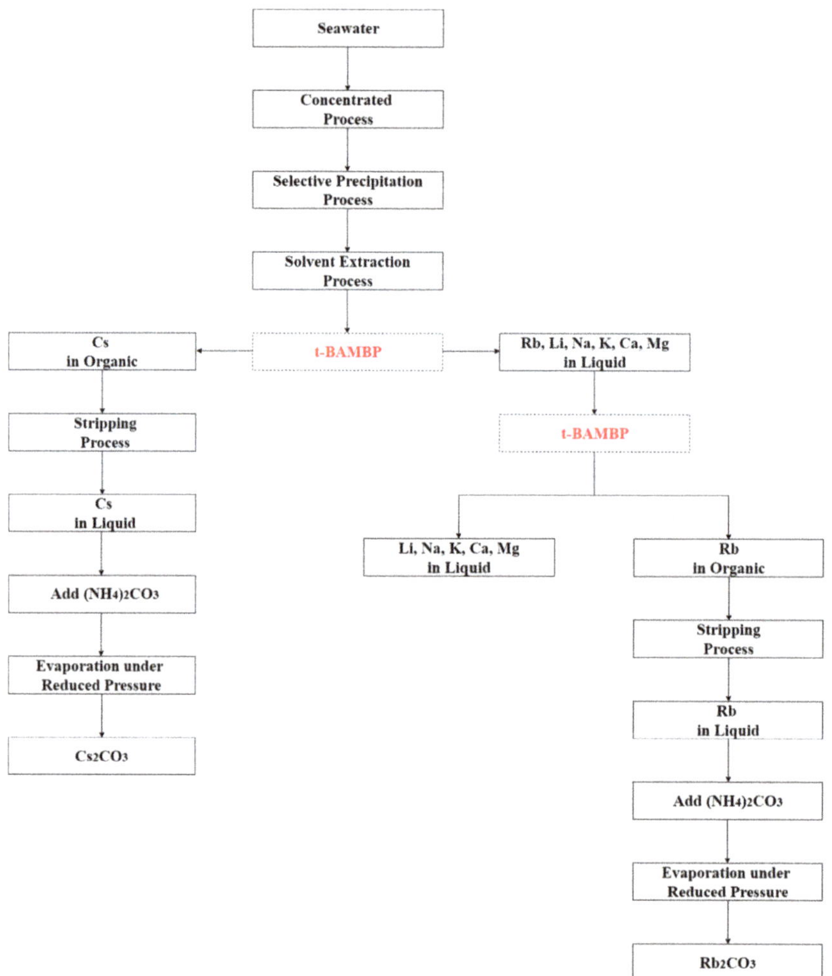

Figure 20. Recommended recovery process of this experiment.

Author Contributions: Conceptualization, W.-S.C. and C.-H.L.; methodology, W.-S.C. and C.-H.L.; validation, W.-S.C. and C.-H.L.; formal analysis, C.-H.L. and Y.-F.C.; investigation, C.-H.L., Y.-F.C. and K.-W.T.; resources, C.-H.L., Y.-F.C. and Y.-J.C.; data curation, C.-H.L., K.-W.T., Y.-J.C. and Y.-A.C.; writing—original draft preparation, C.-H.L.; writing—review and editing, C.-H.L., K.-W.T., Y.-J.C. and Y.-A.C.; visualization, C.-H.L.; supervision, W.-S.C.; project administration, W.-S.C. All authors have read and agreed to the published version of the manuscript.

Funding: This research received no external funding.

References

1. *The United Nations World Water Development Report 2012*; UNESCO: Paris, France, 2012.
2. Wangnick, K. *2004 IDA Worldwide Desalting Plants Inventory*; Wangnick Consulting: Gnarrenburg, Germany, 2004.
3. Kumar, A.; Phillips, K.; Thiel, G.; Schröder, U.; Lienhard, J. Direct electrosynthesis of sodium hydroxide and hydrochloric acid from brine streams. *Nat. Catal.* **2019**, *2*, 106–113. [CrossRef]
4. U.S. Geological Survey. *Mineral Commodity Summaries 2015*; U.S. Geological Survey: Reston, VA, USA, 2015; p. 196. [CrossRef]
5. U.S. Geological Survey. *Mineral Commodity Summaries 2016*; U.S. Geological Survey: Reston, VA, USA, 2016; p. 202. [CrossRef]
6. U.S. Geological Survey. *Mineral Commodity Summaries 2017*; U.S. Geological Survey: Reston, VA, USA, 2017; p. 202. [CrossRef]
7. U.S. Geological Survey. *Mineral Commodity Summaries 2018*; U.S. Geological Survey: Reston, VA, USA, 2018; p. 200. [CrossRef]
8. U.S. Geological Survey. *Mineral Commodity Summaries 2019*; U.S. Geological Survey: Reston, VA, USA, 2019; p. 200. [CrossRef]
9. Li, Z.; Wakai, R.; Walker, T. Parametric modulation of an atomic magnetometer. *Appl. Phys. Lett.* **2006**, *89*, 134105. [CrossRef] [PubMed]
10. Yen, C.; Yano, Y.; Budinger, T.; Friedland, R.; Derenzo, S.; Huesman, R. Brain tumor evaluation using pet and Rb-82. *Clin. Nucl. Med.* **1981**, *6*, 448. [CrossRef]
11. Groeger, S.; Pazgalev, A.; Weis, A. Comparison of discharge lamp and laser pumped cesium magnetometers. *Appl. Phys. B* **2005**, *80*, 645–654. [CrossRef]
12. Downs, J. Drilling and Completing Difficult HP/HT Wells With the Aid of Cesium Formate Brines-A Performance Review. In Proceedings of the IADC/SPE Drilling Conference, Miami, FL, USA, 21–23 February 2006. [CrossRef]
13. Jiayong, C. *Handbook of Hydrometallurgy*; Metallurgical Industry Press Co., Ltd: Beijing, China, 2005.
14. Paschalis, C.; Jenner, F.; Lee, C. Effects of Rubidium Chloride on the Course of Manic-Depressive Illness. *J. R. Soc. Med.* **1978**, *71*, 343–352. [CrossRef] [PubMed]
15. Malek-Ahmadi, P.; Williams, J. Rubidium in psychiatry: Research implications. *Pharmacol. Biochem. Behav.* **1984**, *21*, 49–50. [CrossRef]
16. Canavese, C.; DeCostanzi, E.; Branciforte, L.; Caropreso, A.; Nonnato, A.; Sabbioni, E. Depression in dialysis patients: Rubidium supplementation before other drugs and encouragement? *Kidney Int.* **2001**, *60*, 1201. [CrossRef] [PubMed]
17. Lake, J. *Textbook of Integrative Mental Health Care*; Thieme New York: New York, NY, USA, 2007.
18. O'Neil, M.J. *The Merck Index: An Encyclopedia of Chemicals, Drugs, and Biologicals*; Whitehouse Station, N.J.: Merck: Readington, NJ, USA, 2001.
19. Li, Z.; Pranolo, Y.; Zhu, Z.; Cheng, C. Solvent extraction of cesium and rubidium from brine solutions using 4-tert-butyl-2-(α-methylbenzyl)-phenol. *Hydrometallurgy* **2017**, *171*, 1–7. [CrossRef]
20. Liu, S.; Liu, H.; Huang, Y.; Yang, W. Solvent extraction of rubidium and cesium from salt lake brine with t-BAMBP–kerosene solution. *Trans. Nonferrous Met. Soc. China* **2015**, *25*, 329–334. [CrossRef]
21. Wang, J.; Che, D.; Qin, W. Extraction of rubidium by t-BAMBP in cyclohexane. *Chin. J. Chem. Eng.* **2015**, *23*, 1110–1113. [CrossRef]
22. Xing, P.; Wang, G.; Wang, C.; Ma, B.; Chen, Y. Separation of rubidium from potassium in rubidium ore liquor by solvent extraction with t-BAMBP. *Miner. Eng.* **2018**, *121*, 158–163. [CrossRef]

Article

Removal of Metallic Iron from Reduced Ilmenite by Aeration Leaching

Qiuyue Zhao [1,2,3], Maoyuan Li [1], Lei Zhou [1], Mingzhao Zheng [1] and Ting'an Zhang [1,2,3,*]

[1] School of Metallurgy, Northeastern University, Shenyang 110004, China; zhaoqy@smm.neu.edu.cn (Q.Z.); maoyuan44@gmail.com (M.L.); 1801570@stu.neu.edu.cn (L.Z.); 1871388@stu.neu.edu.cn (M.Z.)

[2] Engineering Research Center of Metallurgy of Non-Ferrous Metal Materials Process Technology of Ministry of Education, Shenyang 110004, China

[3] Key Laboratory of Ecological Utilization of Multi-metal Intergrown Ores of Ministry of Education, Shenyang 110004, China

* Correspondence: zta2000@163.net; Tel.: +86-24-83686283

Received: 24 June 2020; Accepted: 27 July 2020; Published: 29 July 2020

Abstract: Aeration leaching was used to obtain synthetic rutile from a reduced ilmenite. The reduced ilmenite, obtained from the carbothermic reduction of ilmenite concentrate in a rotary kiln at about 1100 °C, contained 62.88% TiO_2 and 28.93% Metallic iron. The particle size was about 200 μm and the size distribution was uniform. The effects of NH_4Cl and HCl concentrations, stirring speed, and aeration leaching time on the extent of removal of metallic iron from the reduced ilmenite were studied at room temperature. The results revealed that aeration leaching is feasible at room temperature. When using the NH_4Cl system, the metallic iron content was reduced to 1.98% in synthetic rutile, but the TiO_2 content only reached 69.16%. Higher NH_4Cl concentration did not improve the leaching. Using 2% NH_4Cl with 3% HCl, we were able to upgrade the synthetic rutile to 75%, with a metallic iron content as low as 0.14% and a total iron content of about 4%. Synthetic rutile could be upgraded to about 90% using HCl solution alone. HCl and NH_4Cl are both effective on the aeration leaching process. However, within the scope of this experiment, hydrochloric acid is more efficient in aeration leaching.

Keywords: reduced ilmenite; synthetic rutile; aeration leaching; Becher process

1. Introduction

Titanium dioxide (TiO_2) is the most widely used titanium product, being employed as pigment, as filler in the paper, plastic, and rubber industries, and as flux in glass manufacture. Synthetic rutile (SR) is one of the major sources of TiO_2 [1–3]. Industrial processes usually involve the initial preparation of titanium dioxide, followed by titanium metal production [4,5]. Several commercial or proposed processes are available to produce SR or high-grade titanium slag from ilmenite which is mainly composed of $FeTiO_3$. These involve a combination of thermal oxidation and reduction by roasting, leaching, and physical separation steps. Iron is converted to soluble ferrous or elemental forms by reduction at a high temperature, followed by acid leaching to obtain a SR product.

Ilmenite generally contains impurities such as iron, which leads to its low grade and cannot be directly used. Synthetic rutile is a kind of titanium rich raw material with the same composition and structural properties as natural rutile by separating most iron components from ilmenite. An industrial process for upgrading ilmenite to SR is typically represented by the Becher process [6–8]. Ilmenite contains 40–65% titanium as TiO_2, with the rest being iron oxide. The Becher process removes the iron oxide, leaving a residue of SR that contains more than 90% TiO_2. The Becher process comprises four major steps: oxidation, reduction, aeration, and acid leaching [9,10]. Oxidation involves heating the ilmenite in a rotary kiln with air to convert the contained iron to iron oxide:

$$4FeTiO_{3(s)} + O_{2(g)} \rightarrow 2Fe_2O_3 \cdot TiO_{2(s)} + TiO_{2(s)} \tag{1}$$

This allows for the use of a wide range of ilmenite materials with various Fe(II) and Fe(III) contents for the subsequent step. Reduction is performed in a rotary kiln with a mixture of pseudobrookite ($Fe_2O_3 \cdot TiO_2$) and coal at about 1200 °C to reduce iron oxide to metallic iron:

$$Fe_2O_3 \cdot TiO_{2(s)} + 3CO \rightarrow 2Fe_{(s)} + 2TiO_{2(s)} + 3CO_{2(g)} \tag{2}$$

Metallic iron is then oxidized and precipitated from the solution as a slime in an aeration or 'rusting' step in large tanks using 1% ammonium chloride solution at 80 °C:

$$4Fe_{(s)} + 3O_{2(g)} \rightarrow 2Fe_2O_3 \tag{3}$$

The finer iron oxide is then separated from the larger SR particles. When most of the iron oxide is removed, the residual portion is leached using 0.5 M sulfuric acid and then separated from the SR. In the aeration leaching step, the removal of metallic iron from the reduced ilmenite (RI) grains is essentially a redox reaction, which can be represented by the following half-cell reactions:

$$2Fe \rightarrow 2Fe^{2+} + 4e^- \text{ (anodic reaction)} \tag{4}$$

$$O_2 + 4H^+ + 4e^- \rightarrow 2H_2O \text{ (cathodic reaction)} \tag{5}$$

The oxidation of ferrous ions is then given by:

$$2Fe^{2+} + 4OH^- + 1/2O_2 \rightarrow Fe_2O_3 \cdot H_2O + H_2O \tag{6}$$

In current industrial practice, the aeration step of the Becher process can take as long as 22 h to complete [11]. Some reports show that the rusting process can be accelerated by improving aeration [12] or by adding a component such as acetic, tartaric, or citric acid [13,14]; a ligand, such as ethylenediammonium dichloride; various phenolic and aldehyde compounds, such as pyrogallol, saccharin, starch, and formaldehyde; sugars, such as glucose and sucrose; and water-soluble redox catalysts, namely, methyl viologen dichloride and diquat dibromide [11,15–18]. These additives differ in effectiveness and cost. Most prior research was carried out at relatively high temperature (70 °C). Other related hydrometallurgical processes include, for example, ultrasonic-assisted acid leaching for iron removal from quartz sand [19–21] and the goethite process for iron removal from hydrochloric acid leaching solution of reduced laterite [22].

In the present work, we report a study of aeration leaching of reduced ilmenite at room temperature. Aeration leaching experiments using the hydrochloric acid system with oxygen injection at room temperature are rarely studied. The effects of hydrochloric acid and ammonia chloride in improving the aeration efficiency were evaluated. The effects of leaching parameters, including stirring speed and NH_4Cl and hydrochloric acid concentrations, were investigated. Through the above research, the method of strengthening the aeration process at room temperature is explored to provide a new way to obtain high-grade SR.

2. Materials and Methods

2.1. Materials

A Chinese source of reduced ilmenite, produced by carbothermic reduction of ilmenite concentrate in a rotary kiln at about 1100 °C, was used. The chemical composition and particle size is reported in Table 1 and Figure 1, respectively. MFe stands for metal iron and TFe stands for all iron in Table 1.

The composition of reduced ilmenite and SR obtained by XRF analysis and MFe was determined by chemical titration. Figure 1 shows that almost 80% of the particles were distributed between 90 and 400 µm, with a mode value of about 200 µm and a uniform distribution.

Table 1. Composition of reduced ilmenite (mass%).

Component	TiO$_2$	MFe	FeO	TFe	CaO	MgO	Mn	Al$_2$O$_3$	SiO$_2$
Content	62.88	28.93	3.69	31.90	0.15	0.23	1.89	1.55	1.84

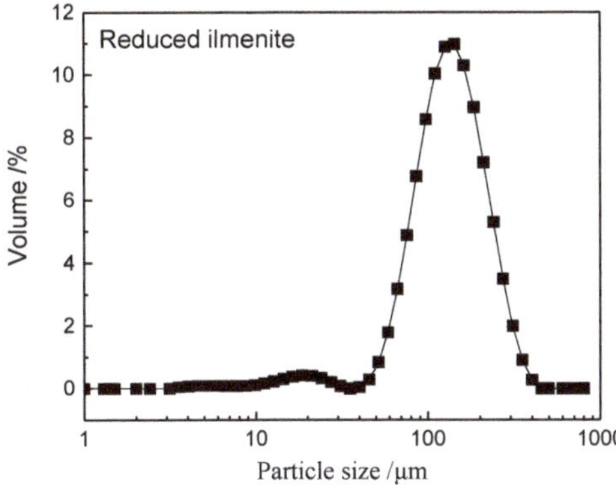

Figure 1. Particle size distribution of reduced ilmenite.

2.2. Aeration Conditions

The aeration leaching experiments were performed in a 1 L stirred reactor. Details of the experimental apparatus are illustrated in Figure 2. The inner diameter of the stirred reactor was 80 mm and the agitator was a four-blade propeller. The blade length as 30 mm.

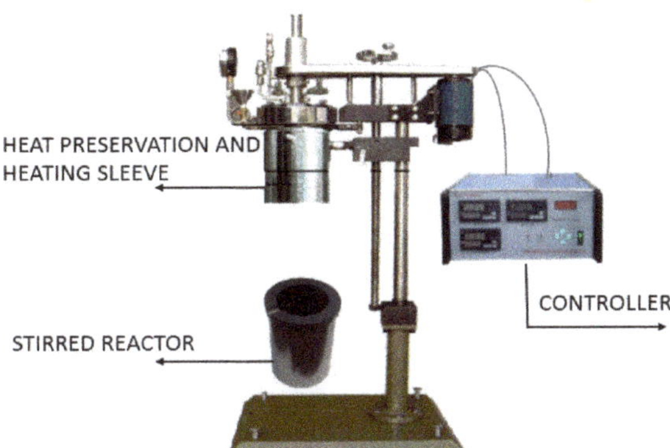

Figure 2. Aeration leaching reactor.

The initial reaction mixture comprised 640 mL solution and 320 g reduced ilmenite, which were added to the stirred reactor. The solution contained different concentrations of ammonium chloride and/or hydrochloric acid. The pulp was stirred by a four-blade agitator. Aeration gas was then introduced and passed through the pulp for the entire duration of the experiment. After 4 h, fine iron oxides were separated from the SR by wet screening. Particles of iron oxides and SR were washed and dried for analysis. We used the content of metallic iron (MFe) remaining in the SR to measure the efficiency of aeration leaching: it was found that the lower the residual iron content, the better the effect.

Aeration leaching is a process of oxygen absorption corrosion of metal iron. Three kinds of corrosion systems are generally selected: ammonium chloride, ammonium chloride plus hydrochloric acid, and hydrochloric acid. It is recognized that the anion provided by hydrochloric acid can destroy the passivation film on the surface of metallic iron in the aeration process [23]. The role of NH_4^+ is to combine with ferrous ions in the ore particles to form a complex that cannot be separated, so as to prevent oxidation and hydrolysis in the ore particles. The complex immediately decomposes when encountering water, and so acts as a carrier. The effect of ammonium chloride was examined using concentrations of 2%, 4%, 6%, and 8% (m/v) NH_4Cl. The stirring speed was 800 rpm. An ambient temperature was employed. Wet separation of the fine iron oxide from the coarse titanium mineral particles was done by using hydrocyclones and spiral classifiers. We measured the MFe and TiO_2 contents in the SR after aeration leaching for 4 h.

The particle size of the samples was analyzed by a laser diffraction particle size analyzer (Bettersize V8.0, Dandong Baite Instrument Co., Ltd., Dandong city, China). The structure and morphology of the reduced ilmenite samples and the product after aeration leaching were characterized by an X-ray diffractometer (BRUKER Inc., Karlsruhe, Germany), applying Cu Kα radiation at 40 kV and 40 mA, with 2θ recording from 10° to 80° with a step size of 0.02° and a counting time of 0.1 s per step. The metallic iron content of the solids was determined by potassium dichromate titration in $FeCl_3$ solution. Other elements were determined by a ZSX PrimusIV X-ray fluorescence spectrum (Japan Neo Confucianism Co., Ltd., Tokyo, Japan).

3. Results and Discussion

3.1. Effect of Solution Composition

The results are shown in Figure 3. With increasing NH_4Cl concentration from 2% to 8%, the metallic iron content in the SR increased from 1.85% to 6.75%, the total iron content decreased from 19.63% to about 16%, and the TiO_2 content increased from 64.97% to about 70%. It can be deduced that an increase in NH_4Cl concentration is not conducive to the aeration leaching process. When the concentration of 2% ammonium chloride was the same as that in Reference [21] and the reaction time was reduced by one hour, the removal rate of MFe in this paper was as high as 98.15%, while the removal rate of iron in Reference [21] was less than 50%. The TFe content was close to 20% and this is not sufficient for the aeration leaching products.

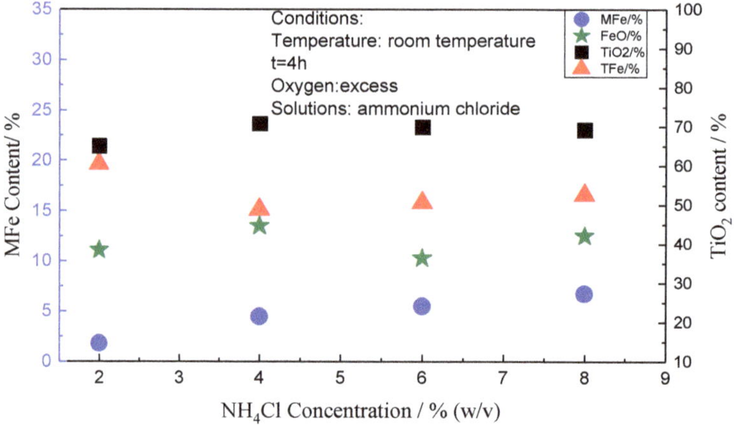

Figure 3. Effect of ammonium chloride concentration on composition of synthetic rutile.

Aeration leaching tests using 2% NH$_4$Cl with 0% to 3% hydrochloric acid were then carried out. The results are shown in Figure 4. For the same reaction time and other conditions, the contents of TFe and MFe in the SR monotonically decreased with an increase in hydrochloric acid concentration from 1% to 3%, while the TiO$_2$ content increased. The addition of hydrochloric acid helped to improve the aeration leaching, but the SR was only upgraded to 75%, which indicated that the reaction needed more time to improve the purity. Similar to Figure 3, the reaction rate of MFe was high but there was over 10% content of TFe in the aeration leaching products.

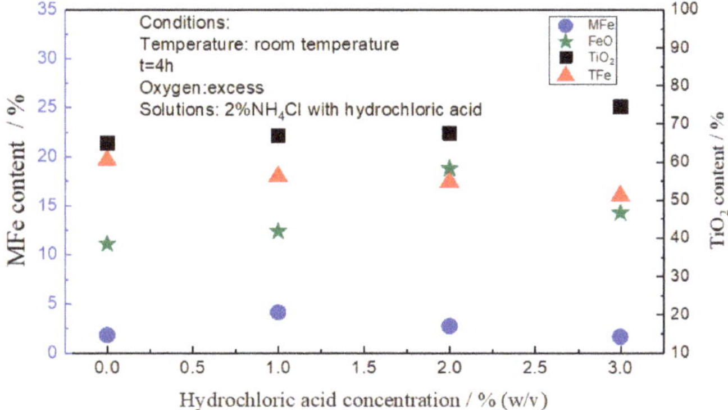

Figure 4. Effect of 2% NH$_4$Cl with hydrochloric acid on composition of synthetic rutile.

The aeration leaching was better with the addition of hydrochloric acid than with the NH$_4$Cl alone, so pure hydrochloric acid was considered for comparative analysis. The hydrochloric acid concentration was selected as 1.5% (m/v). The resulting MFe and TiO$_2$ contents in the SR are shown in Figure 5.

In the hydrochloric acid system, the TFe content in the SR was about 4%, compared with over 10%, and even up to 18%, in the NH$_4$Cl system. This proved that hydrochloric acid is better for aeration leaching than ammonium chloride. However, it is more difficult to store and transport hydrochloric acid, and the accumulation of chloride ion is not conducive to recycling of the corrosion solution. Therefore, comprehensive consideration is needed to select the best aeration leaching solution.

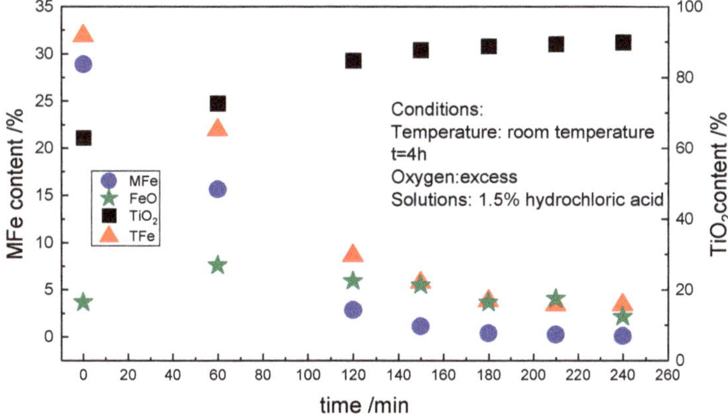

Figure 5. Effect of hydrochloric acid on composition of synthetic rutile.

3.2. Effect of Stirring Speed

Stirring is one of the most important factors in mixing processes in the chemical industry and metallurgy. The purpose is to mix evenly, accelerate the dissolution, or accelerate the reaction process. Generally, too slow a stirring speed will lead to uneven mixing and too fast a stirring speed can damage the product. High-speed mixing consumes more electric energy, which results in an increase in production cost. Selection of an appropriate mixing speed is, therefore, essential [23]. The effect of stirring speed on the removal of metallic iron from the reduced ilmenite was investigated by varying the impeller speed in the range of 400 to 1000 rpm. The concentration of hydrochloric acid was 1.5% m/v, the reaction time was 4 h, and the aeration gas was present in excess.

As shown in Figure 6, the metallic iron remaining in the SR decreased from 6.95% at 400 rpm to 0.49% at 800 rpm, corresponding to a reduction in metallic iron content of 6.46% points. The metallic iron content increased to 2.85% at 1000 rpm and the iron content increased by 2.36% compared with the value of 2.85% at 800 rpm.

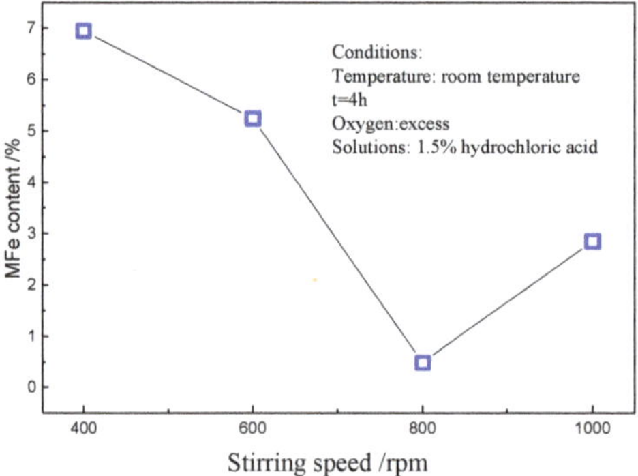

Figure 6. Effect of stirring speed on metallic iron content of synthetic rutile.

The presence of agitation can break up bubbles, increase the specific surface area of bubbles, and accelerate mass transfer from the gas phase to the liquid phase. Agitation can also promote uniform suspension of reduced ilmenite particles, increase the liquid–solid contact area, accelerate the internal diffusion process, and prevent corroded iron ions from reducing ilmenite particles in an in situ reaction. The vortex will be formed at high speed, which will lead to uneven mixing of gas, liquid, and solid.

3.3. Phase Analysis

Figure 7 presents XRD spectra of reduced ilmenite before and after iron removal by aeration under the conditions: room temperature, t = 4 h, excess oxygen, 1.5% hydrochloric acid. The major phases in the reduced ilmenite before iron removal were Fe, TiO_2, and $FeTi_2O_5$. The diffraction peaks of $FeTi_2O_5$ and TiO_2 were strong. Peaks for the metallic iron phase were not to be found in the sample after aeration leaching, but diffraction peaks of FeO(OH) were detected. The diffraction peaks of FeO(OH) and TiO_2 had the same intensity, which indicated that metallic iron transformed into FeO(OH). There were just two main phases in the sample after the aeration leaching process. These results indicate that the transformation of reduced ilmenite into rutile was achieved under these experimental conditions.

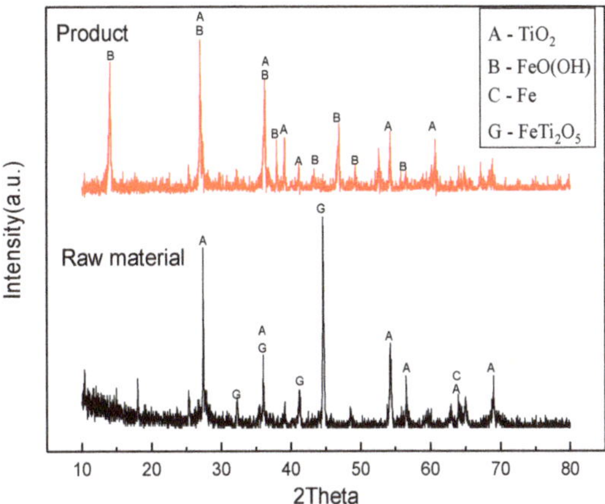

Figure 7. X-ray diffraction patterns of reduced ilmenite before and after iron removal.

Philips ssx-550 scanning electron microscope (SEM) images of the sample before and after aeration leaching are shown in Figure 8. There are obvious differences between the raw material and the product of the aeration leaching process: the sample before aeration leaching was compact and we could not find holes in the surface; after aeration leaching, the interior was full of holes, giving rise to a network structure, which maintained the sample integrity. The holes are attributed to the transformation of metallic iron into iron oxide by the aeration leaching reaction and its removal from the interior of sample.

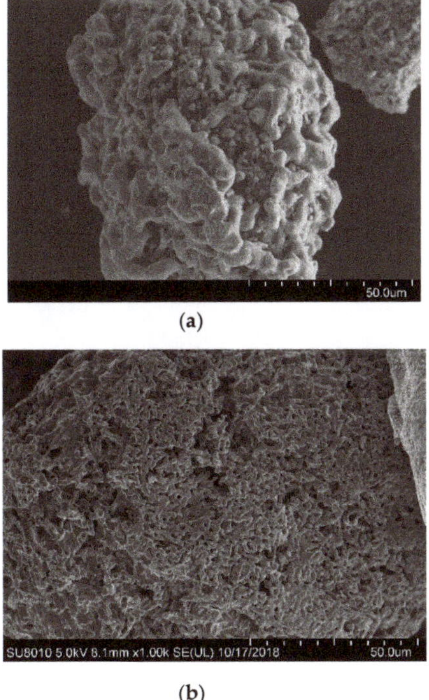

(a)

(b)

Figure 8. *Cont.*

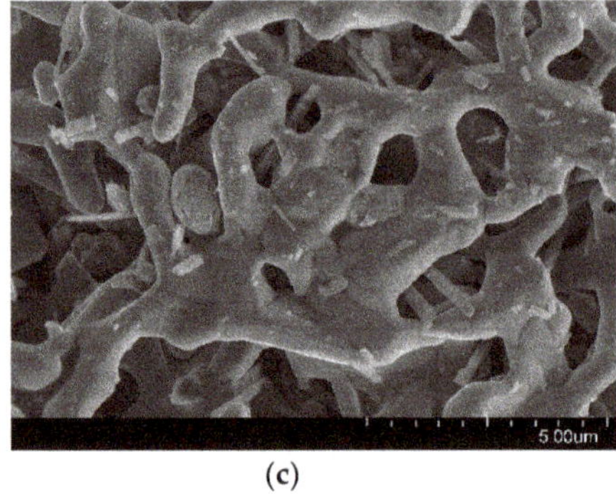

(**c**)

Figure 8. Scanning electron micrographs of sample (**a**) before and (**b**) after aeration leaching process (**c**) after aeration leaching process (enlarged).

4. Conclusions

The following conclusions were drawn on the basis of the results obtained in this work.

Aeration leaching is feasible at room temperature. For a reaction time of 4 h and adequate stirring speed of 800 rpm, the effectiveness of metallic iron removal differed for different solution systems. When using NH_4Cl, the MFe content could be reduced to 1.98% in SR, but the TiO_2 content only reached 69.16%. A higher NH_4Cl concentration did not improve aertion leaching. Using 2% NH_4Cl with hydrochloric acid, the presence of the acid helped to improve the leaching, but the SR was only upgraded to 75%, which indicated that the reaction needed more time to improve the purity. In the hydrochloric acid system, the MFe content was as low as 0.14% and TFe content was about 4%, indicating that the SR could be upgraded to about 90%.

Author Contributions: This is a joint work of the five authors; each author was in charge of their expertise and capability: Q.Z. for writing, formal analysis and original draft preparation, M.L. for data curation, L.Z. for experimental assistance, M.Z. for validation, T.Z. for methodology. All authors have read and agreed to the published version of the manuscript.

Funding: This research was supported by the National Natural Science Foundation of China (NSFC) (Grant No. 51204040); Fundamental Research Funds for the Central Universities (Grant No. N180725023).

Acknowledgments: The authors gratefully acknowledge the Laboratory Center of Northeastern University for chemical analysis. Guangxi Yueqiao New Material Technology Co., Ltd. for providing the reduced ilmenite samples.

Conflicts of Interest: On behalf of all authors, the corresponding author states that there is no conflict of interest.

References

1. Zhang, W.S.; Zhu, Z.W.; Cheng, C.Y. A literature review of titanium metallurgical processes. *Hydrometallurgy* **2011**, *108*, 177–188. [CrossRef]
2. Lakshmanan, V.I.; Bhowmick, A.; Halim, M.A. Titanium dioxide: Production, properties, and applications. *Chem. Phys. Res. J.* **2014**, *7*, 37–42.
3. Que, Y.; Weng, J.; Hu, L. Applications of titanium dioxide in perovskite solar cells. *Prog. Chem.* **2016**, *28*, 40–50.
4. Guo, Y.F.; Liu, S.S.; Jiang, T.; Qiu, G.Z.; Chen, F. A process for producing synthetic rutile from Panzhihua titanium slag. *Hydrometallurgy* **2014**, *134*, 147–148. [CrossRef]
5. Yaraghi, A.; Sapri, M.H.A.; Baharun, N.; Rezan, S.A. Aeration leaching of iron from nitrided Malaysian ilmenite reduced by polystyrene-Coal reductant. *Procedia Chem.* **2016**, *19*, 715. [CrossRef]
6. Becher, R.G. Improved process for the beneficiation of ores containing contaminating iron. *Aust. Patent* **1963**, *247*, 110.
7. Farrow, J.B.; Ritchie, I.M.; Mangono, P. The reaction between reduced ilmenite and oxygen in ammonium chloride solution. *Hydrometallurgy* **1987**, *18*, 21–38. [CrossRef]

8.   Hoecker, W. Process for the Production of Synthetic Rutile. U.S. Patent No. AU19940056301, 22 February 1994.
9.   Sekimoto, H.; Yahaba, S.; Chiba, S.; Yamaguchi, K. New separation technique of titanium and iron for titanium ore upgrading. In Proceedings of the 13th World Conference on Titanium, TMS (The Minerals, Metals & Materials Society), Warrendale, PA, USA, 2 May 2016.
10. Bracanin, B.F.; Clements, R.J.; Davey, J.M. Direct reduction—the Western titanium process for the production of synthetic rutile, ferutil and sponge iron. *Australas. Inst. Min. Metall. Pro.* **1980**, *275*, 33–36.
11. Fletcher, S.; Bruckard, W.J.; Calle, C.; Carey, K.C.; Horne, M.D.; Ruzbacky, R.; Sparrow, G.J. Soluble catalysts for the oxygen reduction reaction, and their application to Becher aeration. *Ind. Eng. Chem. Res.* **2019**, *58*, 10190–10198. [CrossRef]
12. Jayasekera, S.; Marinovich, Y.; Avraamides, J.; Bailey, S.I. Pressure leaching of reduced ilmenite: Electrochemical aspects. *Hydrometallurgy* **1995**, *39*, 183–199. [CrossRef]
13. Nguyen, T.T.; Truong, T.N.; Duong, B.N. Impact of organic acid additions on the formation of precipitated ironcompounds. *Acta Metall. Slovaca* **2016**, *22*, 259–265. [CrossRef]
14. Truong, T.N.; Nguyen, T.T.; Duong, B.N. Acetic acid and sodium acetate mixtures as an aeration catalyst in the removal of metallic iron in reduced ilmenite. *Acta Metall. Slovaca* **2017**, *23*, 371–377. [CrossRef]
15. Bruckard, W.J.; Calle, C.; Fletcher, S.; Horne, M.D.; Sparrow, G.J.; Urban, A.J. The application of anthraquinone redox catalysts for accelerating the aeration step in the becher process. *Hydrometallurgy* **2004**, *73*, 111–121. [CrossRef]
16. Guo, Y.F.; Liu, X.; Qiu, G.Z.; Jiang, T. Strengthening of metallic iron rust in reduced ilmenite. *J. Cent. South Univ. Sci. Technol.* **2012**, *43*, 797–802.
17. Geetha, K.S.; Surender, G.D. ntensification of iron removal rate during oxygen leaching through gas-liquid mass transfer enhancement. *Metall. Mater. Trans. B* **2001**, *3*, 961–963. [CrossRef]
18. Xiang, J.Y.; Pei, G.S.; Lv, W.; Liu, S.L.; Lv, X.W.; Qiu, G.B. Preparation of synthetic rutile from reduced ilmenite through the aeration leaching process. *Chem. Eng. Process.* **2020**, *147*, 107774. [CrossRef]
19. Li, X.; Xing, P.F.; Du, X.H.; Gao, S.B.; Chen, C. Influencing factors and kinetics analysis on the leaching of iron from boron carbide waste-scrap with ultrasound-assisted method. *Ultrason. Sonochem.* **2017**, *38*, 84–91. [CrossRef]
20. Ma, J.Y.; Zhang, Y.F.; Qin, Y.; Wu, Z.K.; Wang, T.L.; Wang, C. The leaching kinetics of K-feldspar in sulfuric acid with the aid of ultrasound. *Ultrason. Sonochem.* **2017**, *35*, 304–312. [CrossRef]
21. Feng, D.; Ren, Q.X.; Ru, H.Q.; Wang, W.; Ren, S.Y.; Jiang, Y.; Liu, B.Y.; Chang, S.X.; Zhang, C.P.; Yang, T. Iron removal from ultra-fine silicon carbide powders with ultrasound-assisted and its kinetics. *Mater. Chem. Phys.* **2020**, *247*, 122860. [CrossRef]
22. Sun, D.L.; Wu, K.M.; Hu, J. Removal of iron from leaching solution of zinc ore by goethite process. *Hydrometall. China* **2015**, *34*, 68–71.
23. Chen, X.; Chen, S.H.; Cai, S.M. Breakdown of the passive film on iron by Cl^- in a acidic solution. *Acta Phys. Chim. Sin.* **1988**, *4*, 3823–3886.

Article

Hydrometallurgical Process and Economic Evaluation for Recovery of Zinc and Manganese from Spent Alkaline Batteries

Lan-Huong Tran *, Kulchaya Tanong, Ahlame Dalila Jabir, Guy Mercier and Jean-François Blais

Institut National de la Recherche Scientifique (Centre Eau, Terre et Environnement), Université du Québec, 490 rue de la Couronne, Québec, QC G1K 9A9, Canada; kulchaya.tanong@ete.inrs.ca (K.T.); ahlame_dalila.jabir@ete.inrs.ca (A.D.J.); guy.mercier@ete.inrs.ca (G.M.); jean-francois.blais@ete.inrs.ca (J.-F.B.)
* Correspondence: lan.huong.tran@ete.inrs.ca; Tel.: +418-654-2550; Fax: +418-654-2600

Received: 17 July 2020; Accepted: 26 August 2020; Published: 1 September 2020

Abstract: An innovative, efficient, and economically viable process for the recycling of spent alkaline batteries is presented herein. The developed process allows for the selective recovery of Zn and Mn metals present in alkaline batteries. The hydrometallurgical process consists of a physical pre-treatment step for separating out the metal powder containing Zn and Mn, followed by a chemical treatment step for the recovery of these metals. Sulfuric acid was used for the first leaching process to dissolve Zn(II) and Mn(II) into the leachate. After purification, Mn was recovered in the form of MnO_2, and Zn in its metal form. Furthermore, during the second sulfuric acid leaching, $Na_2S_2O_5$ was added for the conversion of Mn(IV) to Mn(II) (soluble in the leachate), allowing Mn to be recovered as $MnCO_3$. Masses of 162 kg of Zn metal and 215 kg of Mn (both in the form of MnO_2 and $MnCO_3$) were recovered from one ton of spent alkaline batteries. The direct operating costs (chemicals, labor operation, utilities, energy) and indirect costs (amortization, interest payment) required for a plant treating 8 tons of spent batteries per day was calculated to be \$CAD 726 and \$CAD 534 per ton, respectively, while the total revenue from the sale of the metals was calculated at \$CAD 1359.6 per ton of spent batteries. The development of this type of cost-effective industrial process is necessary for a circular economy, as it contributes to addressing environment- and energy-related issues, and creates opportunities for the economic utilization of metals.

Keywords: spent alkaline battery; recycling; leaching; electrowinning; hydrometallurgy; techno-economic evaluation; metal recovery

---

1. Introduction

The use of electronic compact devices with batteries such as remote controls, watches, electric toys, and pocket lamps has become an integral part of our society. Furthermore, these batteries have a certain lifetime, and the increase in volume of spent batteries over the last few years requires an innovative recycling process. Findings from research into metal recovery in recent years indicate the importance of recycling spent batteries [1–8]. In Canada, Call2recycle collected more than 2.5 kt of batteries for recycling in 2017 and 2.7 kt in 2018, of which 78% consisted of alkaline and Zn–C batteries [9]. However, the collected quantity only represents approximately 20% of all batteries sold in the market [10]. Alkaline batteries consist of a negative zinc metal electrode and a positive manganese dioxide (MnO_2) electrode with an alkaline potassium hydroxide electrolyte, instead of the acidic ammonium chloride electrolyte used in zinc–carbon batteries. After collection, batteries are separated by type and sent to appropriate processing plants. Alkaline and carbon zinc batteries are sent to Retriev (Trail, BC, Canada), Inmetco (Elwood City, PA, USA), Raw Materials Company (Port Colbone, ON, Canada), and Battery Solutions Recovery (Brighton, MI, USA), where the batteries are treated by pyrometallurgical processes [10,11]. These processes separate metals by volatilization and melting, and, therefore, require high-energy consumption and, due to the release of toxic gases, require an additional collection/cleaning system. By contrast, hydrometallurgical processes usually have a lower energy consumption and lower environmental impacts than pyrometallurgical processes. Hydrometallurgical processes consist of metal leaching followed by the separation, purification, and recovery of valuable metals through various techniques including, among others, precipitation, solvent extraction, and electrowinning. Numerous studies have used leaching processes under various conditions for the leaching of Zn and Mn from battery powder (Table 1).

Table 1. Leaching yields of Zn and Mn from alkaline battery by acid leaching.

Leaching Agent	Auxiliary Agent	Conditions	Metal Removal (%)		Ref.
			Zn	Mn	
H_2SO_4 0.54 M	H_2O_2 4% (*v/v*)	2 h, 55 °C, 1/30 *w/v*	100	95.7	[12]
H_2SO_4 0.54 M	H_2O_2 4% (*v/v*)	2 h, 55 °C, 1/10 *w/v*	53.8	42.8	
H_2SO_4 0.5 M	Ascorbic acid 10 g/L		99.5	98.8	
H_2SO_4 0.5 M	Citric acid 10 g/L	3 h, 25 °C, 1/20 *w/v*	99.9	91.6	[13]
H_2SO_4 0.5 M	Oxalic acid 10 g/L		96.4	87.5	
H_2SO_4 1.5 M		3 h, 80 °C, 1/10 *w/v*	90	<20	
H_2SO_4 1.0 M	-	Microwave 1 cycle, 30 s	94	<20	[14]
H_2SO_4 1.0 M		Ultrasonic, 2 min, 0.1 pulse, 20% amplitude	92	<20	
H_2SO_4 1.5 M	Lactose twice of stoichiometry	3 h, 90 °C, 1/10 *w/v*	100	98	[15]
Bio-generated H_2SO_4	H_2O_2 5 vol.% or Na_2SO_3 1 wt. % Calcine 2 h, 750 °C	2 h, 30 °C, 1/25 *w/v*	99	90–98	[16]

Notably, the respective reaction time, temperature, acid concentration, and solid/liquid (S/L) ratio should be compared, as these parameters vary across the different studies. Sulfuric acid is commonly used either singly or in combination with an auxiliary agent. For example, the addition of H_2O_2 (4% *v/v*) in a sulfuric acid solution was used to remove 100% of Zn and 95.7% of Mn from battery powder [12]. In another study, the combination of ascorbic acid with sulfuric acid led to the dissolving of 99.5% of Zn and 98.8% of Mn [13]. Although the use of an auxiliary agent is key to dissolving Mn in sulfuric acid, results from a study where no auxiliary agents were used indicate a 90% removal of Zn and less than 20% removal of Mn in a sulfuric acid medium [14]. Furlani et al. (2009) studied the use of carbohydrates, primarily lactose, as reducing agents for the leaching of manganese from the zinc alkaline battery powder [15]. The carbohydrates reduced Mn(IV) and Mn(III) oxides to acid-soluble Mn(II), and approximately twice the stoichiometric amount of lactose was used for complete leaching [15]. Gallegos et al. (2018) proposed a process using biogenerated sulfuric acid with 5 vol.% H_2O_2 or 1 wt.% Na_2SO_3 for the leaching of Zn and Mn in a single step [16]. In their study, 99% of Zn and 90–98% of Mn was extracted after 2 h of leaching at 30 °C and 0.04 g/mL [16]. Sodium metabisulfite ($Na_2S_2O_5$) was also used as a reducing agent for the dissolution of metals from a mixture of spent batteries. In their research, Tanong et al. (2017) obtained 94 and 99% removal yields for Mn and Zn, respectively, by adding 0.45 g $Na_2S_2O_5$/g to battery powder in H_2SO_4 1.34 M in a single leaching step, with an S/L ratio of 10.9% for 45 min at ambient temperature [6]. In other studies, a thermal pre-treatment was added to increase the efficacy of leaching. For example, Petranikova et al. (2018) investigated the effects of a thermal treatment at 300–950 °C of battery powder on the acid leaching (0.5 M H_2SO_4 at 25 °C for 60 min) [17]. In general, these studies indicate that sulfuric acid leaching allows for the complete dissolving of Zn and partial extraction of Mn (MnO, Mn_2O_3, and Mn_3O_4). The total dissolving of Mn (including MnO_2) demands an auxiliary agent to reduce MnO_2 to MnO, soluble in sulfuric acid. Therefore, the leaching process can dissolve Zn and Mn simultaneously in a single step or via selective leaching in a two-step process. Selective leaching allows for the use of different techniques for recovery.

After leaching, the second challenge in the hydrometallurgical process is the efficient recovery of metals at high purities. The metals present in the leachate can be recovered by precipitation in the form of hydroxides, sulfides, or carbonates according to their respective pH and redox potential. Sobianowska-Turek et al. [18] used NH_4HCO_3 3 M and NH_4OH 1 M to recover almost 100% of manganese, iron, cadmium, and chromium, as well as 98.0% of cobalt, 95.5% of zinc, and 85.0% of copper and nickel from the solution after reductive acidic leaching ($H_2SO_4 + C_2H_2O_4$) [18]. Furthermore, Sayilgan et al. used KOH 2 M and NaOH 2 M for the selective precipitation of manganese and zinc, with complete precipitation obtained for Zn at pH 7–8 and Mn at pH 9–10 [19].

In industry, zinc is usually recovered in metallic form by electrowinning. In a study by Alfantazi and Dreisinger (2003), electrowinning experiments were conducted at an 80 min plating time, 500 A/m^2 current density, and temperature of 38 °C with zinc-containing electrolyte zinc and H_2SO_4 concentrations of 62 and 170 g/L, respectively [20]. In this study, 90% of zinc was recovered by using an Al cathode and Pb anode. Similar results were recorded by Ivanov (2004) with an electrolyte containing 45–55 g Zn/L, 5–6 g Mn/L, and traces of other metals including, among others, Ni, Sb, and Ge [21].

In our study, in addition to developing an efficient battery recycling process, relevant economic factors were also considered. Economic factors generally include processing and operating costs, as well as transport and residue disposal costs. For example, Gasper et al. (2013) evaluated the economic viability for recycling alkaline batteries using mechanical separation [22]. Although the mechanical process developed in their study was cheaper than other reported processes ($US529/ton), it was still not economically feasible, due to the low end-product value. The revenue of the end products was $US 383/ton of batteries that consisted of brass, Zn/ZnO powder, mixed Mn oxides powder, and KOH powder. In our study, we developed and described a complete hydrometallurgical process including the recovery of Zn by electrowinning and Mn by precipitation from alkaline battery powder. The process included dismantling, magnetic separation, leaching, and metal recovery. Furthermore, an economic evaluation was carried out to assess the feasibility on a global scale, especially for applications in Quebec (Canada), where an alkaline battery recycling process is not yet available.

2. Materials and Methods

2.1. Process Description

Figure 1 represents a flow sheet of the entire process for the recovery of Zn and Mn from spent alkaline batteries for this study. The globally relevant process includes two steps, namely, a physical pre-treatment step and a chemical treatment step. After dismantling, a representative sample of batteries is subjected to the attrition process followed by filtration and rinsing to remove any metallic powder attached to the coarse fraction. In addition to the separation of battery products, the removal of alkaline-soluble salts during this process reduces acid consumption in the following step. The coarse fraction is then transferred to the magnetic separator, allowing the magnetic fraction to be recycled as ferrous material and the non-magnetic fraction containing nylon, carton, and plastic to be used as energy sources.

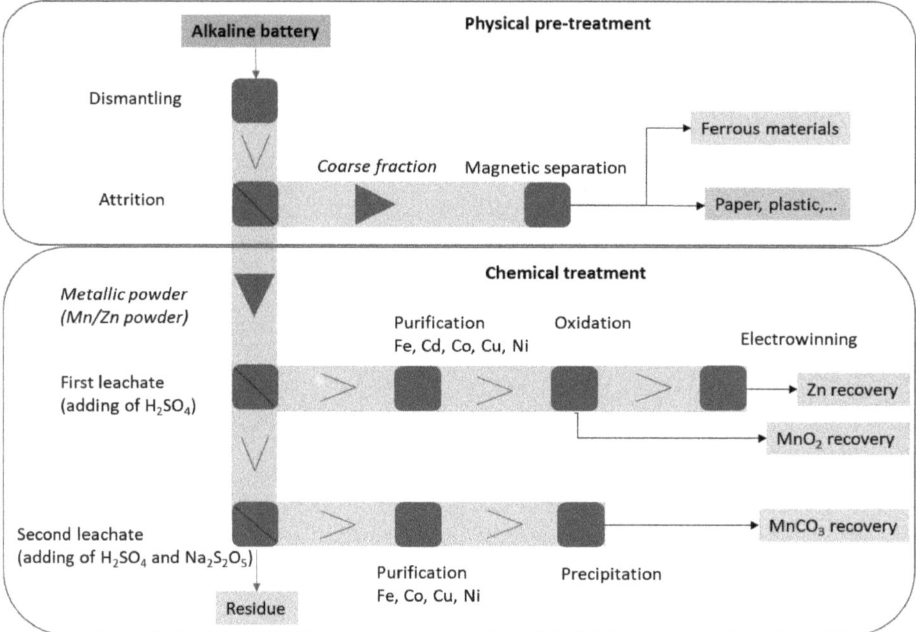

Figure 1. Detailed flow sheet of the hydrometallurgical route to treat the spent alkaline batteries.

The fine fraction containing manganese and zinc described as "metallic powder" is transferred to the chemical treatment, where zinc is dissolved during the first leaching step using sulfuric acid (Equation (1)). A portion of manganese, in the form of MnO, Mn_2O_3, and Mn_3O_4, is also dissolved according to Equations (2) to (4) as follows:

$$ZnO + H_2SO_4 \rightarrow ZnSO_4 \tag{1}$$

$$MnO + H_2SO_4 \rightarrow MnSO_4 + H_2O \tag{2}$$

$$Mn_2O_3 + H_2SO_4 \rightarrow MnSO_4 + MnO_2 + H_2O \tag{3}$$

$$Mn_3O_4 + 2H_2SO_4 \rightarrow 2MnSO_4 + MnO_2 + 2H_2O \tag{4}$$

As MnO_2 is insoluble in sulfuric acid, sodium metabisulfite ($Na_2S_2O_5$) is added during the second leaching step to reduce Mn(IV) to Mn(II) [6]. During this leaching step, MnO_2 in metallic powder is transferred in the leachate solution according to Equation (5):

$$Na_2S_2O_5 + 2MnO_2 + H_2SO_4 \rightarrow 2MnSO_4 + Na_2SO_4 + H_2O \tag{5}$$

After precipitation for the removal of iron, and cementation for the removal of trace metals such as Ni and Cu, the first pregnant leach solution (PLS-1) is treated using sodium persulfate ($Na_2S_2O_8$) to oxidize manganese for MnO_2 recovery. Then, zinc is reduced by electrodeposition, resulting in a high-purity zinc metal. Furthermore, the second pregnant leach solution (PLS-2) is treated for the removal of iron and zinc followed by the precipitation of manganese as $MnCO_3$. While the water used during this process can be reused through a counter-current mode, the final carbon-rich residue can be used for the fabrication of new batteries.

2.2. Physical Pre-Treatment

2.2.1. Sampling and Pre-Treatment

The sample of spent alkaline $Zn-MnO_2$ and Zn-C batteries was obtained from Laurentide Re-Sources Inc. (Victoriaville, QC, Canada). This company receives used batteries from various collection centers located throughout the Province of Quebec (Canada). During physical pre-treatment, spent alkaline batteries were shredded to approximately 1 cm × 4 cm fractions using a mechanical grinder (Muffin Monster, model 30005, JWC Environmental®, Santa Ana, CA, USA).

2.2.2. Attrition Process

The neutral attrition consisted of three 20 min steps with a solid/liquid (S/L) ratio fixed at 30% (*w/w*) and 700 rotations per minute (rpm). The high rotation speed accelerates the separation of fine powder from other parts of the battery including carton pieces, ferrous scraps, and plastic. This step is also useful in removing soluble potassium hydroxide from the battery powder. Attrition experiments were carried out using a 40 L stainless tank reactor equipped with three internal baffles. For each experiment, 3 kg of shredded battery material was combined with 10 L tap water for 20 min at ambient temperature. After each step, S/L separation was carried out using 1.7 mm sieves. The remaining coarser fraction was then rewashed until three attrition stages were complete. The metallic powder was removed from the liquid phase after 2 h of settling. The water used for the washing process was recycled via counter-current mode, the result will show the media after three cycles.

2.3. Chemical Treatment

2.3.1. Leaching Process

After washing, the metal powder was collected and different conditions were used to leach a maximum amount of zinc and manganese from the metallic powder. The challenge within this process is to obtain a high leaching efficiency and a PLS zinc concentration $\geq$40 g/L. This is important for avoiding energy loss during the electrowinning process. The selective leaching process was carried out in two stages in a 40 L stainless steel reactor. The water was recycled three times (LC 1 to 3) as presented in Figure 2. During the first cycle (LC 1), leaching was conducted with 4 kg of metallic powder in 2 M H_2SO_4 and an S/L ratio fixed at 40% (*w/v*) for 45 min at ambient temperature. For the second and third cycles (LC 2 and 3, respectively), leaching was conducted with 2 kg of metallic powder, and an S/L ratio of 20% in 1 M H_2SO_4, using the recycled solution after electrowinning (approximately 35–40 g Zn/L).

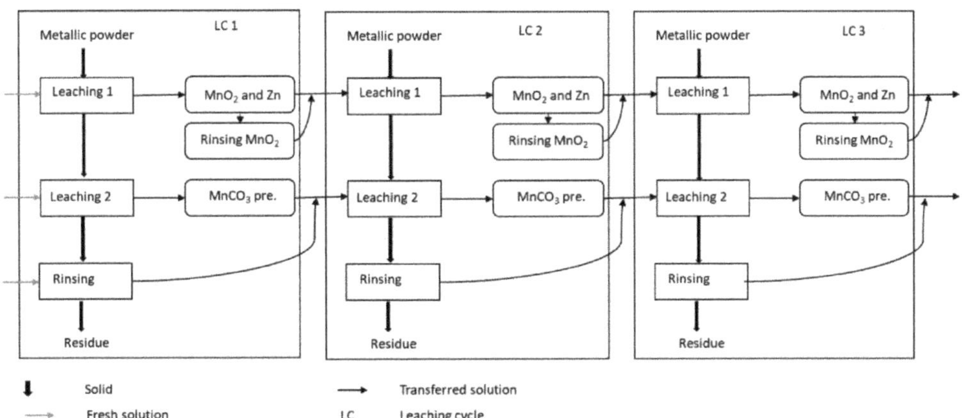

Figure 2. Schematic representation of the counter-current leaching and recovery process including rinsing steps performed on metallic powder.

During the second leaching step, 0.45 g $Na_2S_2O_5$ per gram of metallic powder was added to reduce Mn(IV) to Mn(II), soluble in 1.34 M H_2SO_4, with an S/L ratio of 14% for 45 min at ambient temperature. The PLS-2 obtained from the second leaching step contained a significant proportion of manganese. Following each leaching step, S/L separation was carried out by filtration.

2.3.2. Purification

Both PLS-1 and PLS-2 were purified prior to zinc and manganese recovery. The iron removal from PLS-1 was carried out at pH 4.0–4.5 by the addition of sodium hydroxide (NaOH) and hydrogen peroxide (H_2O_2). The added H_2O_2 dose was 1.5 times that of the stoichiometric molar ratio (SMR) of the total iron concentration (approximately 2 mM for PLS-1 and 30 mM for PLS-2).

After iron removal, the solution was purified by cementation using Zn powder to remove metal traces, such as Ni, Cu, and Co via the following equation:

$$Zn_{(s)} + M^{2+}{}_{(aq)} \rightarrow Zn^{2+}{}_{(aq)} + M_{(s)} \tag{6}$$

where M represents metal traces such as Ni, Cu, and Co.

In our study, a quantity of Zn metal powder equal to 20 times the impurity of metals (around 200 mg/L) was added, and cementation was conducted at 80 °C at pH 4.0–4.5 for 30–120 min.

Using the same principle, iron was also removed from PLS-2. Thereafter, the Zn residual (about 0.1 mol/L) in the PLS-2 was precipitated out using sodium sulfide (Na_2S) at pH 4.0–5.0. The SMR values of 1, 2, and 3 were used during these experiments. The ZnS precipitate was then returned to the first leaching step and the PLS-2 was transferred to the $MnCO_3$ recovery step.

2.3.3. Metal Recovery from PLS-1

Oxidation—MnO_2 Recovery

The recovery of MnO_2 was performed by manganese oxidation using sodium persulfate ($Na_2S_2O_8$) according the following equation:

$$MnSO_4 + Na_2S_2O_8 + 2H_2O \rightarrow MnO_2 + Na_2SO_4 + 2H_2SO_4 \tag{7}$$

The purified solution of $ZnSO_4$ and $MnSO_4$ from the previous step was used to recover MnO_2. The kinetic reaction was carried out with 1 SMR $Na_2S_2O_8$ at 60 °C (nearby 0.6 M of Mn), and the pH increased in the range of 2–4 by adding NaOH. The S/L separation was carried out by filtration and the resulting liquid was transferred to the next step for Zn recovery. The MnO_2 precipitate was washed with an S/L ratio of 10% for 10 min (200 rpm) to obtain high-purity MnO_2.

Electrowinning—Zn Recovery

An acrylic reactor, 8.5 cm (width) × 70 cm (length) × 15 cm (depth), was used for electrodeposition experiments. The electrode sets consisted of 14 aluminum cathodes and 15 Pb/Ag anodes, each with a surface area of 119 cm^2 (4.5 cm (width) × 10 cm (height) × 1 cm (thickness)). Electrodes were placed in parallel, 1 cm apart. The cathodes and anodes were connected to the negative and positive outlets of a DC power supply, respectively, Xantrex XFR (ACA TMetrix, Mississauga, ON, Canada). The working volume was fixed to 5 L at ambient temperature for all electrowinning experiments. Assays were conducted in batch mode with continuous mixing using a water recirculation system in the reactor. The purified solution of $ZnSO_4$ (pH = 2) after MnO_2 precipitate recovery was used in the electrowinning experiments, in which Zn was deposited on the cathode surface with a current density between 250 and 750 A/m^2 and an electrowinning time of 180 min. In this condition, the potential was dropped from 5.6 to 5.2 V. Samples of electrolyte were taken after 15, 30, 45, 60, 90, and 180 min to analyze the residual metal concentration.

2.3.4. Metal Recovery from PLS-2: $MnCO_3$ Precipitation

Sodium carbonate (Na_2CO_3) was used to precipitate manganese from PLS-2 (a quantity of 90 g Na_2CO_3 was added for 1 L of PLS-2). The $MnCO_3$ precipitate was rinsed using an S/L ratio of 10% for 10 min for a moderate-speed mixing (200 rpm) to eliminate dissolved sulfur and sodium from the $MnCO_3$ powder after precipitation. All precipitation experiments were carried out at ambient temperature.

2.4. Analytical Methods

Liquid samples were filtered using G6 glass fiber paper (Fisher brand, Fisher Scientific, Ottawa, ON, Canada, pore size = 1.5 µm) to remove solid particles, and stored in 5% HNO_3 before analysis. Solid samples were dissolved with nitric and hydrochloric acid (HNO_3 and HCl, respectively) before analysis. Metal mobility in the final residual (after the second leaching step) was evaluated using the toxicity characteristic leaching procedure (TCLP)—Method 1311. The metal concentrations in all samples were determined using an inductively coupled plasma-optical emission spectrometer (ICP–OES) (model 725–ES, Agilent Technologies, Santa Clara, CA, USA).

2.5. Techno-Economic Evaluation of the Hydrometallurgical Process

The economic simulation for a battery recycling plant required a large number of parameters (variables) in order to estimate the economic performance of the process. For this study, a computer model was developed to evaluate both direct and indirect costs of the proposed recycling process. This model included more than 260 input variables to define, among others, the various processing steps, capitalization, and operating parameters. The techno-economic analysis was based on parameters and efficiency values obtained from pilot-scale experiments, but increased to a basis of 2800 tons of batteries processed per year ($t \cdot y^{-1}$). This capacity represents the actual amount of batteries collected in Canada. Therefore, the capacity plant was developed for 8 tons per day ($t \cdot d^{-1}$), with a running time of 8 h per day, and annual operation of 350 $d \cdot y^{-1}$. Once the capacity of the plant was established, it was possible to adjust the dimensions of the required equipment, according to the specific process. The total cost was established based on variable equations including dimensions and capacity of equipment, purchase costs and transport of equipment, electric and thermal requirements, as well as energy consumption, as recorded in previous studies [23]. Depreciation and annual interest charges were estimated using a 20-year equipment lifetime, as well as a working capital of 15% of fixed capital costs. In addition, used market parameters were defined as follows: An inflation rate of 2%, an annual interest rate of 5%, and an annual discount rate of 6%. Table 2 presents the basic operating parameters taken from the Canadian market and relevant literature [23–25].

Table 2. Basic operating parameters, market parameters, and capitalization parameters of the techno-economic model for recycling spent alkaline batteries.

Parameters	Values	Units
Basic operating parameters	-	-
Operating period	350	d/yr
Processing capacity of a plant	8	t/d
Daily operation period	8	h/d
Factor of safety (for equipment)	20	%
Market parameters	-	-
Annual inflation rate	2.0	%/yr
Annual interest rate	5.0	%/yr
Annual discount rate	6.0	%/yr
Income tax	30	% of gross income
Exchange rate	1.25	$US/$CAD
Chemical Engineering Plant Cost Index	603.1	dec-18

Table 2. *Cont.*

Parameters	Values	Units
Capitalization parameters	-	-
Amortization period	20	yr
Lifetime of equipment	20	yr
Direct (dir.) costs	-	-
Equipment	-	-
Insulation installation equipment	19	%
Instrumentation and control	3	%
Piping and pipeline systems	7	%
Electrical system	8	%
Building process and services	18	%
Landscaping	3	%
Facilities and services	14	%
Indirect (indir.) costs	-	-
Engineering and supervision	32	%
Construction spending	10	%
Construction management fees	9	% cap. (dir. + indir.)
Contingent fees	26	% cap. (dir. + indir.)
Working capital	15	% fixed capital costs

3. Results and Discussion

3.1. Mass Balance for Physical Pre-Treatment

After three stages of attrition, samples were divided into the following two parts: Coarse fraction (larger than 1.7 mm) and fine fraction (smaller than 1.7 mm). The coarse fraction consisted of a ferrous fraction (22%) and non-ferrous fraction (4%), while the fine fraction contained graphite carbon with metallic powder. Water used for the initial washing contained soluble electrolytes, primarily KOH (4%). According to our results, 1924 g of metallic powder (68% of total mass) was recovered from 2833 g of battery waste with a mass balance of 98.9%. During attrition with water, the pH decreased slightly to 12.8 (stage 1), 11.4 (stage 2), and 10.3 (stage 3).

Table 3 presents the average composition of batteries after physical pre-treatment with recirculated water. This distribution corresponds to results from an investigation on the composition of spent AA alkaline batteries by Almeida et al. [26]. Findings from their study reported alkaline batteries to consist of 2.9% plastic and paper, which is equivalent to the non-ferrous fraction of our study; 21.8% metal and brass, which is equivalent to the ferrous fraction; and 75.3% anode and cathode, which is equivalent to KOH and metallic powder. Although slight differences may be attributed to the sampling method and technical separation, these results indicate the high efficacy of washing for obtaining metallic powder.

Table 3. Composition of the battery after separation by attrition and magnetic separation (physical pre-treatment).

Characteristics	Fraction	Mass (g)	Proportion (%)
Initial		2833 ± 29	
>1.7 mm fraction	Ferrous fraction	639 ± 16	22.4 ± 0.6
	Non-ferrous fraction	124 ± 12	4.4 ± 0.4
<1.7 mm fraction	Metallic powder	1924 ± 16	67.5 ± 0.6
	Soluble fraction (KOH, …)	115 ± 6	4.0 ± 0.2

The chemical composition of the resulting metallic powder from the attrition process is shown in Table 4. The quantities of zinc and manganese are 240 and 326 g per kg of metallic powder, respectively. The high concentration of potassium (25.5 g/kg) is associated with the high quantity of alkaline batteries in comparison to Zn–C batteries in the original sample. The collected metal powder was then transferred to the acid leaching process.

Table 4. Chemical composition of metallic powder.

Elements	Na	Fe	Mn	Zn	K	Other
Concentration (g/kg)	0.10	14.5	326	240	25.5	0.96

3.2. Selective Leaching of Zn and Mn

3.2.1. PLS-1 Leachate

Several groups of researchers have studied the extraction of metals from battery powder using sulfuric acid. In our study, during the first leaching process, metallic powder was leached in 2 M H_2SO_4 with S/L ratios of 10, 20, 40, and 50% for 90 min at ambient temperature. Our results indicate that the leaching efficiency of Zn was related to the amount of total solids and reaction time (Figure 3).

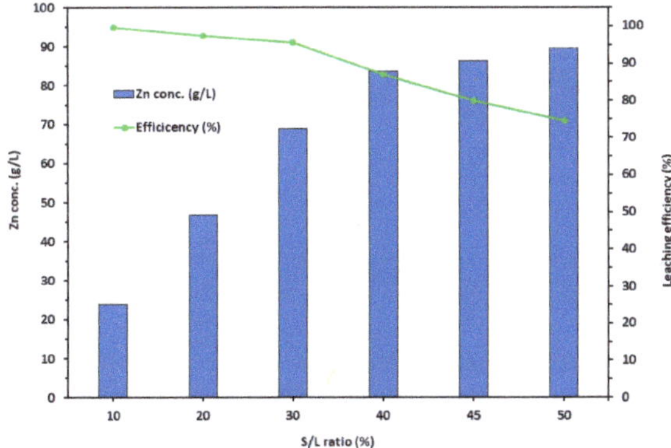

Figure 3. Variation in Zn concentration and the first leaching efficiency (%) with different solid/liquid (S/L) ratios.

The results indicate that an increase in S/L ratio increased leaching up to 90 g Zn/L. Nevertheless, an increase in the S/L ratio also decreased the leaching efficiency. The concentration of Zn in the effluent is an important parameter for electrowinning in this process, and, therefore, an S/L ratio of 40% was chosen for the first cycle of leaching and 20% for the next cycles. A 40% (*w/v*) ratio allowed for the dissolving of 87.1% of Zn with a concentration of 83.6 g/L, after 45 min (Table 5). Under these conditions, the concentration of Mn in PLS-1 was 28.5 g/L (22% of Mn in the battery powder).

Table 5. Composition of metals in different solutions (pregnant leach solution 1 (PLS-1): 2 M H_2SO_4, S/L ratio of 40% and PLS-2: 1.34 M H_2SO_4, S/L ratio of 14.1%, 0.45 g $Na_2S_2O_5$/g metallic powder for 45 min).

Leachate	pH	Conc. of Recovery Metal (g/L)			Conc. of Impurity Metal (mg/L)			
		Mn	Zn	Fe	Cd	Co	Cu	Ni
PLS-1	2.0	28.5 ± 3.5	83.6 ± 0.8	131 ± 13	16.7 ± 2.8	13.4 ± 2.5	42.9 ± 6.1	165 ± 7.4
Fe removal	3.5	29.5 ± 2.3	81.5 ± 1.4	2.99 ± 0.29	14.9 ± 1.7	12.8 ± 1.7	26.6 ± 7.2	153 ± 8.7
Cementation	3.0	31.3 ± 4.0	85.2 ± 2.9	<0.04	<0.04	5.99 ± 0.07	<0.04	21.8 ± 2.1
MnO_2 recovery	2.0	2.03 ± 0.59	83.1 ± 3.0	<0.04	<0.04	<0.04	<0.04	20.7 ± 3.2
Electrowinning	4.0	1.92 ± 0.07	36.1 ± 0.2	<0.04	<0.04	<0.04	<0.04	19.2 ± 5.3
PLS-2	<1	49.1 ± 0.5	5.9 ± 0.3	2754 ± 235	<0.04	2.49 ± 0.29	21.8 ± 2.7	48.4 ± 3.1
Fe removal	4.0	45.2 ± 2.2	5.7 ± 0.5	2.33 ± 0.60	<0.04	2.48 ± 0.31	<0.04	24.1 ± 1.4
Zn removal	4.5	44.6 ± 2.4	<0.04	<0.04	<0.04	<0.04	<0.04	5.50 ± 1.3
$MnCO_3$ recovery	7.3	0.02 ± 0.01	<0.04	<0.04	<0.04	<0.04	<0.04	<0.04

3.2.2. PLS-2 Leachate

Notably, the weight of solids was reduced after the first leaching by the dissolving of metals in PLS-1, with the mass of metal powder decreasing from 400 to 284 g after the first leaching. The addition of $Na_2S_2O_5$ (0.45 g $Na_2S_2O_5$ for 1 g metallic powder) in the second leaching resulted in more than 96% of Mn being dissolved. The S/L ratio had a negative impact on the leaching of Mn with a decrease in the S/L ratio from 35 to 9%, resulting in an increase in Mn extraction rate from 72 to 96%. The S/L ratio of 14.1% made it possible to extract 96.5% of Mn (49.1 g/L). Under these conditions, the concentration of Zn in PLS-2 was 5.87 g/L (a 94.5% solubilization of Zn). Table 5 represents the classification of PLS-1 and PLS-2. Our results confirm the important role of $Na_2S_2O_5$ as the reducing agent for dissolving Mn in the leachate (PLS-2). In an H_2SO_4 solution, the leaching yield of Mn increased from 22 to 96.5% with the addition of $Na_2S_2O_5$. The residue mass decreased from 284 to 140 g after the second leaching.

Table 6 represents the mass distribution for the leaching process for 1 kg of metallic powder and also the global efficiency of the leaching process. The composition of metals (in mg/kg of residue) in residue is also represented in Table 6. The combination of $Na_2S_2O_5$ with sulfuric acid led to the dissolving of not only Zn and Mn, but also other metals (Co, Cu, Ni) of more than 98.4%.

Table 6. Composition of metals in different fractions for each step of leaching.

Samples	Mn	Zn	Fe	Cd	Co	Cu	Ni
Metallic powder (g)	326	240	14.5	0.041	0.045	0.22	0.65
PLS-1 (g)	71.3 ± 8.8	209 ± 2	0.33 ± 0.03	0.042 ± 0.007	0.034 ± 0.006	0.11 ± 0.02	0.41 ± 0.02
PLS-2 (g)	246 ± 3	29.4 ± 1.6	13.8 ± 1.2	<0.0002	0.013 ± 0.002	0.11 ± 0.01	0.24 ± 0.02
Leaching 1 efficiency (%)	21.9 ± 2.7	87.0 ± 0.8	2.3 ± 0.2	101 ± 17	74.3 ± 14.1	48.8 ± 6.9	63.3 ± 2.8
Global leaching efficiency (%)	97.2 ± 2.6	99.3 ± 1.0	97.2 ± 8.3	101 ± 17	102 ± 14	98.4 ± 0.9	101 ± 1.5
Residue (mg/kg)	26.7 ± 2.4	6.8 ± 1.2	4.2 ± 0.5	5.6 ± 3.2	3.2 ± 1.2	20.5 ± 6.9	110 ± 24

3.2.3. Final Residue

A TCLP test was conducted on the final residue in order to validate the effectiveness of the leaching process and residue valorization capacity. Table 7 presents the TCLP results for the residue obtained in our study across three cycles with the recycling of water. All results were below USEPA limits for the potential hazards of waste [27]. The residue obtained after the second leaching is not considered hazardous and could, therefore, be used for other applications, such as a material for the production of batteries.

Table 7. Metal concentration in leachate by toxicity characteristic leaching procedure (TCLP) for final residue.

Elements	US EPA (mg/L)	Residue From		
		LC1	LC2	LC3
As	5.0	<0.01	<0.01	<0.01
Ba	100	0.11	0.19	0.12
Cd	1.0	0.22	0.14	0.47
Cr	5.0	<0.0006	<0.0006	0.01
Pb	5.0	<0.005	<0.005	<0.005
Hg	0.2	0.05	0.01	0.003
Se	1.0	0.07	0.05	0.03
Ag	5.0	0.02	<0.01	<0.01
Ni	250	5.2	3.7	6.9
Cu	250	1.8	0.56	0.19

3.3. Purification and Recovery of Metals

3.3.1. Recovery of MnO_2 from PLS-1 Solution

Purification

Iron precipitation tests were carried out under the same conditions as those used in a study by Blais et al. (2016) with 1.5 times SMR of H_2O_2 for oxidizing Fe^{2+} ions to Fe^{3+} at pH = 4, T = 25 °C, and stirring rate = 200 rpm [28]. Under these conditions, 98% of Fe in the PLS-1 was precipitated. A concentration of 131 mg Fe/L was recorded in the PLS-1 and a 2.99 mg Fe/L was measured after Fe removal (Table 5). In this step, the concentration of Cu decreased from 42.9 to 26.6 mg/L. After iron removal, cementation was used to remove impurities such as Cd, Co, and Ni. The impurities more electropositive than Zn could be removed by adding metal Zn powder and controlling the pH and temperature of the solution [29]. The most efficient removal of Cd (100%), Co (53%), Cu (100%), and Ni (86%) was obtained after 120 min. The low rate of Co removal in our study corresponds to observations made in other studies [29,30]. Furthermore, these studies also mention the difficulty of removing Co due to the small potential difference between Zn and Co–Zn alloys. In addition, the low concentration of cobalt (13.4 mg/L) also makes it more difficult to remove. Notably, Cd and Cu can be removed at lower temperatures (25–40 °C) and at a pH between 2.0 and 2.5. However, a temperature of 80 °C and a pH of 4–5 are necessary for the cementation of Ni [30]. The slight increase in Mn (from 28.5 to 31.3 mg/L) and Zn (83.6 to 85.2 mg/L) concentration can be explained by the evaporation of water during the cementation process at 80 °C.

Precipitation of MnO_2

Across all experiments, 1 SMR of $Na_2S_2O_8$ was used for the oxidation of Mn^{2+} to MnO_2. As suggested by Demopoulos et al. (2002), this reaction was performed at 60 °C and at a controlled pH to investigate the oxidation kinetics of Mn^{2+} ions in MnO_2 at a pH between 2 and 4 [31].

The residual concentration of Mn at pH 2, pH 3, and pH 4 was investigated. According to our results, the oxidation at pH = 3 gave a good oxidation yield (93%). After 120 min, Mn concentration in the solution decreased from 31.3 to 2.03 g/L (Table 8). An amount of 2 g Mn/L in the solution had a positive effect on the next step of Zn electro-deposition [32]. Notably, traces of Co co-precipitated with MnO_2, whereas Ni did not precipitate with MnO_2 (Table 8).

Table 8. Oxidation kinetics of Mn^{2+} to MnO_2 in different pHs.

pH	Temps. (min)	Conc. (g/L)		Conc. (mg/L)	
		Mn	Zn	Co	Ni
	Initial	31.3	85.2	5.99	21.8
2	15	22	85.2	5.87	21.9
	30	19	84.9	5.95	21.8
	60	14	84.9	4.02	21.7
	120	5	83.9	3.09	21.9
3	15	18	84.2	5.13	21.6
	30	16	84.1	4.38	21.9
	60	8	83.6	2.01	21.1
	120	2.03	83.1	ND	20.7
4	15	31.3	84.2	4.35	21.2
	30	24.1	83.7	ND	19.4
	60	3.01	82.5	ND	15.2
	120	ND	79.3	ND	15.1

3.3.2. Electrowinning for Recovery of Zn Metal from PLS-1 Solution

The effect of current density on the electrowinning of Zn was evaluated by measuring the residual concentration of Zn between 250 and 750 A/m^2 for 180 min. The results clearly indicate that the deposition efficiency of Zn increased with an increase in current density. The high concentration of Mn (>5 g/L) in the solution decreased the faradic yield [32], while a lower concentration (between 1 and 5 g/L) had a positive effect on oxidation, thereby protecting the electrodes and increasing the purity of the deposited Zn [32]. At a high current density (750 A/m^2), a decrease in Mn concentration was observed (from 2.03 to 0.22 g/L), which is consistent with the results obtained by Poinsignon et al. [33], where Mn oxidized at the anode and precipitated as MnO_2. The largest Zn deposit was obtained with a current density of 500 A/m^2. Under these conditions, Mn concentration remained stable (from 2.03 to 1.92 g/L).

Figure 4 shows the changes in Zn deposition as a function of time at a current density of 500 mA/m^2. In the figure, two different zones can be distinguished, with the yield of Zn deposits increasing linearly with time until 90 min. After 90 min, the rate of Zn deposits decreased significantly. Notably, at the start of the electrowinning process, Zn concentration was relatively high (83.1 g/L) and, accordingly, the Zn deposit rate was subjected to current control. As the Zn deposit achieved a certain thickness and the Zn concentration was below a certain level (approximately 40 g/L), we conclude that the Zn deposit rate is limited by mass transfer control. This explains why the Zn deposit rate remained constant with time. Figure 4 also presents the change in energy consumption as a function of time, indicating that energy consumption increased linearly to 156 kWh/m^3 after 180 min. It has been established that Zn deposition efficiency is affected by reaction time and the cost of the electrowinning process. In view of reducing power use and increasing Zn deposits, a time of 90 min was selected for the process. Under these conditions, the residual concentration of Zn in the solution was 36.1 g/L. The solution was also recirculating for the first leaching (S/L ratio of 40%) and, consequently, an S/L ratio of 20% was used for the second cycle.

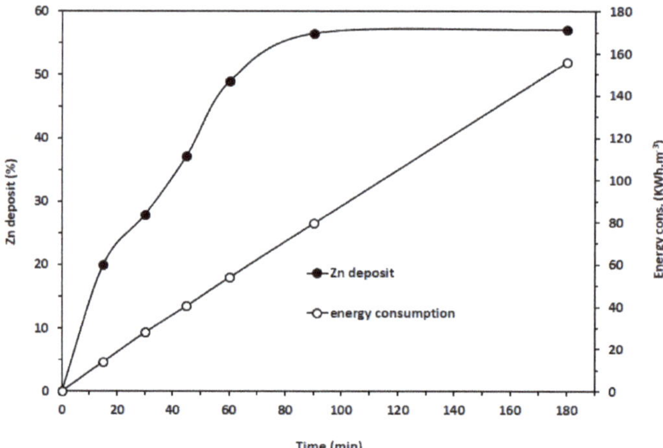

Figure 4. Variation in residual Zn concentration and effect of energy consumption as a function of time applying a current density of 500 A/m^2.

3.3.3. Recovery of $MnCO_3$ from PLS-2 Solution

Using the same principle, iron precipitation assays were carried out with 1.5 SMR of H_2O_2 for the oxidation of Fe^{2+} ions to Fe^{3+} at pH = 4 and ambient temperature. Within this context, nearly 99.9% of Fe in PLS-2 was precipitated. A concentration of 2754 mg Fe/L was recorded in PLS-2 and a concentration of 2.33 mg Fe/L was measured after Fe removal (Table 5). In this step, the concentration of Co remained stable, while all Cu was removed and the concentration of Ni decreased from 48.4 to 24.1 mg/L.

To obtain a pure manganese solution, zinc precipitation was achieved by adding Na_2S (2 SMR of Na_2S at pH 4–5). Under these conditions, up to 100% of zinc was precipitated (Table 5) and the obtained ZnS could be added to the first leaching stage for PLS-1 rich in Zn.

The precipitation of $MnCO_3$ was achieved by the addition of Na_2CO_3. The quantity of Na_2CO_3 was calculated for 1 SMR, with pH 4.5–5.0 and at ambient temperature. After rinsing, the collected $MnCO_3$ had a purity of 95%.

3.4. Economic Evaluation

Table 9 presents the direct and indirect costs, as well as profit for a basic plant that recycles 8 tons of batteries per day. In this simulation, the cost of shredding was not considered. However, the non-magnetic fraction (plastic, carton, etc.) and the final residue were considered waste products that would need to be transported over 50 km with a disposal cost of $CAD 75 per ton of residue. This cost included loading, transport, and landfill calculated at $CAD 28.2 per ton of batteries. In addition to chemical products, labor costs and utilities costs were the two other important parameters, from an economic point of view, which accounted for 18.5 and 10.0% of the total direct costs, respectively.

A total cost of $CAD 1260 per ton was calculated, of which the direct and indirect costs were calculated as $CAD 725.7 and $CAD 534.5 per ton, respectively. Regarding the cost simulation, cost distribution was dominated by indirect costs including amortization (9.6%), financing (15.8%), and marginal social benefits (1.5%). The revenue was calculated as $CAD 1359.6 per ton, whereas Zn, MnO_2, and $MnCO_3$ values were estimated as $CAD 527.4, $CAD 451.6, and $CAD 357.8 per ton, respectively.

Figure 5 illustrates the variation in cost as a function of the recovery capacity of the plant. Indirect costs showed an influence on total cost, and costs decreased with the increase in recovery capacity. Total costs were evaluated to be $CAD 1015 and $CAD 913 per ton for treatment capacities of 16 and 24 t/day, respectively, equivalent to 40 and 60% of batteries sold in the Canadian market.

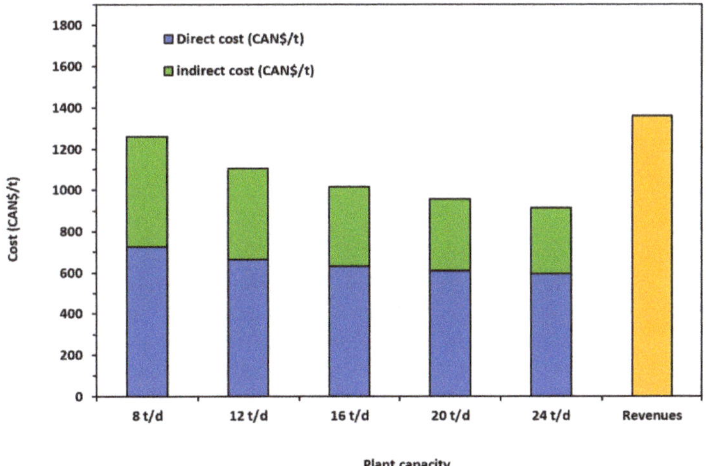

Figure 5. Exploitation costs and revenues depending on the processing capacity of the recovery plant.

Table 9. Parameters related to hydrometallurgical recovery in spent alkaline battery.

Parameters	Values	Units	Cost/profit (\$CAD/t)
Direct operating costs	-	-	−725.7
Chemicals	-	-	-
H_2SO_4	0.21	\$CAD/kg	−129.0
NaOH	0.35	\$CAD/kg	−89.3
Na_2CO_3	0.21	\$CAD/kg	−14.7
H_2O_2	1.60	\$CAD/kg	−15.6
$Na_2S_2O_8$	0.30	\$CAD/kg	−18.6
$Na_2S_2O_5$	0.31	\$CAD/kg	−86.9
Na_2S	0.90	\$CAD/kg	−45.7
Labor	-	-	-
Unit cost	25.0	\$CAD/h	−115.9
Supervision	20.0	% (labor cost)	−23.2
Utilities			
Unit cost of electricity	0.07	\$CAD/kWh	−27.5
Unit cost of water process	0.50	\$CAD/m^3	−2.6
Unit cost of fuel	3.50	\$CAD/M Btu	−20.7
Residue management	-	-	
Loading and transport cost	2.50	\$CAD/t	−3.3
Transportation cost (estimate of 50 km)	0.15	\$CAD/t/km	-
Unit cost of landfill or treatment	75.0	\$CAD/t	−24.9
Maintenance and repairs	2.00	% fixed capital costs/yr	−66.1
Current materials	0.75	% fixed capital costs/yr	−24.8
Laboratory charges	10	% operating labor	−11.6
Patents and royalties	5.00	\$CAD/t	−5.0
Indirect and General costs			−534.5
Marginal social benefits	22.0	% operating labor + supervision	−30.6
Amortization	-	-	−189.9
Financing (interest redemption)	-	-	−314.0
Metal revenues	-	-	1359.6
Fe	0.10	\$CAD/kg	22.8
134 kg of MnO_2 (85.6 kg Mn)	3.37	\$CAD/kg	451.6
286 kg of $MnCO_3$ (136.5 kg Mn)	1.25	\$CAD/kg	357.8
162 kg of Zn metal	3.25	\$CAD/kg	527.4
Profit	-	-	98.3

4. Conclusions

In this study, we established a pilot process for the recycling of metals in spent alkaline batteries, including physical separation techniques and hydrometallurgical steps. Spent batteries were crushed and attrited to separate out and collect the metallic powder. Metallic powder was then treated using sulfuric acid leaching. The first leaching for zinc and a part of manganese included the use of 2 M H_2SO_4 for 45 min at ambient temperature and a solid content of 40% (*w/v*). Under these conditions, up to 87% of zinc was solubilized. In the second leaching step, 0.45 g $Na_2S_2O_5$ was added per 1 g metallic powder in 1.34 M H_2SO_4, resulting in up to 97% of manganese being dissolved. The dissolved manganese was recovered by precipitation in the form of MnO_2 and $MnCO_3$, whereas zinc was recovered in metallic form by electrowinning. Notably, the leaching efficacy remained high, even after three cycles of processing with recirculated water. Furthermore, results from our technical and economic simulation indicate that the suggested process for recycling spent alkaline batteries is economically competitive and feasible. It is necessary to highlight that the profits from the process greatly depend on the market price of zinc, manganese (IV) oxide, and manganese carbonate. However, despite economic challenges, the process of recycling alkaline batteries reduces negative environmental impacts.

Author Contributions: Conceptualization, J.-F.B., G.M., L.-H.T. and K.T.; methodology, K.T., A.D.J.; software, J.-F.B.; validation, J.-F.B., G.M., L.-H.T. and K.T.; formal analysis, K.T. and A.D.J.; investigation, K.T. and A.D.J.; data curation, A.D.J. and K.T.; writing—original draft preparation, L.-H.T.; writing—review and editing, L.-H.T, K.T., J.-F.B. and G.M.; supervision, J.-F.B. and G.M.; project administration, J.-F.B.; funding acquisition, J.-F.B. All authors have read and agreed to the published version of the manuscript.

Funding: This research was funded by Natural Sciences and Engineering Research Council of Canada, grant number RGPIN-2019-05767.X.

Acknowledgments: The authors thank Call2recycle for samples supporting.

Conflicts of Interest: The authors declare no conflict of interest.

References

1. Leite, D.D.S.; Carvalho, P.L.G.; de Lemos, L.R.; Mageste, A.B.; Rodrigues, G.D. Hydrometallurgical recovery of Zn(II) and Mn(II) from alkaline batteries waste employing aqueous two-phase system. *Sep. Purif. Technol.* **2019**, *210*, 327–334. [CrossRef]

2. Ye, G.; Magnusson, M.; Väänänen, P.; Tian, Y. Recovery of Zn and Mn from spent alkaline batteries. In *The Minerals, Metals and Materials Series*; Springer International Publishing: Berlin, Germany, 2018; Volume F10, pp. 329–341.

3. Abid Charef, S.; Affoune, A.M.; Caballero, A.; Cruz-Yusta, M.; Morales, J. Simultaneous recovery of Zn and Mn from used batteries in acidic and alkaline mediums: A comparative study. *Waste Manag.* **2017**, *68*, 518–526. [CrossRef]

4. Yuliusman; Amiliana, R.A.; Wulandari, P.T.; Ramadhan, I.T.; Kusumadewi, F.A. Selection of organic acid leaching reagent for recovery of zinc and manganese from zinc-carbon and alkaline spent batteries. *IOP Conf. Ser. Mater. Sci. Eng.* **2018**, *333*, 012041. [CrossRef]

5. Sayilgan, E.; Kukrer, T.; Civelekoglu, G.; Ferella, F.; Akcil, A.; Veglio, F.; Kitis, M. A review of technologies for the recovery of metals from spent alkaline and zinc-carbon batteries. *Hydrometallurgy* **2009**, *97*, 158–166. [CrossRef]

6. Tanong, K.; Coudert, L.; Chartier, M.; Mercier, G.; Blais, J.F. Study of the factors influencing the metals solubilisation from a mixture of waste batteries by response surface methodology. *Environ. Technol.* **2017**, *38*, 3167–3179. [CrossRef]

7. Formánek, J.; Jandová, J.; Sís, J. A review of hydromet-allurgical technologies for the recovery of Zn and Mn from spent alkaline and zinc batteries. *Chem. Listy* **2012**, *106*, 350–356.

8. Tanong, K.; Tran, L.-H.; Mercier, G.; Blais, J.-F. Recovery of Zn (II), Mn (II), Cd (II) and Ni (II) from the unsorted spent batteries using solvent extraction, electrodeposition and precipitation methods. *J. Clean. Prod.* **2017**, *148*, 233–244. [CrossRef]

9. Canada, C.R. Canadians Are Going Green: Call2Recycle Canada Diverts Record Number Of Batteries in 2018. 2019. Available online: https://www.call2recycle.ca/canadians-are-going-green-call2recycle-canada-diverts-record-number-of-batteries-in-2018/ (accessed on 14 February 2020).

10. Recyc-Quebec. Piles et batteries. In Fiches D'information 2019. Available online: https://www.recyc-quebec.gouv.qc.ca/sites/default/files/documents/Fiche-info-piles.pdf (accessed on 14 February 2020).

11. Canada, C.R. Explore the Secret Life of Batteries. 2018. Available online: https://www.call2recycle.ca/explore-the-secret-life-of-batteries/ (accessed on 10 February 2018).

12. Veloso, L.R.S.; Rodrigues, L.E.O.C.; Ferreira, D.A.; Magalhães, F.S.; Mansur, M.B. Development of a hydrometallurgical route for the recovery of zinc and manganese from spent alkaline batteries. *J. Power Sources* **2005**, *152*, 295–302. [CrossRef]

13. Chen, W.-S.; Liao, C.-T.; Lin, K.-Y. Recovery Zinc and Manganese from Spent Battery Powder by Hydrometallurgical Route. *Energy Procedia* **2017**, *107*, 167–174. [CrossRef]

14. Maryam Sadeghi, S.; Vanpeteghem, G.; Neto, I.F.F.; Soares, H.M.V.M. Selective leaching of Zn from spent alkaline batteries using environmentally friendly approaches. *Waste Manag.* **2017**, *60*, 696–705. [CrossRef] [PubMed]

15. Vellingiri, K.; Tsang, D.C.W.; Kim, K.H.; Deep, A.; Dutta, T.; Boukhvalov, D.W. The utilization of zinc recovered from alkaline battery waste as metal precursor in the synthesis of metal-organic framework. *J. Clean. Prod.* **2018**, *199*, 995–1006. [CrossRef]

16. Gallegos, M.V.; Peluso, M.A.; Sambeth, J.E. Preparation and Characterization of Manganese and Zinc Oxides Recovered from Spent Alkaline and Zn/C Batteries Using Biogenerated Sulfuric Acid as Leaching Agent. *JOM* **2018**, *70*, 2351–2358. [CrossRef]

17. Furlani, G.; Moscardini, E.; Pagnanelli, F.; Ferella, F.; Vegliò, F.; Toro, L. Recovery of manganese from zinc alkaline batteries by reductive acid leaching using carbohydrates as reductant. *Hydrometallurgy* **2009**, *99*, 115–118. [CrossRef]

18. Petranikova, M.; Ebin, B.; Mikhailova, S.; Steenari, B.M.; Ekberg, C. Investigation of the effects of thermal treatment on the leachability of Zn and Mn from discarded alkaline and Zn–C batteries. *J. Clean. Prod.* **2018**, *170*, 1195–1205. [CrossRef]

19. Sobianowska-Turek, A.; Szczepaniak, W.; Maciejewski, P.; Gawlik-Kobylińska, M. Recovery of zinc and manganese, and other metals (Fe, Cu, Ni, Co, Cd, Cr, Na, K) from Zn-MnO2 and Zn-C waste batteries: Hydroxyl and carbonate co-precipitation from solution after reducing acidic leaching with use of oxalic acid. *J. Power Sources* **2016**, *325*, 220–228. [CrossRef]

20. Sayilgan, E.; Kukrer, T.; Yigit, N.O.; Civelekoglu, G.; Kitis, M. Acidic leaching and precipitation of zinc and manganese from spent battery powders using various reductants. *J. Hazard. Mater.* **2010**, *173*, 137–143. [CrossRef]

21. Ivanov, I. Increased current efficiency of zinc electrowinning in the presence of metal impurities by addition of organic inhibitors. *Hydrometallurgy* **2004**, *72*, 73–78. [CrossRef]

22. Gasper, P.; Hines, J.; Miralda, J.P.; Bonhomme, R.; Schaufeld, J.; Apelian, D.; Wang, Y. Economic feasibility of a mechanical separation process for recycling alkaline batteries. *J. New Mater. Electrochem. Syst.* **2013**, *16*, 297–304. [CrossRef]

23. Metahni, S.; Coudert, L.; Blais, J.-F.; Tran, L.H.; Gloaguen, E.; Mercier, G.; Mercier, G. Techno-economic assessment of an hydrometallurgical process to simultaneously remove As, Cr, Cu, PCP and PCDD/F from contaminated soil. *J. Environ. Manag.* **2020**, *263*, 110371. [CrossRef]

24. Ulrich, G.D. *A Guide to Chemical Engineering Process Design and Economics*; John Wiley & Sons: New York, NY, USA, 1984.

25. Peters, M.S.; Timmerhaus, K.D. *Plant Design and Economics for Chemical Engineers*, 4th ed.; McGraw-Hill: New York, NY, USA, 1991.

26. Almeida, M.F.; Xará, S.M.; Delgado, J.; Costa, C.A. Characterization of spent AA household alkaline batteries. *Waste Manag.* **2006**, *26*, 466–476. [CrossRef]

27. USEPA. Hazardous Waste Charateristics. 2009. Available online: https://www.epa.gov/sites/production/files/2016-01/documents/hw-char.pdf (accessed on 10 February 2018).

28. Blais, J.F.; Mercier, G.; Tanong, K.; Tran, L.H.; Coudert, L. Method for recycling Valuable Metals from Spent Batteries. U.S. Patent 20170170532A1, 15 June 2017.

29. Bøckman, O.; Østvold, T. Products formed during cobalt cementation on zinc in zinc sulfate electrolytes. *Hydrometallurgy* **2000**, *54*, 65–78. [CrossRef]

30. Safarzadeh, M.S.; Moradkhani, D.; Ashtari, P. Recovery of zinc from Cd–Ni zinc plant residues. *Hydrometallurgy* **2009**, *97*, 67–72. [CrossRef]

31. George, P.; Demopoulos, L.R.; Wang, Q. Method for Removing Manganese from Acidic Sulfate Solution. U.S. Patent 6391270B1, 21 May 2002.

32. Zhang, Z.-Y.; Zhang, F.-S.; Yao, T. An environmentally friendly ball milling process for recovery of valuable metals from e-waste scraps. *Waste Manag.* **2017**, *68*, 490–497. [CrossRef]

33. Poinsignon, C.-J.-L.; Tedjar, F. Method for Electrolytical Processing of Used Batteries. European Patent 0620607A1, 19 October 1994.

 *metals* 

Article

Selective Recovery of Molybdenum over Rhenium from Molybdenite Flue Dust Leaching Solution Using PC88A Extractant

Ali Entezari-Zarandi [1,2,*], **Dariush Azizi** [1,3], **Pavel Anatolyevich Nikolaychuk** [4], **Faïçal Larachi** [1] **and Louis-César Pasquier** [3]

[1] Department of Chemical Engineering, Université Laval, Québec, QC G1V 0A6, Canada; Dariush.Azizi@ete.inrs.ca (D.A.); faical.larachi@gch.ulaval.ca (F.L.)

[2] Centre Technologique des Résidus Industriels (CTRI), Rouyn-Noranda, QC J9X 0E1, Canada

[3] Center Eau Terre Environnement, Institut National de la recherche scientifique (INRS), Québec, QC G1K 9A9, Canada; louis-cesar.pasquier@inrs.ca

[4] Lehrstuhl für Thermodynamik und Energietechnik, Universität Paderborn, Warburger Straße 100, 33098 Paderborn, Germany; npa@csu.ru

* Correspondence: Ali.Entezarizarandi@cegepat.qc.ca; Tel.: +1-819-762-0931 (ext. 1737)

Received: 31 August 2020; Accepted: 23 October 2020; Published: 26 October 2020

Abstract: Selective solvent extraction of molybdenum over rhenium from molybdenite (MoS_2) flue dust leaching solution was studied. In the present work, thermodynamic calculations of the chemical equilibria in aqueous solution were first performed, and the potential–pH diagram for the $Mo–Re–SO_4^{2-}–H_2O$ system was constructed. With the gained insight on the system, 2-ethylhexyl phosphonic acid mono-(2-ethylhexyl)-ester (PC88A) diluted in kerosene was used as the extractant agent. Keeping constant the reaction temperature and aqueous-to-organic phase ratio (1:1), organic phase concentration and pH were the studied experimental variables. It was observed that by increasing the acidity of the solution and extractant concentration, selectivity towards Mo extraction increased, while the opposite was true for Re extraction. Selective Mo removal (+95%) from leach solution containing ca. 9 g/L Mo and 0.5 g/L Re was achieved when using an organic phase of 5% PC88A at pH = 0. No rhenium was coextracted during 10 min of extraction time at room temperature. Density functional theory (DFT) calculations were performed in order to study the interactions of organic extractants with Mo and Re ions, permitting a direct comparison of calculation results with the experimental data to estimate selectivity factors in Mo–Re separation. For this aim, PC88A and D2EHPA (di-(2-ethylhexyl) phosphoric acid) were simulated. The interaction energies of D2EHPA were shown to be higher than those of PC88A, which could be due to its stronger capability for complex formation. Besides, it was found that the interaction energies of both extractants follow this trend considering Mo species: $MoO_2^{2+} > MoO_4^{2-}$. It was also demonstrated through DFT calculations that the interaction energies of D2EHPA and PC88A with species are based on these trends, respectively: $MoO_2^{2+} > MoO_4^{2-} > ReO_4^-$ and $MoO_2^{2+} > ReO_4^- > MoO_4^{2-}$, in qualitative agreement with the experimental findings.

Keywords: rhenium; molybdenum; solvent extraction; separation; hydrometallurgy

---

1. Introduction

Molybdenum (Mo) is a strategic metal that has an extensive demand in different branches of the industry. Rhenium (Re) is also a strategic metal, although less common with wide applications in the oil industry (e.g., production of reforming catalyst) and in heat-resistant alloys (e.g., aerospace). Typically, in Mo sulphide concentrates (molybdenite, MoS_2), Re coexisting with varying concentrations ranging from 0.001% to 0.1% is identified [1]. Upon roasting, molybdenite transforms into technical-grade Mo oxide, while Re content escapes the reactor in the form of Re_2O_7, which partly deposits in the filters alongside the flue dust. Scrubbing the flue gases and leaching flue dusts are effective methods to recover Re values. Such solutions typically have 5–10 g/L Mo and 0.4–0.9 g/L Re. Due to similar chemical properties, Mo–Re separation is a challenge, and a number of studies have been devoted to introducing ad hoc technologies in this regard. Ion exchange and solvent extraction are the most used separation methods [2]. However, the separation of Mo and Re from liquors obtained from leaching molybdenite roasting flue dust is an essential step in order to produce high-purity final products, such as ammonium perrhenate and ammonium (para) molybdate. The following hydrolysis and acidic reactions illustrate the reactants/products involved in the leaching of flue dust [3,4]:

$$MoO_3 \ (s) + H_2O \ (l) \ \rightleftharpoons \ H_2MoO_4 \ (aq) \quad \Delta_r G° = -31.9 \ kJ/mol \tag{1}$$

$$H_2MoO_4 \ (aq) \ + 2H^+ \ (aq) \ \rightleftharpoons \ MoO_2^{2+} \ (aq) \ + 2H_2O \ (l) \quad \Delta_r G° = -2.79 \ kJ/mol \tag{2}$$

$$Re_2O_7 \ (s) \ + \ H_2O \ (l) \ \rightleftharpoons \ 2ReO_4^- \ (aq) \ + \ 2H^+ \ (aq) \quad \Delta_r G° = -64.4 \ kJ/mol \tag{3}$$

$$ReO_4^- \ (aq) \ + \ H^+ \ (aq) \rightleftharpoons \ HReO_4 \ (aq) \quad \Delta_r G° = -1465.8 \ kJ/mol \tag{4}$$

At high acid concentration, the anionic oxyspecies of rhenium (i.e., perrhenate ion (ReO_4^-)) is highly stable, and other possible short-life species readily hydrolyse to it. On the other hand, molybdenum forms cationic oxyspecies (MoO_2^{2+}) at high acid concentration, evolving into neutral and anionic species with increasing pH [5].

Many attempts have been made on Mo and Re separation. Solvating extractant TBP (tributyl phosphate) is used for Re removal in near-zero pH, followed by Mo removal employing commercial extractant LIX984N [6]. Stepwise removal of Mo with TBP was conducted at pH near 2, and afterwards, Re was removed at pH lower than 0 [7]. Similarly, the coextraction of molybdenum and rhenium by N235 (tri-octyl amine) and their separation from stripping solution by using D201 ion-exchange resin (containing quaternary ammonium group [N-$(CH_3)_2C_2H_4OH$]) has been reported [8]. Selective extraction of rhenium over molybdenum from alkaline solutions has also been studied by employing an organic phase composed of 20% N235 and 30% TBP diluted in kerosene [9]. An organic phase composed of 5 vol% N235 (in kerosene) was found to selectively recover Re over Mo at equilibrium pH 0.0 [10]. Table 1 lists recent attempts to separate Re and Mo from their aqueous mother liquor.

The use of appropriate solvents is a challenging problem. The commercially available extractant Cyanex 923 has been shown to be an appropriate choice for the purpose of rhenium recovery [11]. The lower solubility of Cyanex 923 in water compared with that of TBP (0.05 to 0.4 g/L at 25 °C, respectively) and its complete miscibility with diluents at low temperature are mentioned as some of its advantages over other solvents such as TOPO (trioctylphosphine oxide) and Aliquat 336. Pathak et al. [12] studied the extraction behaviour of molybdenum from acidic radioactive wastes using PC88A. They found that by increasing HNO_3 concentration in the aqueous phase, Mo extraction decreases, while increasing the organic concentration until 0.15 M causes an increase in metal extraction.

In the present work, the application of PC88A (2(ethylhexyl)phosphonic acid mono-2(ethylhexyl)-ester) extractant was studied on Mo–Re separation in solutions obtained from leaching molybdenite flue dust under various conditions of organic concentration and aqueous solution acidity. Next, the performance of PC88A was compared with that of D2EHPA using both experimentation and density functional theory simulations.

Table 1. Literature on Mo and Re solvent extraction.

Extractant/Diluent	Acidity	Selectivity	Ref.
N235 + TBP/kerosene	pH = 9	Re over Mo 15 g/L Mo + 0.1 g/L Re 97.6%: 1.6%	[9]
N235/kerosene	pH = 0	Re over Mo	[10]
Cyanex 923/kerosene	pH = 0 HCl	Re	[11]
N235 + isooctanol/kerosene	pH = 0 HNO_3	Re over Mo	[13]
LIX 63/kerosene	pH = 2–6 H_2SO_4	Mo over W	[14]
D2EHPA/kerosene	pH = 3–4 H_2SO_4	Mo over W	[15]
TBP/kerosene	pH = 2 pH = 0 H_2SO_4	Mo over ReRe over Mo	[7]
TBP/kerosene	pH = 0 <3.0 M HCl	Re over Mo and V Re and Mo over V	[16]
TBP/kerosene	pH = 1.5 pH = −0.3 H_2SO_4	Mo over Re Re over Mo	[1]
Alamine 304-1/Anysol-150	pH = 2–3 H_2SO_4	Re over Mo 260–280 mg/L Re + 80–90 mg/L Mo	[17]
PC88A/*n*-dodecane	0.1–4.0 M HNO_3	Mo 0.01 mol/L	[12]
PC88A/Sulfonated kerosene	pH = −0.2 ~ 0.5 HCl	Mo	[18]

N235 = tri-octyl amine, TBP = tributyl phosphate, Cyanex 923 = trialkylphosphine oxide, LIX 63 = 5,8-diethyl-7-hydroxy-6-dodecanone oxime, D2EHPA = di-(2-ethylhexyl)phosphoric acid, Alamine 304-1 = tri-n-dodecyl amine, PC88A = 2(ethylhexyl)phosphonic acid mono-2(ethylhexyl)-ester.

2. Experimentation

2.1. Materials and Reagents

Babakan Ferromolybdenum Co. (Kerman, Iran) kindly provided molybdenite flue dust. Table 2 lists the main chemical composition of the flue dust sample. Deionised water (industrial grade) was used as the leaching agent. D2EHPA and PC88A extractants were analytical-grade products kindly provided by Farapoyan Isatis Co., Yazd, Iran. Kerosene (Tehran Refinery, Tehran, Iran) was used as diluent, and sulphuric acid and sodium hydroxide (Merck) were the pH-adjusting agents used in our protocols. An iron-rich copper solvent extraction raffinate sample (NICICO, Tehran, Iran) was also used to investigate the possibility of selective separation of Mo from present metal impurities.

Table 2. Composition of flue dust before leaching (ppm).

Component	As	Ca	Cu	Fe	Mg	Mo	Na	Pb	Re	S	Se	Zn
Content	785	3670	3450	1720	775	36.6%	2380	1260	3980	25.2%	6650	242

2.2. Flue Dust Leaching

Molybdenite flue dust leaching was carried out in a 5 L glass reactor equipped with a mechanical agitation system (600 rpm) and a water jacket. The pulp density and reaction temperature were 20 wt.% and 80 °C, respectively. A mixture of water and alcohols (e.g., ethyl or methyl alcohol) was found to help in the selective leaching of rhenium over molybdenum [19]. Selective leaching has the advantage of separation of desired values from the very first steps of hydrometallurgical treatment [20]. However, for the sake of simplicity in the current work, only deionised water of industrial quality was used to eliminate the probable effect of additives on the solvent extraction step. It is worth noting that the application of acidic medium for leaching is unfavourable due to excessive introduction of Mo and other impurities, such as Fe, into the solution [21]. The leach liquor samples (5 mL without compensation) were withdrawn at predetermined intervals and immediately filtered through a medium quantitative filter paper to be analysed for Re and Mo content after required dilution with deionised water. The pH was monitored over the filtered samples in order to understand its variations during the leaching process (Figure 1a). After 360 min of contact, the remaining hot solution was filtered under gentle vacuum and cooled down to room temperature in order to obtain the stock solution. The solution underwent a series of colour changes in the course of leaching (Appendix A, Figure A1). It first became pale yellow, then turned into green, and finally dark blue. Metal recovery and pH variation profiles are presented in Figure 1b. Stock solution composition is listed in Table 3.

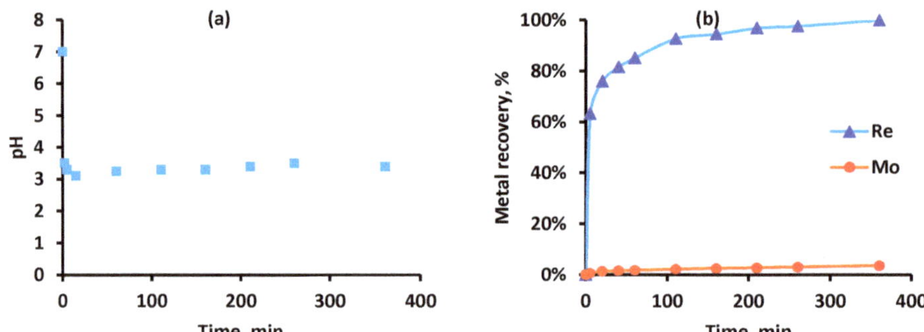

Figure 1. Dissolution of metal oxides from flue dust: (**a**) pH variations and (**b**) metal recovery over time. T = 85 °C, t = 360 min, 600 rpm.

Table 3. Chemical composition of stock solution.

Element	Mo	Re	Cu	Se	Fe
Concentration (ppm)	9150	455	2.5	6.2	2000

2.3. Experimental Procedure

Different concentrations of PC88A diluted in kerosene (i.e., 5, 15, and 30 vol%) were used to test the influence of the extractant concentration on Re–Mo extraction and separation. Likewise, various pH classes of stock solution (i.e., −1, 1, 0, 3, 7, and 9) were prepared using sulphuric acid and sodium hydroxide. All the solvent extraction experiments were performed at room temperature. For all extraction experiments, an organic-to-aqueous ratio of 1:1 (25 mL for each) was imposed. Good contacting between phases was achieved during 10 min of mixing under magnetic stirring (600 rpm) with a cross-shaped magnet (3 cm in diameter) in 250 mL capacity beakers. Agitated mixtures were then transferred to a separation funnel and retained there for another 10 min before aqueous phase separation and chemical analysis. However, the phase separation time was measured to be in the order of 20 to 30 s. It is worth noting that for the case of pH values of 7 and 9, the solution was filtered before the experiment to separate formed iron precipitates.

2.4. Chemical Analysis

An inductively coupled plasma optical emission spectrometer (ICP-OES) was used to determine the Re and Mo concentrations in all aqueous solutions. The Cu and Fe contents were measured by atomic absorption spectrometry (AAS) where necessary. The samples were treated prior to analysis by the addition of appropriate amounts of nitric acid, followed by dilutions to a predetermined volume.

2.5. Thermodynamic Analysis of Equilibria in Aqueous Solution

Thermodynamic calculations were performed in order to identify the equilibrium aqueous species in the leaching stock solutions. According to Table 3, the total contents of molybdenum and rhenium in the solution are equal to ~9 g/L (9.4×10^{-2} mol/L) and ~0.5 g/L (2.7×10^{-3} mol/L), respectively. The calculations were performed at pH values of 2, 1, 0, and −1, which correspond to the total content of sulphuric acid ranging from 6.5×10^{-3} to 10.09 mol/L. The concentrations of different dissociation products of sulphuric acid were calculated as presented in Appendix B and used to calculate the ionic strength of the solutions.

The activity coefficients of both ReO_4^- and MoO_2^{2+} ions were calculated from the extended Debye–Hückel theory [22]. The values of the effective radii of the ions and the values for water dielectric constant were taken from references [23,24], respectively. The calculation details are presented in Appendix B. As can be seen, the average activities of molybdenum species are ~0.01 mol/L, and those of rhenium species are ~0.001 mol/L. The activities of sulphur species have no effect on the chemical equilibria.

The chemical and electrochemical equilibria in the leaching stock solutions were presented in the form of convenient potential–pH diagrams. The diagrams for molybdenum [4] and sulphur [25] were constructed earlier. The thermodynamic characteristics of the reactions for rhenium were calculated using data from [26].

The potential–pH diagram for the $Mo-Re-SO_4^{2-}-H_2O$ system is plotted at 25 °C, air pressure of 1 bar and the activities of the molybdenum species 0.01 mol/L, the activities of the rhenium species 0.001 mol/L, and the activities of the sulphur species 0.1 mol/L, and presented in Figure 2.

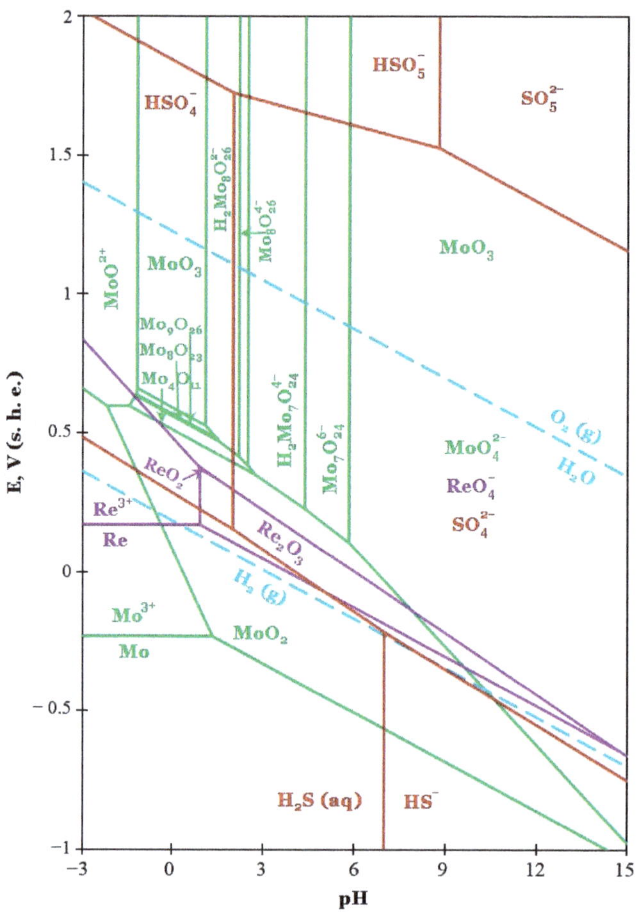

Figure 2. The potential–pH diagram for the Mo–Re–SO$_4^{2-}$–H$_2$O system is plotted at 25 °C, air pressure of 1 bar and the activities of molybdenum, rhenium, and sulphur species equal to 0.01, 0.001, and 0.1 mol/L, respectively.

2.6. DFT Computational Details

Density functional theory (DFT)-based simulations were conducted using Dmol3 module implemented in Material Studio 2016 software package. The structures of nonsolvated and water-solvated forms of MoO$_4^{2-}$, MoO$_2^{2+}$, and ReO$_4^-$ were optimised in the aqueous pregnant solution. For the interactions in the organic phase, only the optimised structures of MoO$_4^{2-}$, MoO$_2^{2+}$, and ReO$_4^-$, deprived of their inner-sphere water molecules, were considered [26]. Similarly, the structures of PC88A and D2EHPA were optimised in the organic phase. The complexation (or interaction) energies in the aqueous (E$_{C/w}$) and organic (E$_{C/o}$) phases of the complex species were determined as follows [27]:

$$E_{C/w} = E_{Species + Reagent/w} - E_{Species/w} - E_{Reagent/o} \tag{5}$$

$$E_{C/o} = E_{Species + Reagent/o} - E_{Species/o} - E_{Reagent/o} \tag{6}$$

where E$_{Species+Reagent/w}$ is the total energy after complex formation in the aqueous phase between water-solvated species aqua-complexes and two reagent (PC88A and D2EHPA) species; E$_{Species+Reagent/o}$

is the total energy after complex formation between species and two reagent (PC88A and D2EHPA) species in the organic phase; $E_{Species/w}$ is the energy of optimised solvated MoO_4^{2-}, MoO_2^{2+}, and ReO_4^{-} aqua-complexes; $E_{Species/o}$ is the energy of optimised MoO_4^{2-}, MoO_2^{2+}, and ReO_4^{-} species in the organic phase; and $E_{reagent/o}$ is the energy of solvated PC88A and D2EHPA optimised in the organic phase. Note the "w" and "o" index solvent parameterizations in Equations (5) and (6) for the conductor-like screening model (COSMO) used as an implicit solvation model to account for the aqueous, interfacial, and organic environments of the simulated structures. In this regard, dielectric constant was considered to be 78.54 (water) [24] and 1.8 (kerosene) for simulations in the aqueous phase and the organic phase, respectively.

The generalised gradient approximation (GGA) with Perdew-Burke-Ernzerhof exchange-correlation density functional (PBEsol) was used to describe the exchange correlation interactions. The double numerical plus polarization (DNP) basis was selected. The self-consistent field (SCF) convergence was fixed to 2×10^{-6} (0.005 kJ/mol), and the convergence criteria for the energy, maximum force, and maximum displacement were set to 2×10^{-5} Ha (0.05 kJ/mol), 0.05 Ha/Å, and 0.002 Å, respectively. No special treatment of core electrons was considered, and all the electrons were included in the calculations. In addition, a smearing value was fixed at 5×10^{-3} through calculation. In the spin-unrestricted condition, the calculation was performed by the use of various orbitals for different spins. Besides, the initial value for the number of unpaired electrons for each atom was taken from the formal spin introduced for each atom. In this situation, the starting value can be subsequently optimised throughout the calculations. Maximum SCF interactions and calculation interactions were set at 2000 and 1000, respectively, and calculation of the interactions step was set to 0.3 Å. It is worth mentioning that different initial positions were considered for all cases during DFT simulations, and only the most stable configuration and results have been reported. Besides, it is worthy to mention that similar studies in rhenium/ molybdenum solvent extraction have not been conducted based on the authors' best of knowledge.

3. Results and Discussion

3.1. Effects of pH and Organic Phase Concentration

A potential–pH equilibrium diagram for the $Mo–Re–SO_4^{2-}–H_2O$ system is presented in Figure 2. It can be inferred that soluble perrhenate (ReO_4^{-}) ion is the dominant species over the whole pH range. However, for the case of Mo, several cationic and anionic species may be present depending on the solution pH. In a neutral to alkaline region, MoO_4^{2-} is the predominant species, while moving towards the acidic region, complex anionic species will form.

The precipitation of molybdenum oxide MoO_3 in the acidic media was not experimentally observed because of the formation of complex compounds of molybdenum (VI) with sulphate anions [28,29].

Variation of oxidation states of Mo over pH could simply be presented as follows [30]:

$$MoO_4^{2-} \xrightarrow{pH < 6} Mo_7O_{24}^{6-} \xrightarrow{pH < 5} Mo_8O_{26}^{4-} \xrightarrow{pH < 1}$$
$$\xrightarrow{pH < 1} Mo_2O_5(SO_4)_2^{2-} \xrightarrow{c\,(H_2SO_4) > 0.3M} MoO_2^{2+} \tag{7}$$

These complexes of molybdenum (VI) with sulphate were not included in the thermodynamic calculations due to lack of information on their stability constants.

PC88A, a close analog of D2EHPA, is an acidic organophosphorus extractant that is typically present as a dimer with a noticeable potential to extract cationic molybdenum species through ion exchange. The highly hydrophobic organic anion forms an organic neutral complex with the metal ions that are present in the aqueous phase:

$$Me^{n+}\,(aq)\,+m\,\overline{(HX)_2}\,(org)\,\rightarrow\,\overline{Me(HX)_{2m-n}}\,(org)\,+n\,H^+\,(aq) \tag{8}$$

where $\overline{(HX)}_2$ represents the dimeric form of PC88A. Due to the formation of several oxidation states in the aqueous phase, molybdenum chemistry is rather complicated. Depending on the acidity of the aqueous solution, the size and type of molybdenum ion change [31]. However, the cationic molybdenum may form at a low pH range that describes the high Mo extraction values in Figure 3a. Conversely, little amounts of Re are extracted at the same low pH range, which is correlated to the existence of Re in its ReO_4^- form that is not extractable by PC88A (Figure 3b). It can be inferred from Figure 3a,b that by keeping the concentration of PC88A constant and changing proton concentration, selectivity increases towards Mo extraction over Re.

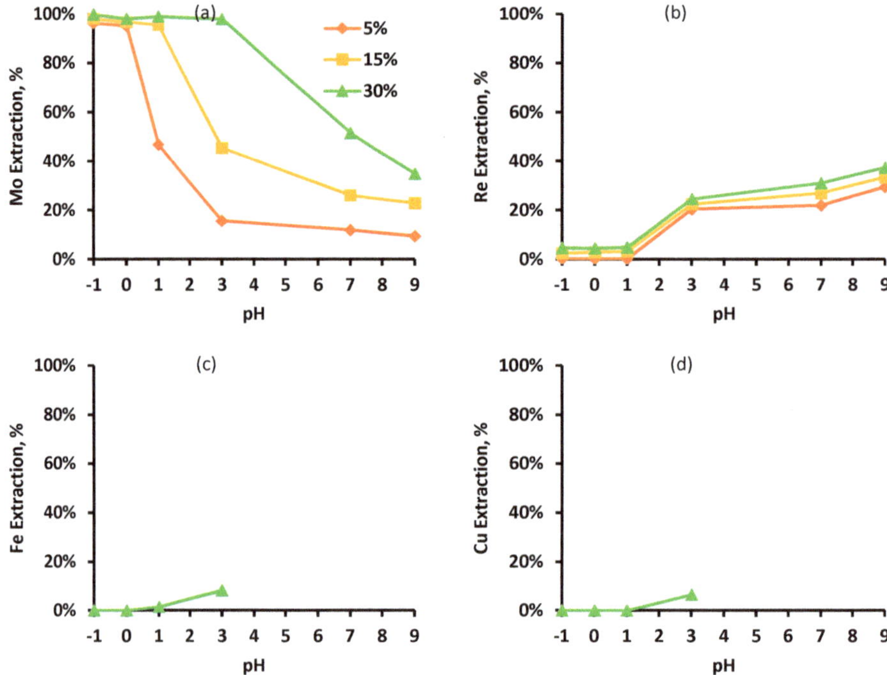

Figure 3. Metal extraction profiles as a function of pH and organic phase concentration: (**a**) Re, (**b**) Mo, (**c**) Fe, and (**d**) Cu.

Likewise, the effect of PC88A concentration on the extraction of Mo and Re was studied. However, for the case of Fe and Cu extraction, this effect was not studied (one-time test using 30% PC88A). It is clear that in the pH range of 1 to 3, increasing extractant concentrations does effectively change the Mo–Re extraction help in a better separation, notably near pH = 1.

3.2. Effect of Extractant

The organophosphoric acid reagents PC88A and D2EHPA are close analogs. Metal extraction at pH = 1, 3, and 7 was performed using an organic phase composed of 15% D2EHPA diluted in kerosene to be compared with PC88A. Experimental results show that PC88A is highly selective in the extraction of Mo over Re at pH = 1, while D2EHPA is more capable of providing a noticeable separation at neural pH (Table 4).

Table 4. Comparison of PC88A and D2EHPA capability of separation between Re and Mo.

Extractant	pH	Org. Conc.	Mo Recovery (%)	Re Recovery (%)
PC88A	1	15	95.48	3.19
D2EHPA	1	15	81.51	15.35
PC88A	3	15	45.32	22.45
D2EHPA	3	15	57.23	4.3
PC88A	7	15	25.94	27
D2EHPA	7	15	90.08	14.1

Such observed behaviours could be attributed to the Re and Mo species properties and the properties of the extractants (PC88A and D2EHPA) over the entire range of studied pH. To disclose the effects of these parameters on the separation of Mo over Re, first, the properties of the two extractants were studied via DFT calculations. Figures 4 and 5 display the converged structures of the two extractants along with the distributions of their highest occupied molecular orbitals (HOMOs) and lowest unoccupied molecular orbitals (LUMOs).

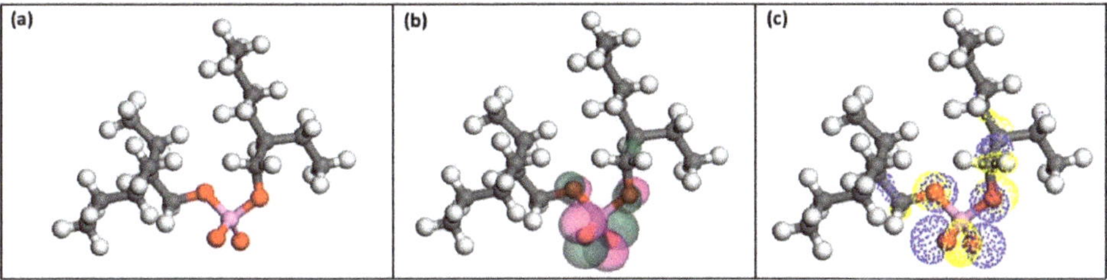

Figure 4. (a) Optimised structure of D2EHPA, (b) distribution of highest occupied molecular orbitals (HOMOs) of D2EHPA, and (c) distribution of lowest unoccupied molecular orbitals (LUMOs) of D2EHPA.

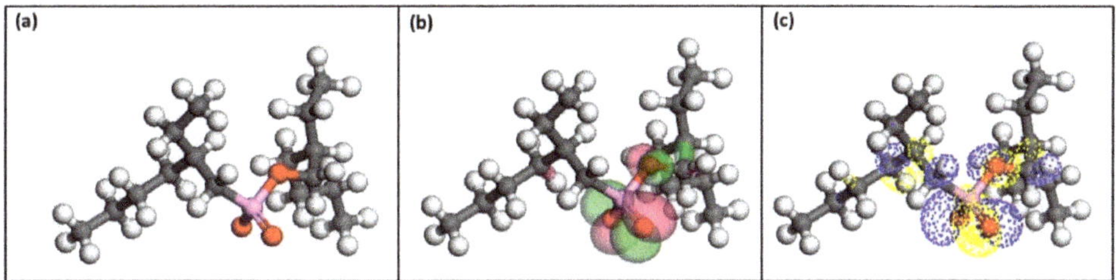

Figure 5. (a) Optimised structure of PC88A, (b) distribution of highest occupied molecular orbitals (HOMOs) of PC88A, and (c) distribution of lowest unoccupied molecular orbitals (LUMOs) of PC88A.

As seen, HOMOs and LUMOs are mostly located around their $P = O$ and $P - O$ groups, which are considered polar electron donor groups in the structure of these extractants. Furthermore, a Hirshfeld charge analysis was performed to assess the electronic charge of O atoms in $P = O$ and $P - O$ groups as presented in Table 5.

Table 5. Charge properties of PC88A and D2EHPA.

Extractants	Active Group	Charge (e)	Average Charge (e)
D2EHPA	P = O P – O	−0.551 −0.553	−0.552
PC88A	P = O P – O	−0.54 −0.53	−0.545

As illustrated in the Table 5, more electronic charges accumulate in both polar groups of D2EHPA, which are responsible for the complex formation through solvent extraction. This means that D2EHPA can be considered a stronger extractant, while PC88A has a potential for being a more selective reagent through solvent extraction. Such behaviour of these two extractants has already been reported in the literature [32]. Considering that the pK_a of D2EHPA and that of PC88A are around 3.01 and 4.21, respectively [33], this implies that these extractants will lie in their molecular forms for $pH \leq pK_a$ and will be dissociated into their ionic forms outside this pH range. In addition to extractant properties, Re and Mo species and properties are varied at studied pH as demonstrated in Figure 2. Variations in pH should play a critical role through complex formations in the present solvent extraction system. In this regard, as seen in Figure 2, while ReO_4^- can be considered a dominated Re species over the entire range of studied pH, MoO_2^{2+} at pH = 1, $Mo_7O_{24}^{6-}$ and $Mo_8O_{26}^{4-}$ at pH = 3, and MoO_4^{2-} at natural pH are regarded as dominated Mo species. Water-solvated forms of ReO_4^-, MoO_4^{2-}, and MoO_2^{2+} were studied by DFT calculations (Figure 6) to disclose more details about these Re and Mo species. It was found that ReO_4^-, MoO_4^{2-}, and MoO_2^{2+} are solvated with five, four, and seven water molecules, respectively. It was also realised that bond lengths of water-ReO_4^- are less than others (Table 6), indicating stronger interactions between water molecules and ReO_4^- as compared with those involved with the Mo species. Besides, Hirshfeld charge analysis was performed to compare charges of Re and Mo in these three species after solvation. It was realised that changes in the charge of Re are higher compared with those of Mo after solvation (water) of these three species as presented in Table 6, indicating higher interaction between water molecules and ReO_4^- as already established through bond length assessments.

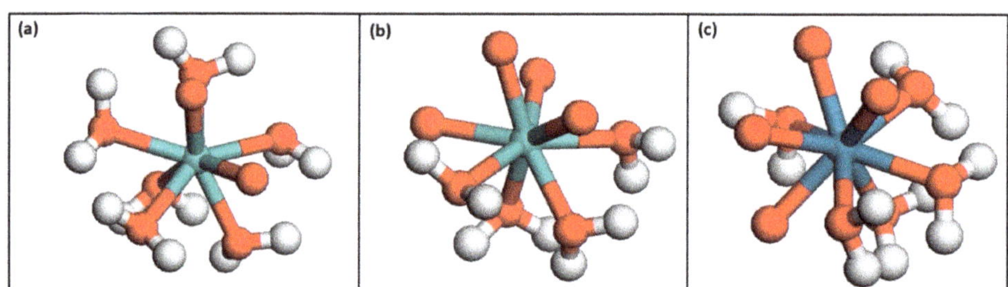

Figure 6. Optimised structures of (**a**) MoO_2^{2+}, (**b**) MoO_4^{2-}, and (**c**) ReO_4^-.

Table 6. Charge analysis and bond length of solvated Re and Mo species in the aqueous phase.

Species	Charge e (Re or Mo) Nonsolvated	Charge e (Re or Mo) Solvated	Changes (e)	Water Bond Length (Å)
MoO_4^{2-}	0.83	0.61	0.22	2.3
MoO_2^{2+}	0.89	0.6	0.29	2.3
ReO_4^-	0.74	0.4	0.34	2.2

As seen in Table 4, PC88A performance in Mo extraction decreased once pH increased. At pH = 1, PC88A is in molecular form, but positively charged Mo species, MoO_2^{2+}, with relatively low interaction with water molecule is the prevailing species. Therefore, making bond between PC88A and MoO_2^{2+} occurred. On the other hand, at pH = 3, relatively bulky species of Mo, including $Mo_7O_{24}^{6-}$ and $Mo_8O_{26}^{4-}$, are predicted to be the dominant forms in the solution, while the extractant is in its molecular form. It may be speculated that hindrance effects prevent bond formations between Mo species and PC88A, thus the declining recovery of Mo. Finally, at pH = 7, negatively charged MoO_4^{2-} forms along with PC88A dissociated into its ionic moieties. In this situation, it is suggested that electrostatic repulsions hinder the bond formation. Such effect can be more pronounced when it is noticed that PC88A is not a strong extractant based on DFT calculations. However, the case of Re is different as its recovery increased with increasing pH. Over the entire range of studied pH, Re is in the form of ReO_4^-, showcasing strong interactions with water molecules. PC88A dissociates in its ionic constituents with a potential for bond formation only at pH 7. Hence, it can be speculated that at pH 7, PC88A was more likely to create bonds with the Re species. Even in this situation, it is worthy to mention that PC88A is not a strong extractant based on DFT calculations; thus, although enhancement in Re recovery occurred through pH evolution, generally, its recovery is relatively lower than that of Mo in all ranges of pH. This could be due to the strength degree of PC88A in bond formation and the stability of ReO_4^- in interaction with water molecules. In the case of D2EHPA, hindrance effect at pH = 3 and repulsion at pH = 7 are less pronounced through Mo separation due to its stronger capability in complex formation (according to DFT calculations).

In order to confirm these explanations for the solvent extraction system, the interaction energy of the extractants with the Re and Mo species were obtained and are reported in Table 7. Interactions in both the organic and aqueous phases were taken into consideration since complexations could occur in both environments. In the case of aqueous phase calculations, the water-solvated forms of the species were considered. As seen in Table 7, the interaction energies of D2EHPA are higher than those of PC88A, which could be due to its stronger capability for complex formation. Besides, it can be seen that the interaction energies of both extractants follow this trend considering the Mo species: $MoO_2^{2+} > MoO_4^{2-}$. This trend is in line with the observation in Table 4. Besides, the interaction energies of D2EHPA and PC88A with the species are based on these trends, respectively: $MoO_2^{2+} > MoO_4^{2-} > ReO_4^-$ and $MoO_2^{2+} > ReO_4^- > MoO_4^{2-}$. These are also in line with observed recovery at different pH values.

Table 7. Interaction energy (kJ/mol) species with extractants in the organic and aqueous phases.

Species	PC88A		D2EHPA	
	Organic	Aqueous	Organic	Aqueous
MoO_4^{2-}	−178.1	−138.1	−321.3	−237.5
MoO_2^{2+}	−411.4	−321.1	−372.1	−221.8
ReO_4^-	−189.1	−155.3	−195.3	−157.1

4. Conclusions

PC88A was found to be a suitable solvent extraction candidate for selectively removing Mo over Re from molybdenite flue dust leach solutions. Acidifying the leach solution with H_2SO_4 at pH = 0–1 and employing an organic phase composed of 10–15% PC88A diluted in kerosene led to an ultimate separation between Mo and Re, transferring ca. 97% of Mo to the organic phase and leaving ca. 98% of Re in the leach solution. DFT simulations also indicated that the interaction energies of metals with D2EHPA were stronger than those with PC88A, thus explaining a different capability for complex formation. Besides, considering the Mo species, it was found that the interaction energies of both extractants followed the trend $MoO_2^{2+} > MoO_4^{2-}$ in line with the experimental observations. It was also confirmed via DFT simulations that the interaction energies of D2EHPA and PC88A with the metal oxyspecies follow the trends $MoO_2^{2+} > MoO_4^{2-} > ReO_4^-$ and $MoO_2^{2+} > ReO_4^- > MoO_4^{2-}$, respectively. These are also in line with observed recoveries at different pH values.

Author Contributions: Conceptualization, A.E.-Z.; data curation, A.E.-Z., D.A., P.A.N., F.L., and L.-C.P.; formal analysis, A.E.-Z. and D.A.; methodology, A.E.-Z.; project administration, F.L.; software, P.A.N.; writing—original draft, A.E.-Z.; writing—review and editing, D.A., P.A.N., F.L., and L.-C.P. All authors have read and agreed to the published version of the manuscript.

Funding: This research received no external funding.

Acknowledgments: The authors are grateful to the Babakan Ferromolybdenum Co. and Ali Amirarmand (Yazd Science and Technology Park) for providing the materials.

Conflicts of Interest: The authors declare no conflict of interest.

Appendix A Appearance of the Leaching Solution

Figure A1. The solution colour changes in the course of molybdenite flue dust leaching. It first became pale yellow (**right**), then turned into green, and finally dark blue (**left**).

Appendix B Details of the Thermodynamic Calculations

In order to construct the potential–pH diagram (Figure 2), the thermodynamic activities of ionic species in the solution should be calculated. To do it, the composition of the leaching solution should be estimated first. The leaching reactions of molybdenite flue dust are described by Equations (1)–(4). According to them, the primary aqueous species for Mo and Re are MoO_2^{2+} and ReO_4^-, respectively. Sulphuric acid is used to maintain the desired pH value. However, it is dibasic; it does not dissociate completely and may form sulphate and hydrosulphate ions and undissociated H_2SO_4 in the solution according to the following equations:

$$H_2SO_4 \ (aq) \ \rightleftharpoons \ HSO_4^- \ (aq) \ + \ H^+ \ (aq), \ K_1 \tag{A1}$$

$$HSO_4^- \ (aq) \ (aq) \ \rightleftharpoons \ SO_4^{2-} \ (aq) \ + \ H^+ \ (aq), \ K_2 \tag{A2}$$

The dissociation constants are presented in Table A1.

Table A1. Dissociation constants of sulphuric acid at 25 °C.

Step i	K_i, mol/L	Reference
1	1000	[34]
2	0.012	[35]

Let $c^o_{H_2SO_4}$ be the initial concentration of sulphuric acid and $c_{H_2SO_4}$, $c_{HSO_4^-}$, and $c_{SO_4^{2-}}$ the equilibrium concentrations of different species. Then the following equations for the equilibrium constants may be written:

$$K_2 = \frac{c_{SO_4^{2-}} \cdot c_{H^+}}{c_{HSO_4^-}} \tag{A3}$$

$$K_1 = \frac{c_{HSO_4^-} \cdot c_{H^+}}{c_{H_2SO_4}} \tag{A4}$$

$$c^o_{H_2SO_4} = c_{H_2SO_4} + c_{HSO_4^-} + c_{SO_4^{2-}} \tag{A5}$$

Rearranging Equations (A3) and (A4) and substituting them into Equation (A5) yields

$$c_{HSO_4^-} = \frac{c_{SO_4^{2-}} \cdot c_{H^+}}{K_2} \tag{A6}$$

$$c_{H_2SO_4} = \frac{c_{HSO_4^-} \cdot c_{H^+}}{K_1} = \frac{\frac{c_{SO_4^{2-}} \cdot c_{H^+}}{K_2} \cdot c_{H^+}}{K_1} = \frac{c_{SO_4^{2-}} \cdot c_{H^+}^2}{K_1 \cdot K_2} \tag{A7}$$

$$c^o_{H_2SO_4} = \frac{c_{SO_4^{2-}} \cdot c_{H^+}^2}{K_1 \cdot K_2} + \frac{c_{SO_4^{2-}} \cdot c_{H^+}}{K_2} + c_{SO_4^{2-}} \tag{A8}$$

Rearranging Equation (A8) gives

$$c^o_{H_2SO_4} = c_{SO_4^{2-}} \cdot \left(\frac{c_{H^+}^2}{K_1 \cdot K_2} + \frac{c_{H^+}}{K_2} + 1 \right) \tag{A9}$$

$$c^o_{H_2SO_4} = c_{SO_4^{2-}} \cdot \left(\frac{c_{H^+}^2}{K_1 \cdot K_2} + \frac{K_1 \cdot c_{H^+}}{K_1 \cdot K_2} + \frac{K_1 \cdot K_2}{K_1 \cdot K_2} \right) \tag{A10}$$

$$c^o_{H_2SO_4} = c_{SO_4^{2-}} \cdot \frac{c_{H^+}^2 + K_1 \cdot c_{H^+} + K_1 \cdot K_2}{K_1 \cdot K_2} \tag{A11}$$

$$K_1 \cdot K_2 \cdot c^o_{H_2SO_4} = c_{SO_4^{2-}} \cdot \left(c_{H^+}^2 + K_1 \cdot c_{H^+} + K_1 \cdot K_2 \right) \tag{A12}$$

$$c_{SO_4^{2-}} = \frac{K_1 \cdot K_2 \cdot c^o_{H_2SO_4}}{c_{H^+}^2 + K_1 \cdot c_{H^+} + K_1 \cdot K_2} \tag{A13}$$

The equilibrium concentrations of hydrosulphate ions and undissociated sulphuric acid might be obtained by substituting Equation (A13) into Equations (A6) and (A7):

$$c_{HSO_4^-} = \frac{c_{SO_4^{2-}} \cdot c_{H^+}}{K_2} = \frac{\frac{K_1 \cdot K_2 \cdot c^o_{H_2SO_4}}{c_{H^+}^2 + K_1 \cdot c_{H^+} + K_1 \cdot K_2} \cdot c_{H^+}}{K_2} = \frac{K_1 \cdot c_{H^+} \cdot c^o_{H_2SO_4}}{c_{H^+}^2 + K_1 \cdot c_{H^+} + K_1 \cdot K_2} \tag{A14}$$

$$c_{H_2SO_4} = \frac{c_{SO_4^{2-}} \cdot c_{H^+}^2}{K_1 \cdot K_2} == \frac{\frac{K_1 \cdot K_2 \cdot c^o_{H_2SO_4}}{c_{H^+}^2 + K_1 \cdot c_{H^+} + K_1 \cdot K_2} \cdot c_{H^+}^2}{K_1 \cdot K_2} = \frac{c_{H^+}^2 \cdot c^o_{H_2SO_4}}{c_{H^+}^2 + K_1 \cdot c_{H^+} + K_1 \cdot K_2} \tag{A15}$$

Let us introduce the mole fractions of the three species in the solution:

$$x_{SO_4^{2-}} = \frac{n_{SO_4^{2-}}}{n_{H_2SO_4} + n_{HSO_4^-} + n_{SO_4^{2-}}} \overset{V = const}{=} \frac{c_{SO_4^{2-}}}{c_{H_2SO_4} + c_{HSO_4^-} + c_{SO_4^{2-}}} = \frac{c_{SO_4^{2-}}}{c^o_{H_2SO_4}} \tag{A16}$$

$$x_{HSO_4^-} = \frac{n_{HSO_4^-}}{n_{H_2SO_4} + n_{HSO_4^-} + n_{SO_4^{2-}}} \overset{V = const}{=} \frac{c_{HSO_4^-}}{c_{H_2SO_4} + c_{HSO_4^-} + c_{SO_4^{2-}}} = \frac{c_{HSO_4^-}}{c^o_{H_2SO_4}} \tag{A17}$$

$$x_{H_2SO_4} = \frac{n_{H_2SO_4}}{n_{H_2SO_4} + n_{HSO_4^-} + n_{SO_4^{2-}}} \overset{V = const}{=} \frac{c_{H_2SO_4}}{c_{H_2SO_4} + c_{HSO_4^-} + c_{SO_4^{2-}}} = \frac{c_{H_2SO_4}}{c^o_{H_2SO_4}} \tag{A18}$$

Consequently,

$$x_{SO_4^{2-}} = \frac{c_{SO_4^{2-}}}{c_{H_2SO_4}^o} = \frac{\dfrac{K_1 \cdot K_2 \cdot c_{H_2SO_4}^o}{c_{H+}^2 + K_1 \cdot c_{H+} + K_1 \cdot K_2}}{c_{H_2SO_4}^o} = \frac{K_1 \cdot K_2}{c_{H+}^2 + K_1 \cdot c_{H+} + K_1 \cdot K_2} \tag{A19}$$

$$x_{HSO_4^-} = \frac{c_{HSO_4^-}}{c_{H_2SO_4}^o} = \frac{\dfrac{K_1 \cdot c_{H+} \cdot c_{H_2SO_4}^o}{c_{H+}^2 + K_1 \cdot c_{H+} + K_1 \cdot K_2}}{c_{H_2SO_4}^o} = \frac{K_1 \cdot c_{H+}}{c_{H+}^2 + K_1 \cdot c_{H+} + K_1 \cdot K_2} \tag{A20}$$

$$x_{H_2SO_4} = \frac{c_{H_2SO_4}}{c_{H_2SO_4}^o} = \frac{\dfrac{c_{H+}^2 \cdot c_{H_2SO_4}^o}{c_{H+}^2 + K_1 \cdot c_{H+} + K_1 \cdot K_2}}{c_{H_2SO_4}^o} = \frac{c_{H+}^2}{c_{H+}^2 + K_1 \cdot c_{H+} + K_1 \cdot K_2} \tag{A21}$$

It is worth noting that

$$x_{H_2SO_4} + x_{HSO_4^-} + x_{SO_4^{2-}} = 1 \tag{A22}$$

If the pH value is predetermined and the values of the equilibrium constants K_1 and K_2 (see Equations (A3) and (A4)), are known, the mole fractions of the three species in the solution might be unambiguously calculated. The dependency of the mole fractions of the species on pH value is called the speciation diagram. The speciation diagram for the sulphuric acid is presented in Figure A2.

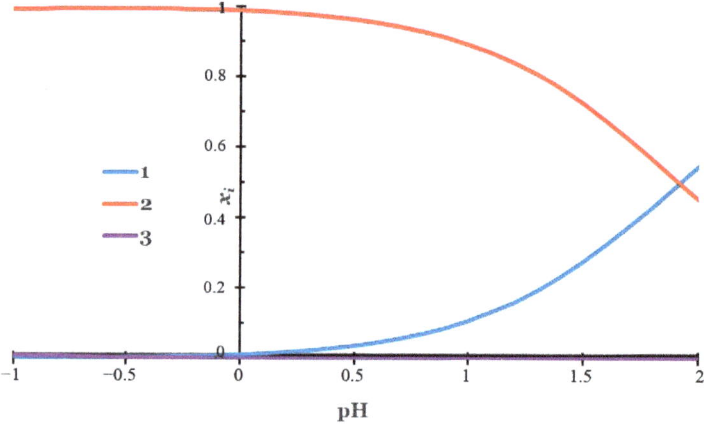

Figure A2. The speciation diagram for the sulphuric acid in the pH range from −1 to 2. (1)—SO_4^{2-} (aq), (2)—HSO_4^- (aq), (3)—H_2SO_4 (aq).

Because pH is not an independent variable and is determined by the addition of sulphuric acid, one needs to estimate the relationship between the total content of sulphuric acid in the solution and the pH value. To do it, the material balance equations should be considered.

In the first dissociation step (Equation (A1)), let the initial concentration of sulphuric acid be $c_{H_2SO_4}^o$. Let x moles of free acid per litre of solution dissociate to hydrosulphate ions. Consequently, the equilibrium concentration of the hydrosulphate ions is equal to x. From the reaction stoichiometry, it follows that the equilibrium concentration of hydrogen ions according to that dissociation step is also equal to x.

In the second dissociation step (Equation (A2)), let y moles of hydrosulphate ions per litre of solution dissociate further to sulphate ions. Consequently, the equilibrium concentration of the sulphate ions is equal to y. From the reaction stoichiometry, it follows that the equilibrium concentration of hydrogen ions according to that dissociation step is also equal to y.

Considering both dissociation steps, one might notice that the equilibrium concentration of hydrogen ions according to both dissociation steps equals $x + y$, the equilibrium concentration of hydrosulphate ions equals $x - y$, and the equilibrium concentration of free acid equals $c^o_{H_2SO_4} - x$ (see Table A2).

Table A2. Material balance of sulphuric acid dissociation.

	H_2SO_4 (aq)	$\rightleftharpoons$	HSO_4^- (aq)	+	H^+ (aq)
Initial state	$c^o_{H_2SO_4}$		0		0
Equilibrium state	$c^o_{H_2SO_4} - x$		x		x
	HSO_4^- (aq)	$\rightleftharpoons$	SO_4^{2-} (aq)	+	H^+ (aq)
Initial state	x		0		x
Equilibrium state	$x - y$		y		$x + y$

If the pH value is known, the ratio of equilibrium concentrations of sulphate and hydrosulphate ions is constant. The following system of equations may be written:

$$c_{H^+} = 10^{-pH} = x + y \tag{A23}$$

$$\frac{c_{HSO_4^-}}{c_{SO_4^{2-}}} = \frac{x - y}{y} \tag{A24}$$

Rearranging Equation (A24) yields

$$\frac{x - y}{y} = \frac{x}{y} - 1 = \frac{c_{HSO_4^-}}{c_{SO_4^{2-}}} \tag{A25}$$

$$\frac{x}{y} = \frac{c_{HSO_4^-}}{c_{SO_4^{2-}}} + 1 \tag{A26}$$

$$x = y \cdot \left(\frac{c_{HSO_4^-}}{c_{SO_4^{2-}}} + 1 \right) \tag{A27}$$

$$c_{H^+} = x + y = y \cdot \left(\frac{c_{HSO_4^-}}{c_{SO_4^{2-}}} + 1 \right) + y = y \cdot \left(\frac{c_{HSO_4^-}}{c_{SO_4^{2-}}} + 2 \right) \tag{A28}$$

$$y = \frac{c_{H^+}}{\dfrac{c_{HSO_4^-}}{c_{SO_4^{2-}}} + 2} \tag{A29}$$

Obviously, the ratio of equilibrium concentrations of sulphate and hydrosulphate ions is equal to the ratio of their mole fractions:

$$\frac{c_{HSO_4^-}}{c_{SO_4^{2-}}} = \frac{x_{HSO_4^-}}{x_{SO_4^{2-}}} = \frac{K_1 \cdot c_{H^+}}{c_{H^+}^2 + K_1 \cdot c_{H^+} + K_1 \cdot K_2} : \frac{K_1 \cdot K_2}{c_{H^+}^2 + K_1 \cdot c_{H^+} + K_1 \cdot K_2} = \frac{K_1 \cdot c_{H^+}}{K_1 \cdot K_2} = \frac{c_{H^+}}{K_2}. \tag{A30}$$

Substituting Equation (A30) into Equation (A29) gives

$$y = c_{SO_4^{2-}} = \frac{c_{H^+}}{\dfrac{c_{HSO_4^-}}{c_{SO_4^{2-}}} + 2} = \frac{c_{H^+}}{\dfrac{c_{H^+}}{K_2} + 2} = \frac{K_2 \cdot c_{H^+}}{c_{H^+} + 2 \cdot K_2} \tag{A31}$$

Substituting Equation (A31) into Equations (A6) and (A7) yields

$$c_{HSO_4^-} = \frac{c_{SO_4^{2-}} \cdot c_{H^+}}{K_2} = \frac{\frac{K_2 \cdot c_{H^+}}{c_{H^+} + 2 \cdot K_2} \cdot c_{H^+}}{K_2} = \frac{c_{H^+}^2}{c_{H^+} + 2 \cdot K_2} \tag{A32}$$

$$c_{H_2SO_4} = \frac{c_{SO_4^{2-}} \cdot c_{H^+}^2}{K_1 \cdot K_2} = \frac{\frac{K_2 \cdot c_{H^+}}{c_{H^+} + 2 \cdot K_2} \cdot c_{H^+}^2}{K_1 \cdot K_2} = \frac{c_{H^+}^3}{c_{H^+} \cdot K_1 + 2 \cdot K_1 \cdot K_2} \tag{A33}$$

Therefore, by using Equations (A5) and (A31)–(A33), the equilibrium concentrations of different species and the total content of sulphuric acid at the given pH value might be calculated straightforwardly. The calculated concentrations for pH values equal to 2, 1, 0, and −1 are presented in Table A3.

Table A3. Equilibrium concentrations of various aqueous sulphuric acid species and total concentration of sulphuric acid at different pH values.

PH	$c_{H_2SO_4}$, mol/L	$c_{HSO_4^-}$, mol/L	$c_{SO_4^{2-}}$, mol/L	$c_{H_2SO_4}^0$, mol/L
2	2.94×10^{-8}	0.00294	0.00353	0.00647
1	8.06×10^{-6}	0.0806	0.00968	0.0903
0	0.000977	0.977	0.0117	0.989
−1	0.0998	9.976	0.0119	10.088

The leaching reactions of molybdenite flue dust are given by reactions (1) through (4). According to them, the aqueous species to molybdenum and rhenium in the leaching liquor are MoO_2^{2+} and ReO_4^-, respectively. According to Table 3, the total content of molybdenum in a solution equals:

$$c_{[Mo]} = 9\frac{g}{L} : 95.94 \frac{g}{mol} = 0.094 \frac{mol}{L} \tag{A34}$$

The total content of rhenium equals:

$$c_{[Re]} = 0.5\frac{g}{L} : 186.207 \frac{g}{mol} = 0.0027 \frac{mol}{L} \tag{A35}$$

The parameters of the extended Debye–Hückel equation are:

$$T = 298.15 \text{ K} \tag{A36}$$

$$\varepsilon = 87.74 - 0.4008 \cdot (T - 273.15) + $$
$$+ 9.398 \cdot 10^{-4} \cdot (T - 273.15)^2 + 1.41 \cdot 10^{-6} \cdot (T - 273.15)^3 = 78.3294 \tag{A37}$$

$$A = \frac{1.825 \cdot 10^6}{(\varepsilon \cdot T)^{\frac{3}{2}}} = \frac{1.825 \cdot 10^6}{(78.3294 \cdot 298.15)^{\frac{3}{2}}} = 0.5114 \; \frac{L^{\frac{1}{2}}}{mol^{\frac{1}{2}}} \tag{A38}$$

$$B = \frac{5.029 \cdot 10^{11}}{(\varepsilon \cdot T)^{\frac{1}{2}}} = \frac{5.029 \cdot 10^{11}}{(78.3294 \cdot 298.15)^{\frac{1}{2}}} = 3.291 \cdot 10^9 \; \frac{L^{\frac{1}{2}}}{m \cdot mol^{\frac{1}{2}}}. \tag{A39}$$

The electrostatic radii of the individual ions are presented in Table A4.

Table A4. Electrostatic radii of individual ions.

Ion	a_i, Å
ReO_4^-	4.5
MoO_2^{2+}	4.5

The ionic strength of the solution is calculated as follows:

$$I = \frac{C_{ReO_4^-} \cdot z^2_{ReO_4^-} + C_{MoO_2^{2+}} \cdot z^2_{MoO_2^{2+}} + C_{SO_4^{2-}} \cdot z^2_{SO_4^{2-}} + C_{HSO_4^-} \cdot z^2_{HSO_4^-} + C_{H^+} \cdot z^2_{H^+}}{2} \tag{A40}$$

The activity coefficients of molybdenum and rhenium ions in a solution are calculated according to the extended Debye–Hückel equation. The thermodynamic activities are calculated straightforwardly:

$$\lg \gamma_{ReO_4^-} = -A \cdot z^2_{ReO_4^-} \cdot \frac{\sqrt{I}}{1 + B \cdot a_{ReO_4^-} \cdot \sqrt{I}} \tag{A41}$$

$$a_{ReO_4^-} = c_{[Re]} \cdot 10^{\gamma_{ReO_4^-}} \tag{A42}$$

$$\lg \gamma_{MoO_2^{2+}} = -A \cdot z^2_{MoO_2^{2+}} \cdot \frac{\sqrt{I}}{1 + B \cdot a_{MoO_2^{2+}} \cdot \sqrt{I}} \tag{A43}$$

$$a_{MoO_2^{2+}} = c_{[Mo]} \cdot 10^{\gamma_{MoO_2^{2+}}} \tag{A44}$$

The calculated values are presented in Table A5.

Table A5. Activity coefficients and thermodynamic activities of molybdenum and rhenium species in leaching stock solution at different pH values.

pH	I, mol/L	$\lg \gamma_{ReO_4^-}$	$a_{ReO_4^-}$, mol/L	$\lg \gamma_{MoO_2^{2+}}$	$a_{MoO_2^{2+}}$, mol/L
2	0.20288	−0.138	0.00196	−0.553	0.0263
1	0.299	−0.155	0.00189	−0.618	0.0226
0	1.096	−0.214	0.00165	−0.855	0.0131
−1	10.201	−0.285	0.00140	−1.140	0.0068

As can be seen, the average activities of the molybdenum species are ~0.01 mol/L, and those of the rhenium species are ~0.001 mol/L. The activities of the sulphur species have no influence on the position of the lines on the diagram.

The potential–pH diagram for the Mo–Re–SO$_4^{2-}$ system (Figure 2) is plotted at 25 °C, air pressure of 1 bar and the activities of the molybdenum species 0.01 mol/L, the activities of the rhenium species 0.001 mol/L, and the activities of the sulphur species 0.1 mol/L.

References

1. Habashi, F. *Handbook of Extractive Metallurgy*; Wiley-VCH: Weinheim/Heidelberg, Germany, 1997.
2. Virolainen, S.; Laatikainen, M.; Sainio, T. Ion exchange recovery of rhenium from industrially relevant sulfate solutions: Single column separations and modeling. *Hydrometallurgy* **2015**, *158*, 74–82. [CrossRef]
3. Srivastava, R.R.; Lee, J.-C.; Kim, M.-S. Complexation chemistry in liquid–liquid extraction of rhenium. *J. Chem. Technol. Biotechnol.* **2015**, *90*, 1752–1764. [CrossRef]
4. Cheema, H.A.; Ilyas, S.; Masud, S.; Muhsan, M.A.; Mahmood, I.; Lee, J.-C. Selective recovery of rhenium from molybdenite flue-dust leach liquor using solvent extraction with TBP. *Sep. Purif. Technol.* **2018**, *191*, 116–121. [CrossRef]
5. Nikolaychuk, P.A.; Tyurin, A.G. Utočnënnaâ diagramma Purbe dlâ molibdena. *Butlerovskie Soobŝeniâ* **2011**, *24*, 101–105.
6. Khoshnevisan, A.; Yoozbashizadeh, H.; Mohammadi, M.; Abazarpoor, A.; Maarefvand, M. Separation of rhenium and molybdenum from molybdenite leach liquor by the solvent extraction method. *Miner. Metall. Process.* **2013**, *30*, 53–58. [CrossRef]

7.   Alamdari, E.K.; Darvishi, D.; Haghshenas, D.F.; Yousefi, N.; Sadrnezhaad, S.K. Separation of Re and Mo from roasting-dust leach-liquor using solvent extraction technique by TBP. *Sep. Purif. Technol.* **2012**, *86*, 143–148. [CrossRef]

8.   Cao, Z.F.; Zhong, H.; Jiang, T.; Liu, G.Y.; Wang, S. Selective electric-oxidation leaching and separation of Dexing molybdenite concentrates. Zhongguo Youse Jinshu Xuebao/Chin. *J. Nonferrous Met.* **2013**, *23*, 2290–2295.

9.   Zhan-fang, C.; Hong, Z.; Zhao-hui, Q. Solvent extraction of rhenium from molybdenum in alkaline solution. *Hydrometallurgy* **2009**, *97*, 153–157. [CrossRef]

10. Kang, J.; Kim, Y.U.; Joo, S.H.; Yoon, H.S.; Kumar, J.R.; Park, K.H.; Shin, S.M. Behavior of Extraction, Stripping, and Separation Possibilities of Rhenium and Molybdenum from Molybdenite Roasting Dust Leaching Solution Using Amine Based Extractant Tri-Otyl-Amine (TOA). *Mater. Trans.* **2013**, *54*, 1209–1212. [CrossRef]

11. Srivastava, R.R.; Kim, M.-S.; Lee, J.-C.; Ilyas, S. Liquid–liquid extraction of rhenium (VII) from an acidic chloride solution using Cyanex 923. *Hydrometallurgy* **2015**, *157*, 33–38. [CrossRef]

12. Pathak, S.K.; Singh, S.; Mahtele, A.; Tripathi, S.C. Studies on extraction behaviour of molybdenum (VI) from acidic radioactive waste using 2 (ethylhexyl) phosphonic acids, mono 2 (ethylhexyl) ester (PC-88A)/n-dodecane. *J. Radioanal. Nucl. Chem.* **2010**, *284*, 597–603. [CrossRef]

13. Wang, Y.; Jiang, K.; Zou, X.; Zhang, L.; Liu, S. Recovery of metals from molybdenite concentrate by hydrometallurgical technologies. In *TT Chen Honorary Symposium on Hydrometallurgy, Electrometallurgy and Materials Characterization*; John Wiley & Sons, Inc.: Hoboken, NJ, USA, 2012; pp. 315–322.

14. Nguyen, T.H.; Lee, M.S. Separation of molybdenum (VI) and tungsten (VI) from sulfate solutions by solvent extraction with LIX 63 and PC 88A. *Hydrometallurgy* **2015**, *155*, 51–55. [CrossRef]

15. Morís, M.A.A.; Díez, F.V.; Coca, J. Solvent extraction of molybdenum and tungsten by Alamine 336 and DEHPA in a rotating disc contactor. *Sep. Purif. Technol.* **1999**, *17*, 173–179. [CrossRef]

16. Truong, H.T.; Lee, M.S. Separation of rhenium (VII), molybdenum (VI), and vanadium (V) from hydrochloric acid solution by solvent extraction with TBP. *Geosyst. Eng.* **2017**, *20*, 224–230. [CrossRef]

17. Kim, H.S.; Park, J.S.; Seo, S.Y.; Tran, T.; Kim, M.J. Recovery of rhenium from a molybdenite roaster fume as high purity ammonium perrhenate. *Hydrometallurgy* **2015**, *156*, 158–164. [CrossRef]

18. Xiao, C.; Zeng, L.; Xiao, L.; Zhang, G. Solvent Extraction of Molybdenum (VI) from Hydrochloric Acid Leach Solutions Using P507. Part I: Extraction and Mechanism. *Solvent Extr. Ion Exch.* **2017**, *156*, 130–144. [CrossRef]

19. Entezari, A.; Karamoozian, M.; Eskandari Nasab, M. Investigation on selective rhenium leaching from molybdenite roasting flue dusts. *J. Min. Environ.* **2013**, *4*, 77–82.

20. Entezari-Zarandi, A.; Larachi, F. Selective dissolution of rare-earth element carbonates in deep eutectic solvents. *J. Rare Earths* **2019**, *37*, 528–533. [CrossRef]

21. Keshavarz Alamdari, E. Selective Leaching-Recovery of Re and Mo from Out-Gas Dust of Molybdenite Roasting Furnace. *Trans. Indian Inst. Met.* **2017**, *70*, 1995–1999. [CrossRef]

22. Debye, P.; Hückel, E. Zur Theorie der Elektrolyte. I. Gefrierpunktserniedrigung und verwandte Erscheinungen. *Phys. Z* **1923**, *24*, 185–206.

23. Kielland, J. Individual activity coefficients of ions in aqueous solutions. *J. Am. Chem. Soc.* **1937**, *59*, 1675–1678. [CrossRef]

24. Coolidge, W. Dielektrische Untersuchungen und elektrische Drahtwellen. *Ann. Phys.* **1899**, *305*, 125–166. [CrossRef]

25. Nikolaychuk, P. Das revidierte pourbaix-diagramm für schwefel. In *Materialien Zum Wissenschaftlichen Seminar der Stipendiaten der Programme „Mikhail Lomonosov" und „Immanuel Kant"*; Deutscher Akademischer Austausch Dienst und Ministerium für Bildung und Wissenschaft der RF: Moscow, Russia, 2015; pp. 72–76.

26. Wagman, D.D.; Evans, W.H.; Parker, V.B.; Schumm, R.H.; Halow, I. The NBS tables of chemical thermodynamic properties. Selected values for inorganic and C_1 and C_2 organic substances in SI units, National Standard Reference Data System. *J. Phys. Chem. Ref. Data* **1982**, *11*, 37–38.

27. Azizi, D.; Larachi, F. Behavior of bifunctional phosphonium-based ionic liquids in solvent extraction of rare earth elements-quantum chemical study. *J. Mol. Liq.* **2018**, *263*, 96–108. [CrossRef]

28. Palant, A.; Iatsenko, N.; Petrova, V. Solvent extraction of molybdenum (VI) by diisododecylamine from sulphuric acid solution. *Hydrometallurgy* **1998**, *48*, 83–90. [CrossRef]

29. Loewenschuss, A.; Shamir, J.; Ardon, M. Vibrational spectra of binuclear molybdenum sulfate complexes of high bond order. *Inorg. Chem.* **1976**, *15*, 238–241. [CrossRef]
30. Alamdari, E.K.; Sadrnezhaad, S.K. Thermodynamics of extraction of from aqueous sulfuric acid media with TBP dissolved in kerosene. *Hydrometallurgy* **2000**, *55*, 327–341. [CrossRef]
31. Kholmogorov, A.G.; Kononova, O.N.; Panchenko, O.N. A Review of the Use of Ion Exchange for Molybdenum Recovery in Russia. *Can. Metall. Q.* **2004**, *43*, 297–304. [CrossRef]
32. Shi, Q.; Zhang, Y.; Huang, J.; Liu, T.; Liu, H.; Wang, L. Synergistic solvent extraction of vanadium from leaching solution of stone coal using D2EHPA and PC88A. *Sep. Purif. Technol.* **2017**, *181*, 1–7. [CrossRef]
33. Cheng, C.Y.; Barnard, K.R.; Zhang, W.; Robinson, D.J. Synergistic solvent extraction of nickel and cobalt: A review of recent developments. *Solvent Extr. Ion Exch.* **2011**, *29*, 719–754. [CrossRef]
34. Sue, K.; Uchids, M.; Adschiri, T.; Arai, K. Determination of sulfuric acid first dissociation constants to 400 °C and 32 MPa by potentiometric pH measurements. *J. Supercrit. Fluids* **2004**, *31*, 295–299. [CrossRef]
35. Wu, Y.C.; Feng, D. The second dissociation constant of sulfuric acid at various temperatures by the conductometric method. *J. Solut. Chem.* **1995**, *24*, 133–144. [CrossRef]

Publisher's Note: MDPI stays neutral with regard to jurisdictional claims in published maps and institutional affiliations.

Review

Separation of Radioactive Elements from Rare Earth Element-Bearing Minerals

Adrián Carrillo García [1], Mohammad Latifi [1,2], Ahmadreza Amini [1] and Jamal Chaouki [1,*

[1] Process Development Advanced Research Lab (PEARL), Chemical Engineering Department, Ecole Polytechnique de Montreal, C.P. 6079, Succ. Centre-ville, Montreal, QC H3C 3A7, Canada; adrian.carrillo-garcia@polymtl.ca (A.C.G.); mohammad.latifi@polymtl.ca (M.L.); ahmad-reza.amini@polymtl.ca (A.A.)

[2] NeoCtech Corp., Montreal, QC H3G 2N7, Canada

* Correspondence: jamal.chaouki@polymtl.ca

Received: 8 October 2020; Accepted: 13 November 2020; Published: 17 November 2020

Abstract: Rare earth elements (REE), originally found in various low-grade deposits in the form of different minerals, are associated with gangues that have similar physicochemical properties. However, the production of REE is attractive due to their numerous applications in advanced materials and new technologies. The presence of the radioactive elements, thorium and uranium, in the REE deposits, is a production challenge. Their separation is crucial to gaining a product with minimum radioactivity in the downstream processes, and to mitigate the environmental and safety issues. In the present study, different techniques for separation of the radioactive elements from REE are reviewed, including leaching, precipitation, solvent extraction, and ion chromatography. In addition, the waste management of the separated radioactive elements is discussed with a particular conclusion that such a waste stream can be employed as a valuable co-product.

Keywords: rare earth elements; thorium; uranium; separation methods; precipitation; solvent extraction; leaching; membrane

---

1. Introduction

The REE are fifteen lanthanide elements in the periodic table with atomic numbers of 57 to 71, including lanthanum (La), cerium (Ce), praseodymium (Pr), neodymium (Nd), promethium (Pm), samarium (Sm), europium (Eu), gadolinium (Gd), terbium (Tb), dysprosium (Dy), holmium (Ho), erbium (Er), thulium (Tm), ytterbium (Yb), and lutetium (Lu) as well as scandium (Sc) and yttrium (Y) with atomic numbers of 21, and 39, respectively. There is a fast growth in new applications and demand for the REE, especially in energy, environment, and high technology fields with durability, high efficiency and low carbon emissions [1–5]. These elements are called "rare" owing to their difficult extraction from deposits that is attributed to the similarity in the physical and chemical properties of REE and gangue minerals and to the difficulty to find concentrated deposits. Another challenge for REE production is the heterogeneity of these elements in the deposits [4,6–9], which plays a vital role in configuring the unit operations regarding the geology, versatility, and composition of the minerals [4,6]. The production of REE requires several steps of magnetic, gravity, and electrostatic separations in addition to flotation to efficiently separate the REE from associated gangues with similar physical properties. The individual production of the REE is very challenging owing to their similar chemical properties, and specific extraction techniques are required to recover REE; however, europium and cerium are exceptions, i.e., cerium can be formed as, either trivalent or tetravalent during hydrometallurgical process where the tetravalent cerium can be separated from the trivalent REE [10,11]. Figure 1 shows a typical REE production process, including geology, mining, physical beneficiation, hydrometallurgy, and separation of individual elements [6].

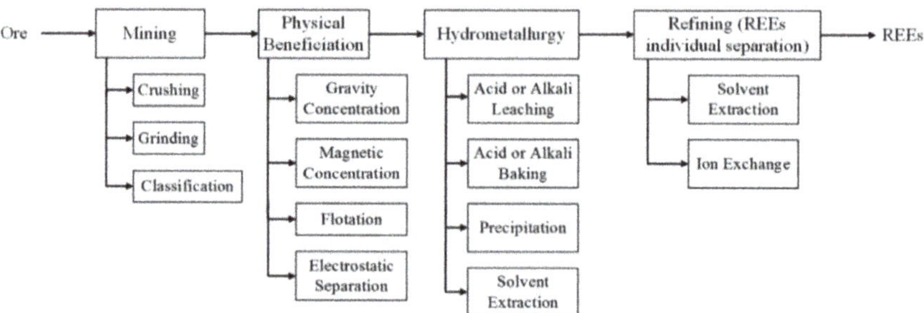

Figure 1. The REE general processing plant.

Since mining and refining of low-grade REE-bearing minerals are technically infeasible owing to a lithophilic nature of the REE [12,13], high-grade REE-bearing minerals, such as bastnäsite, monazite, and xenotime, are used for economic extraction of these elements [4,12,14–16]. Table 1 summarizes both the low-grade and high-grade REE-bearing minerals.

Among various gangues, radioactive elements are a serious challenge in the REE production process, regarding specific regulations for safety management in the processing units (for more details refer to [17–19]). Thorium (Th) and uranium (U) are naturally occurring radioactive materials (NORM), which can be found in the REE deposits (Table 1). Monazite and xenotime are the most known REE-bearing minerals that contain radioactive elements. For instance, the REE-bearing ore at Mountain Pass, i.e., a major bastnäsite resource in California with rare earth oxides (REO) of 8.5 wt.%, contains thorium (Th) and uranium (U) of 0.02, and 0.002 wt.%, respectively [19]. In addition, the Bayan Obo bastnäsite and monazite deposit in China contains minerals such as fluorite, magnetite, barite, calcite and quartz with magnetic susceptibility, specific gravity, electrical conductivity, or floatability similar to REE-bearing minerals [20,21].

The low concentration of radioactive elements in the upstream rare earth ore processing units, e.g., in physical beneficiation, results in quite low emission of radioactivity, whereas the higher concentration of the radioactive elements in downstream separation lines requires safety measurements to be carefully considered. For example, an exposure of a worker to an ore containing 500 ppm thorium and 50 ppm uranium, staying 1 m away from a large mass of the ore for an entire working year, leads to a total exposure of 2.4 mSv that is below the dose limit for a NORM worker, 20 mSv [17]. This exposure is mainly caused by ore dust inhalation (at 1 mg/m^3) and incidentally ore ingestion (at 100 mg/day). Therefore, a step for separation of thorium and uranium is required to minimize risks associated with REE production in terms of safety, environmental hazards, and quality of the final product [22].

In the present paper, various hydrometallurgical techniques applied during REE production process for separating thorium (Th) and uranium (U) are reviewed, including leaching, precipitation, solvent extraction, ion chromatography, and membrane to understand the advantages and limitations of each technique. In addition, the process selection with regards to the feed properties, waste management of the separated radioactive elements, and how they can be treated to produce valuable co-products are also discussed.

Table 1. REE-bearing minerals and gangue minerals from deposits.

Type	Mineral	Formula	Average Composition (wt.%)			Ref.
			REE Oxide	ThO_2	UO_2	
Carbonate	Ancylite	$Sr(Ce,La)(CO_3)_2OH \cdot H_2O$	46	0–0.4	0.1	[6]
	Bastnäsite	**$(Ce,La)CO_3F$**	**74**	**0–0.3**	**<0.9**	[6]
	Parisite	$Ca(REE)_2(CO_3)_3F_2$	59	0–0.5	0–0.3	[6]
Phosphate	Apatite	$Ca_5(PO_4)_3(F,Cl,OH)$	19	-	-	[4,6]
	Britholite	$(REE,Ca)_5(SiO_4,PO_4)_3(F,OH)$	56	1.5	-	[4,6]
	Monazite	**$(REE,Th)PO_4$**	**35–71**	**0–20**	**0–16**	[6,23]
	Xenotime	**YPO_4**	**61**	-	**0–5**	[4,6]
Oxide	Brannerite	$(U,REE,Ca)(Ti,Fe)_2O_6$	6	-	-	[6]
	Perovskite	$(Ca,REE)TiO_3$	<37	0–2	<0.05	[6]
Silicate	Allanite	$(REE,Ca)_2(Al,Fe)_3(SiO_4)_3(OH)$	30	0.3	-	[4,6]
	Cheralite	$(REE,Th,Ca)(P,Si)O_4$	5	<30	-	[4,6]

It is worthy of mentioning that recovery of REE from secondary sources such as electronic wastes [24,25], red mud (Bauxite) [26–29], and coal [30] has also been recently investigated. The separation of radioactive elements during these processes is out of the scope of the present review article and requires further study.

2. Separation by Leaching

Leaching is a process based on the different solubility of elements in a leach liquor. To separate the radioactive elements from REE, leaching process is typically applied on, either fresh or concentrated ore [4], in order to maximize the solubility of radioactive elements [31–34] or REE [35] in the liquor.

According to the literature, a one-step leaching process faces technical issues during separation of thorium and uranium from REE owing to the occurrence of undesired reactions and leaching of un-wanted components [36]. For instance, Lapidus and Doyle [33] applied a one-step leaching process to separate radioactive elements from a monazite concentrate using oxalate reagent for leaching out thorium oxalate in the liquid form while rare earth oxalate remains in the solid form. They observed that either $Th(HPO_4)_2$ or $Th_3(PO_4)_4$ is re-precipitated in the liquor at a pH < 3, and oxalate reagent forms stable complexes with other metal impurities, instead of reaction with radioactive elements. To overcome the drawbacks of this process, a two-step cracking-leaching process was proposed. In the first step (cracking), an alkaline reagent, e.g., NaOH, cracks the concentrate of the REE-bearing mineral, either monazite or xenotime, to produce a hydroxide cake containing the REE, thorium, uranium, and some other impurities, reactions 1 and 2 [31,33,35]. This step eliminates the re-precipitation of thorium phosphate resulting in the formation of hydroxide forms of the REE and thorium, which can be separated in the second leaching step [34,37,38],

$$2REPO_4 + 6NaOH \rightarrow 2RE(OH)_3(\downarrow) + 2Na_3PO_4 \tag{1}$$

$$Th_3(PO_4)_4 + 12NaOH \rightarrow 3Th(OH)_4(\downarrow) + 4Na_3PO_4 \tag{2}$$

After the cracking step, acid leaching of the produced hydroxide cake is performed wherein the acidic oxalate reagents are employed to leach the radioactive elements while the REE oxalate is insoluble [34]. If the REE is preferred to be in the liquor solution, other acids such as nitric acid (reaction 3) [35] or hydrochloric acid (reaction 4) [37] can be used instead of oxalate reagents; however, the uranium also tends to leach out with the REE if the pH is not properly controlled,

$$RE(OH)_3 + 3HNO_3 \rightarrow RE(NO_3)_{3(aq)} + 3H_2O \tag{3}$$

$$RE(OH)_3 + 3HCl \rightarrow RECl_{3(aq)} + 3H_2O \tag{4}$$

The two-step approach can be improved by employing high-pressure leaching to stabilize products that are unstable at atmospheric conditions. Figure 2 shows an increase in the recovery of thorium by ammonium carbonate (250%), from 13% at atmospheric pressure to 98% at 6.5–10 atm at 70 °C, while no significant change occurs in the recovery of REE [31].

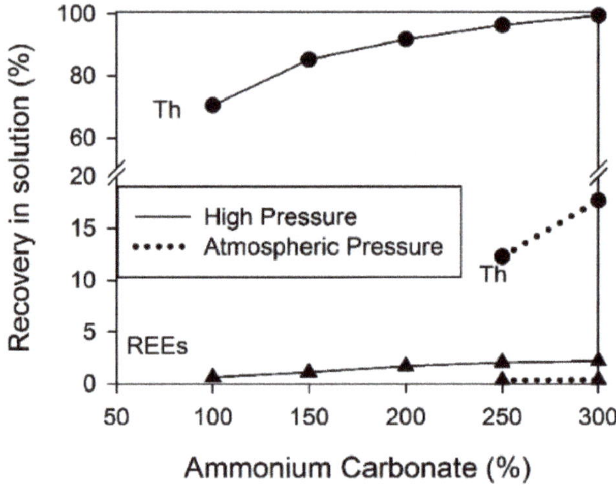

Figure 2. Effect of the pressure on the recovery of thorium (Th) and REE at 70 °C, adapted from [31]. (Reproduced with permission from ref. [31], copyright (2002), Elsevier)

In such a high-pressure leaching process, the ammonium carbonate leaches both the thorium, reaction 5, and uranyl hydroxides, reaction 6, while it produces a solidus complex after reaction with REE, reaction 7 [31,32]. Typically, an excess amount of reagent is employed in either atmospheric or high-pressure leaching owing to the consuming part of the reagent in the secondary reactions,

$$Th(OH)_4 + 5(NH_4)_2CO_3 \rightarrow (NH_4)_6[Th(CO_3)_5] + 4NH_4OH \tag{5}$$

$$UO_2(OH)_2 + 3(NH_4)_2CO_3 \rightarrow (NH_4)_4[UO_2(CO_3)_3] + 2NH_4OH \tag{6}$$

$$2RE(OH)_3 + 4(NH_4)_2CO_3 \rightarrow RE_2(CO_3)_3(NH_4)_2CO_3(\downarrow) + 6NH_4OH \tag{7}$$

Figure 3 illustrates a flowsheet summarizing the cracking-leaching approaches (two-step leaching process) wherein, either REE or radioactive elements can be extracted in the leach liquor, depending on the leaching reagent. Table 2 summarizes the experimental conditions and the overall recovery of REE, thorium, and uranium in leach liquor for the two-step cracking-leaching process.

Table 2. Uranium and thorium separation from REE by a two-step cracking-leaching process.

REE-Bearing Mineral	Reagent	Operating Conditions			Overall Recovery in Leach Liquor (%)			Ref.
		Pressure (atm)	Time (h)	T (°C)	Th	U	REE	
Xenotime	Nitric acid	1	-	60	<1	-	>98	[35]
Concentrated monazite (87 wt.% REE)	Ammonium carbonate	6.5–10	1–2	80	99	95	2.5	[31]
Monazite (18.5 wt.% REE)	Ammonium oxalate	1	<2	40	100	>40	<1	[33]

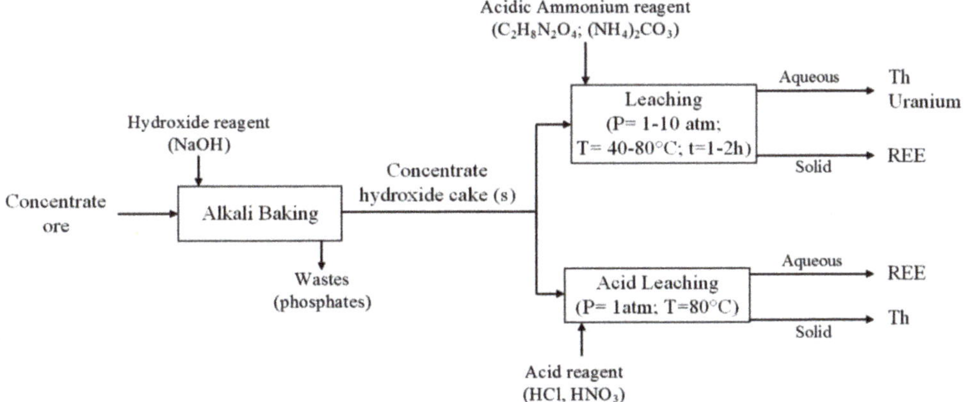

Figure 3. Leaching-cracking approaches for separation of REE from radioactive elements.

3. Separation by Precipitation

The REE can be separated from other elements in a liquor via precipitation method using an appropriate reagent such as oxalic acid, reaction 8 [39–42],

$$2RE^{3+} + 3H_2C_2O_4 + 10H_2O \rightarrow RE_2(C_2O_4)_3 \cdot 10H_2O(\downarrow) + 6H^+ \tag{8}$$

To separate the REE from the radioactive elements through precipitation, the most important parameters, include the type of the liquor feed, REE concentration in the feed, the concentration of the precipitation reagent, and the mass ratio of the reagent to REE. The pH is another parameter that determines the performance of selective precipitation of uranium, thorium, and REE. The ideal case is to find distinct pH values to separate radioactive elements from REE. Uranium and REE precipitate at close pH values with a risk of co-precipitation, whereas a high recovery of thorium is achievable since its precipitation requires a different pH [43–46]. Table 3 presents the pH ranges required for the precipitation of uranium, thorium, and REE in both chloride and sulfate liquors. In an acid liquor, some of alkali precipitation reagents are preferred owing to higher selectivity towards radioactive elements [46–49]. For instance, ammonium hydroxide (NH₄OH) at a pH close to 5 [50], and sodium hydroxide [47] precipitate thorium with a small loss of REE. Whereas, potassium iodate (KIO₃) is inefficient in the precipitation of thorium due to the co-precipitation of REE [50]. In addition, ammonium hydroxide (NH₄OH) precipitates uranium at a pH close to 4.5, which is far enough from the REE precipitation pH of about 6 [51].

Table 3. The pH ranges for precipitation of thorium, uranium and the REE in different liquors.

Elements	Precipitation pH (Approx.)	
	Chloride Liquor	**Sulfate Liquor**
Th	4.8–5.8	1–2
U	5.5–7	6
REE	6.8–8	3–5.5

3.1. Types of the Liquor and Reagent

Typically, the main liquors produced through the industrial hydrometallurgical processes are chloride and sulfate [4]. In the chloride liquor, precipitation of the radioactive elements is usually achieved by adding an alkali reagent. If the pH is kept close to or below 5.5, the total thorium and a part of uranium are likely precipitated and recovered while the loss of the REE is about 2% [44,45].

A lower pH for precipitation of radioactive elements has been reported in a sulfate liquor compared to the chloride liquor. Table 3 presents the pH range required for precipitating thorium and uranium from chloride and sulfate liquors. For example, 100% thorium is precipitated from sulfate liquor at a pH close to 1 using ammonium hydroxide. However, REE (44.7% La, 63.5% Ce, and 63.2% Nd) are also co-precipitated under such a highly acidic condition [43,46].

In some cases, nitrate liquor has also been studied for separation of uranium from REE [51,52] wherein the uranium was precipitated by 90% using hydroxide reagents at a pH of 4.5 [51]. Further, a higher efficiency has also been obtained for a precipitation pH between 6 to 8, at which almost total recovery of uranium is achieved in the form of β–$UO_2(OH)_2$ [52].

The type of reagent is another critical parameter that determines which elements, e.g., thorium, uranium, or REE, are precipitated, while the others remain in the aqueous phase. A double sulphate reagent produces the REE precipitate in the form of $MRE(SO_4)_2$ with M representing sodium, potassium or ammonium, leaving radioactive elements in the aqueous phase, reaction 9 [53]. Precipitation of the trivalent REE is attributed to the low solubility of the REE sulfate in water, resulting in the separation from the radioactive elements and tetravalent REE, i.e., cerium (IV):

$$RE(SO_4)_3 + Na_2SO_4 + 2H_2O \rightarrow Na_2SO_4RE_2(SO_4)_3 \cdot 2H_2O \downarrow \tag{9}$$

3.2. Multi- or Single-Step Precipitation

The precipitation for separation of radioactive elements and REE can be conducted in either multi-steps or a single step, regarding the types of the liquor and reagents. A secondary purification step is necessary when the reagent is selective to one of the radioactive elements, while the other one remains with REE. For instance, a 30% concentrated oxalic acid precipitates 99% thorium along with 98% REE at 30 °C, leaving uranium in the solution according to reactions 10 to 13 [35,39,46,54]. Next, a mixed alkali solution of Na_2CO_3 and $NaHCO_3$ selectively leaches and recovers 99% thorium from the oxalate cake, reaction 14, while the REE remains in the solid form as carbonates [39]:

$$RE_2(SO_4)_3 + 3H_2C_2O_4 \rightarrow RE_2(C_2O_4)_3 \downarrow +3H_2SO_4 \tag{10}$$

$$UO_2SO_4 + H_2C_2O_4 \rightarrow UO_2(C_2O_4) + H_2SO_4 \tag{11}$$

$$UO_2(C_2O_4) + 3H_2SO_4 \rightarrow UO_2SO_4 + 2CO_2 + 3H_2O \tag{12}$$

$$Th(SO_4)_2 + 2H_2C_2O_4 \rightarrow Th(C_2O_4)_2 \downarrow +2H_2SO_4 \tag{13}$$

$$Th(C_2O_4)_2 + 4Na_2CO_3 + 2NaHCO_3 \rightarrow Na_6Th(CO_3)_5 + 2Na_2C_2O_4 + CO_2 + H_2O \tag{14}$$

Vijayalakshmi et al. [46] also reported a multi-step separation of radioactive elements from REE wherein the thorium is initially precipitated from a sulfate liquor by adding ammonium hydroxide (NH_4OH). Then, REE was separated from uranium in a secondary precipitation step in the form of REE oxalate. They suggested employing an excess amount of oxalic acid to lower the pH of the solution, and adjust it between 1 and 2 to avoid co-precipitation of impurities, e.g., aluminum and iron, with REE [40]. In such a multi-step separation of radioactive elements and REE, control of pH is more comfortable during the process, and the efficiency of the recovery is high. However, a large amount of reagent is required, which is economically infeasible.

To overcome the drawbacks of the multi-step precipitation, various studies were conducted to recover either the REE or both thorium and uranium in a single-step process. Kul et al. [41] applied a double-salt single-step approach and reported 98% recovery of REE in the precipitate while only 15% thorium is co-precipitated. The hydroxide reagents have the potential for a single-step recovery of the radioactive elements from a chloride liquor [48,49,55]. For instance, the hydrated lime, $Ca(OH)_2$, precipitates thorium at a pH of 2.5, while the loss of REE through co-precipitation is minimized [48]. Yu et al. [49] employed the hydrated lime to extract the thorium and uranium from a chloride liquor which was produced from the processing of monazite and REE carbonatite minerals from the Montviel deposit in the North West of Quebec in Canada. They demonstrated that the addition of hydrogen peroxide (H_2O_2) is necessary to oxidize the iron and facilitate its precipitation. They reported recovery of 99% Th, 80% U, and over 90% iron and phosphate impurities, while minimizing the co-precipitation of REE to less than 2%. The advantage of the single-step separation process is a small amount of reagent required for the precipitation, whereas adjusting a proper pH for selective precipitation is difficult.

The flowsheet in Figure 4 [39,46,48], illustrates various potential precipitation pathways to separate radioactive elements from REE in a sulfate or chloride liquor that would be produced in the upstream hydrometallurgical processes. Table 4 summarizes the recovery yield of the radioactive elements obtained from different precipitation processes.

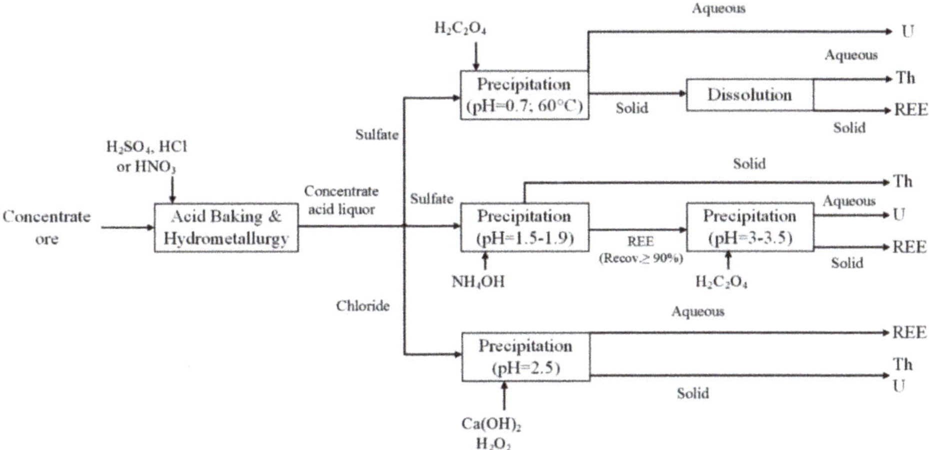

Figure 4. Process flowsheet for separation of Th and U from REE by precipitation.

Table 4. Uranium and thorium separation from REE by precipitation technique.

Original Ore	Upstream Liquor Feed	Precipitation Reagent	Final pH	Th	U	REE	Ref.
Monazite	Sulfate	1st step: Oxalic acid	0.7	98	-	99	[39]
		2nd step: Alkali leaching	-	>99	-	<1	
Synthesized solution	Nitrate	KOH and NH₄OH	4.5	-	90	low	[51]
Monazite, Apatite	Chloride	Hydrated lime and NH₄OH	2.5	>99	-	5	[48]
Monazite, REE carbonate	Chloride	Lime and H_2O_2	-	99	>80	<2	[49]
Bastnäsite, Monazite	Hydroxide cake	HCl	5.8	>99	>99	2.3	[56]
Xenotime	Sulfate	1st step: NH₄OH	1.5–1.9	>99	-	<6	[46]
		2nd step: Oxalic acid	-	-	-	>98	

4. Separation by Solvent Extraction

The solvent extraction technique has received significant attention for separation of thorium and uranium from REE by using appropriate extractants, which can be dissolved into an organic solvent to provide an immiscible phase and enough interface with the aqueous liquor of the REE.

This process is typically conducted by two main groups of extractants, including amine and organophosphorus extractants. The amine group can be divided into primary and tertiary types where the primary amine is highly selective towards the thorium in either sulfate or chloride liquors, and the tertiary amine is selective towards the uranium in sulfate liquors. The organophosphorus extractants are usually divided into acid and neutral types, which are applicable to nitrate, chloride and sulfate liquors.

4.1. Solvent Extraction with Amine Extractants

Amine extractants have been employed in the AMEX process, which was developed in the late 1950s, for extracting radioactive elements from REE-bearing minerals. In this process, the amine extractants are mixed with a sulfate leach liquor produced from monazite sands [57]. Thorium is first extracted with a primary amine (reaction 15) followed by a nitric acid stripping step. Then, uranium is extracted with a tertiary amine (reaction 16) and stripped with sodium carbonate,

$$4RNH_2H^+HSO_4^-{}_{(org)} + [Th(SO_4)_4]^{4-}_{(aq)} \rightarrow \left(RNH_2H^+\right)_4[Th(SO_4)_4]^{4-}_{(org)} + 4HSO_4^-{}_{(aq)} \tag{15}$$

$$4R_3NH^+HSO_4^-{}_{(org)} + [UO_2(SO_4)_3]^{4-}_{(aq)} \rightarrow \left(R_3NH^+\right)_4[UO_2(SO_4)_3]^{4-}_{(org)} + 4HSO_4^-{}_{(aq)} \tag{16}$$

Table 5 lists some amines employed for solvent extraction of radioactive elements from the REE [58–60].

Table 5. Performance of amine extractants for uranium and thorium separation from REE.

Amine		Experimental Conditions				Extraction (%)			Ref.
Type	Name	Liquor	Ore	pH	Time (min)	Thorium	Uranium	REE	
Primary	Primene JM-T	Sulfate	Monazite	-	5	95.4	8.8	0.32	[58]
	N$_{1923}$	Sulfate	Bastnäsite	-	-	>97	-	E$_{Ce}$: 3–8	[60]
Secondary	Octylamine	Sulfate	Monazite	4	15	70–80	50–60	-	[59]
	N-methylaniline	Sulfate	Monazite	4	15	70–80	5–10	0	[59]
Tertiary	Alamine 336	Sulfate	Monazite	-	5	3.1	82.4	0.02	[58]
	N,N-dimethylaniline	Sulfate	Monazite	4	15	70	15–20	E$_{Ce, Eu, Y}$: 0	[59]
				7	30	0	55	E$_{Ce, Eu, Y}$: 0	[59]
Mixture	Primene JM-T and Alamine 336	Sulfate	Monazite	-	5	45.5	58.8	0.04	[58]

The Primene JM-T is the most applied primary amine to extract thorium, and the Alamine 336 is a tertiary amine that has been employed for uranium extraction from the REE-bearing minerals [58,61,62]. The N$_{1923}$, i.e., (C$_n$H$_{2n+1}$)$_2$CHNH$_2$ ($n = 9$–11), is an alternative primary amine for this process, which has been employed in Baotou and Sichuan in China [60,63–65]. The N$_{1923}$ extractant is characterized by low solubility in water and a high separation factor between thorium and the REE, especially in sulfate liquors [66]. For instance, the selectivity of the N$_{1923}$ for thorium is 600 times higher than for cerium in a sulfate liquor produced from the bastnäsite and monazite concentrates [60,66]. The reaction between the N$_{1923}$ and thorium takes place in the interfacial zone, and it is controlled by the thorium mass transfer [67]. Figure 5 [60] shows a flowsheet for the recovery of thorium from REE-bearing minerals, where a thorium recovery of over 99% is achievable in an organic phase using a primary amine in a multi-step process, including solvent extraction, scrubbing, and precipitation [60].

Employing a mixture of primary and tertiary amines is a promising method for the simultaneous extraction of thorium and uranium from the REE, which is economically and technically favored due to a decrease in the number of steps in the solvent extraction process. For instance, simultaneous separation of Th and U has been reported from sulfate liquor during the processing of monazite using a mixture of Primene JM-T (i.e., a primary amine) and Alamine 336 (i.e., a tertiary amine), wherein the optimized process conditions (concentration of amines, contact time, and pH) resulted in the extraction of 99.9 and 99.5% thorium and uranium, respectively, while the extraction of REE was less than 0.1% [58].

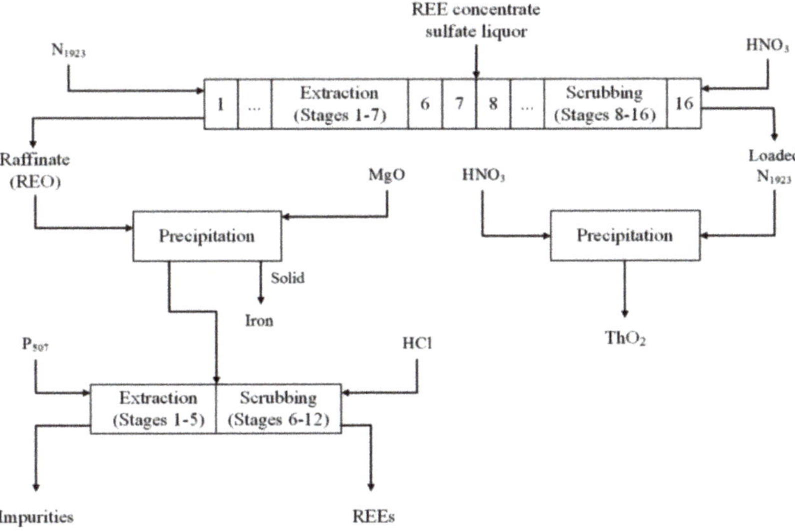

Figure 5. Process of separating Th from REE of Baotou bastnäsite leaching.

4.2. Solvent Extraction with Phosphorous Extractants

Phosphorous extractants are alternative for amines to extract the radioactive elements [4,68–72]. This group of extractants includes phosphate esters and amines. Table 6 presents the performance of different types of organophosphorus extractants during the solvent extraction process for separating Th and U from different REE-bearing liquors.

Table 6. Phosphorous extractants for uranium and thorium separation from REE.

Type of Extractant	Feed Liquor	Phosphorous Extractant	Extraction (%)			Ref.
			Th	U	REE	
	Acidic	Cyanex 272	83	-	12	[69]
Acid	Nitrate	DEHEHP	20		Ce: 95	[65]
	Nitrate	Mixture of HEH and EHP in kerosene	95	-	Ce: 99	[73]
	Nitrate	TiAP (2 solvent extraction steps)	99	95	<2	[74]
Neutral	Nitrate	p-phosphorylated calixarene	99	-	5	[75]
	Nitrate	TEHP in n-paraffin	50–77	2.5	Y: 0.17	[68]
Other	Nitrate	Aliquat 336	97	54	<3	[32,76]
	Nitrate	polyaramide	90	-	>48	[77]

4.2.1. Acid Organophosphorus Extractants

Acidic organophosphorus extractants such as di-(2-ethylhexyl)-phosphoric acid (DEHP), (2-ethylhexyl) 2-ethylhexyl-phosphonic acid (EHEHP), and di-(2-ethylhexyl) 2-ethylhexyl phosphonate (DEHEHP), have been proposed as promising reagents to separate radioactive elements from REE [64,65,78,79]. Reaction 17 shows absorption of thorium during treatment with DEHEHP [64],

$$\text{Th(NO}_3)^{3+} + 3\text{NO}_3^- + \text{L}_{(aq)} \rightarrow \text{Th(NO}_3)_4 \cdot \text{L}_{2(org)} \tag{17}$$

where L is the ligand or extractant. The acid organophosphorus extractants are efficient, especially for the extraction of radioactive elements from highly acidic sulfate solutions. In addition, DEHP, EHEHP, and DEHEHP can be employed for individual separation of the REE [80], since these extractants have a low affinity to extract trivalent REE while cerium(IV) is simultaneously extracted with thorium and uranium [65]. Nonetheless, to avoid a high loss of the tetravalent cerium, it would be individually recovered in a downstream stripping. Moreover, cerium(IV) can be recovered by applying a second solvent extraction process to the organic phase, owing to a high separation factor of cerium (IV) over thorium, which is 36.

The Cyanex is another type of acid organophosphorus extractant that is widely employed for thorium and uranium separation from REE [69,81–83]. The mechanism of metals extraction by Cyanex is the cation exchange [69], where the strength of the acid determines the performance of the Cyanex extractants. The commonly used Cyanex extractants are the Cyanex 272 (di-2,4,4-trimethyl phosphonic acid), Cyanex 301 (bis-2,4,4-trimethylprntyl phosphonic acid) [84], and Cyanex 302 [69]. The Cyanex 272 shows higher efficiency compared to the Cyanex 302 for separating thorium from lanthanides due to the higher strength of the latter acid [83]. In addition, applying a mixture of extractants, e.g., a mixture of HEH and EHP in kerosene, is a promising approach to increase the efficiency of the solvent extraction for separating thorium from REE (Table 6).

4.2.2. Neutral Organophosphorus Extractants

The tri-n-butyl phosphate (TBP) is the most common neutral organophosphorus extractant for separating thorium and uranium from REE that is especially efficient in a nitrate medium. Tris(2-ethylhexyl) phosphate (TEHP), and tri-iso-amyl phosphate (TiAP) are other neutral organophosphorus extractants, which are usually dissolved in n-paraffin or in xylene [68,74,85–87]. In terms of efficiency, the TEHP has a special affinity to uranium, which leads to the loss of some thorium in the aqueous phase. Therefore, the TEHP should be employed in a two-step solvent extraction (Figure 6) to recover uranium and then thorium [68].

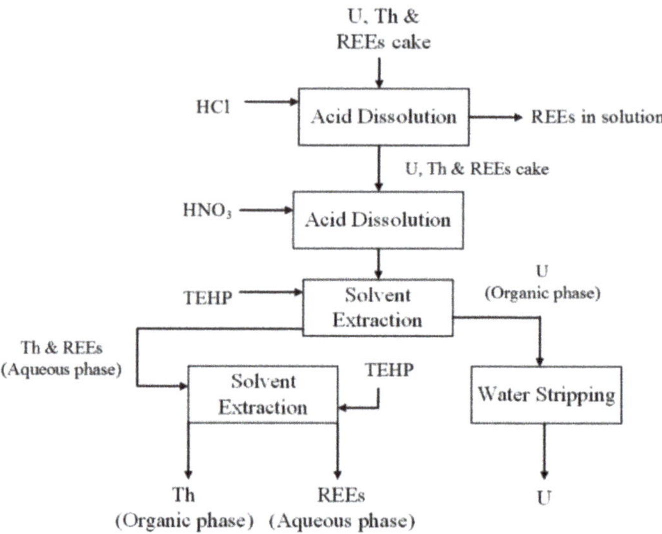

Figure 6. Separation of thorium and uranium from REE by solvent extraction method with TEHP in n-paraffin.

Calixarene is another group of neutral organophosphorus extractants appropriate for recovering radioactive elements from REE [71,73,75,88]. For instance, 5,11,17,23-tetra(diethoxyphosphoryl)-25,26,27,28-tetrapropyloxy calixarene and p-phosphorylated calixarene have been employed, either as a resin in chromatography to concentrate thorium and uranium [88] or as a solvent extractant to recover thorium and uranium from the REE nitrate liquors [71,75]. The efficiency of thorium extraction (reaction 18) would vary with the concentration of nitric acid used in the upstream process. It is reported that around 85% of thorium is extracted using 2 mol/L nitric acid while the extraction of REE (Gd, La, and Yb) is less than 10% [75],

$$Th^{4+} + 4NO_3^- + 3Calixarene \rightarrow Th(NO_3)_4 \cdot 3Calixarene. \tag{18}$$

4.2.3. Improvement of Solvent Extraction Process

Although various types of extractants have been applied in the solvent extraction process to separate radioactive elements from REE [32,76,77,89], the selection of an appropriate extractant is a function of its cost, selectivity towards one of the radioactive elements, and the feed liquor (sulfate, nitrate or chloride). For instance, Cyanex extractant shows poor metal extraction, is costly and requires high acidic environments, which make it infeasible for industrial applications.

Employing a mixture of acid and neutral organophosphorus extractants potentially improves the solvent extraction process that would result in a higher separation efficiency than using each extractant separately. For instance, although the selectivity in uranium extraction by acid organophosphorus PC88A (mono (2-ethyl hexyl) ester) showed better performance than the neutral organophosphorus TBP, the best result was obtained with a mixture of both, where the distribution of uranium was 0.1, 1.1, and 2.3 after 20 min treatment with TBP, PC88A, and their mixture (0.15 M TBP and 0.15 M PC88A in xylene), respectively [86,90]. This enhancement in performance is attributed to the role of TBP, which dehydrates uranium, resulting in an enhancement in the hydrophobic nature of the species and improving the extraction by the mixture.

In addition, solvent extraction is typically followed by a stripping step to separate the recovered radioactive element(s) from the accompanying reagent. Recovery of the radioactive elements is usually over 90% after stripping [74,82]. For the stripping step, an appropriate acidic reagent is employed depending on the feed of the solvent extraction [69,74,85,86,90]. For instance, for a feed of sulfate solution, nitric or sulfuric acid is employed as the stripping reagent [58]. For a feed of nitrate liquor, the stripping reagent is usually hydrochloric acid or sulfuric acid [69,76,82].

5. Separation by Ion Chromatography

5.1. Cation Exchange Resin

Cation exchange column has been employed to investigate chromatography principles on the separation of radioactive elements from REE [91–94]. In ion chromatography, it is possible to employ an active stationary phase or an active mobile phase, circulating in a neutral chromatography column. A configuration of a silica gel resin impregnated with ammonium nitrate (NH_4NO_3) in mobile phases of HDEHP (di-(2-ethylhexyl)-dithiophosphoric acid) and HDiBDTP (di(iso-butyl)dithiophosphoric acid) resulted in the separation of metal ions and the individual REE [94,95]. This method is based on different retention times of thorium, uranium, and REE in the column, which is affected by the employed resin [91]. For this purpose, various resins have been employed such as transUranic-Element Specific (TRU Spec), i.e., a resin composed of TBP and octyl(phenyl)-N,Ndiisobutylcarbamoylmethylphosphine oxide (CMPO) and supported on an inert polymer substrate [93], and the Dionex, i.e., a column for the separation of transition and lanthanide metals from contaminated water [92].

These columns adsorb all the elements of interest and separate them using different eluent. Once adsorbed, the order of elements elution can vary in function of the column resin. By employing appropriate eluents, thorium and uranium can thus be recovered individually with an efficiency of over 90% [91,93]. For instance, Figure 7 [91] shows that the elements of interest, i.e., thorium, uranium, and the REE, are adsorbed in the Ion Pac CS 5 column, then recovered separately with a mixture of hydrochloric acid and ammonium sulfate as eluent. Accordingly, uranium is initially recovered in the form of UO_2Cl^-, then thorium is recovered as $ThSO_4^{2+}$ [91]. Another chromatographic investigation showed over 95% recovery of the light REE, which was adsorbed in a transuranic-element specific resin (TRU Spec), using a LN Spec resin column by a nitric acid elution; later, thorium and uranium were recovered, respectively, by hydrochloric acid and a mixture of hydrochloric and hydrofluoric acids [93,96].

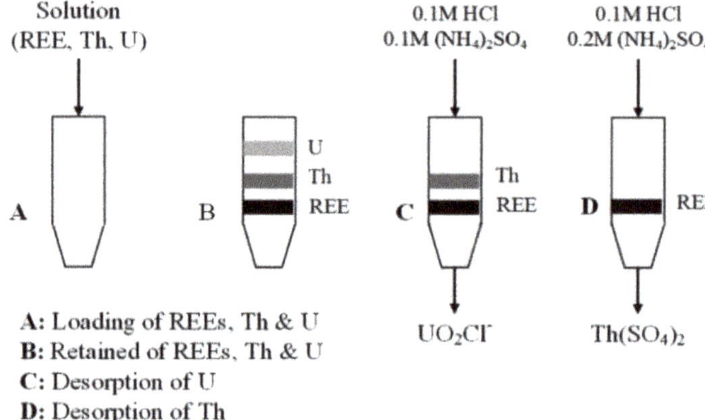

Figure 7. Cation exchange extraction with Dionex Ion Pac CS5 column.

In addition, a combination of different chromatographic columns with different resins can achieve a higher recovery and purity of the products, in addition to the separation of REE from one to another [97,98]. However, the low flow rate of the eluent, e.g., 0.25 to 1.5 mL/min, in the cation exchange resin columns hinder their industrial application. This technique might be suitable in a downstream step where high purity is required.

5.2. Anion Exchange Resin

The anion exchange column is an alternative process for separating metal ions from the contaminating elements where chelating resins such as Dowex [99] and Amberlite [100–106] are employed. The efficiency of the separation depends on both the anion exchange resin and the type of acidic eluent [99–101,106]. A malonic acid eluent in methanol is employed to circulate impurities through the Dowex ion-exchange column, resulting in the adsorption of uranium with an efficiency of over 99%. However, thorium recovery was inefficient, and it remained with REE in the eluate. The Amberlite XAD-4 is another anion exchange resin that is applicable in a wide range of pH [100] and could achieve 99% recovery of uranium [107]. Compared to the XAD-4, the Amberlite XAD-2 has less surface area and larger pore diameter, and extracts uranium when impregnated with Cyanex 302; however, it also partially co-extract thorium (separation factor U/Th = 1.2×10^4) [101]. In some cases, methanol is employed to elute the collected uranium in the column [101,107].

The anion exchange resin column is effective for uranium separation from REE, whereas it is inefficient for thorium separation [108]. Therefore, it can be used as the last step of REE purification. The Amberlite IRA 402 Cl resin was applied as a final purification step in a successive separation of thorium and uranium by precipitation. The low concentration uranium remained with the REE was then separated by anion chromatography, where REE recovery of 99% was achieved by the elution with NaCl. Next, 99% of uranium was recovered by water elution, Figure 8 adapted from [106].

In terms of the potential for the scale-up, in anion exchange resins, similar to cation exchange resins, the small flow rate of the feed solution, e.g., around 1 mL/min, is a significant limitation for an industrial-scale application [99,100,106].

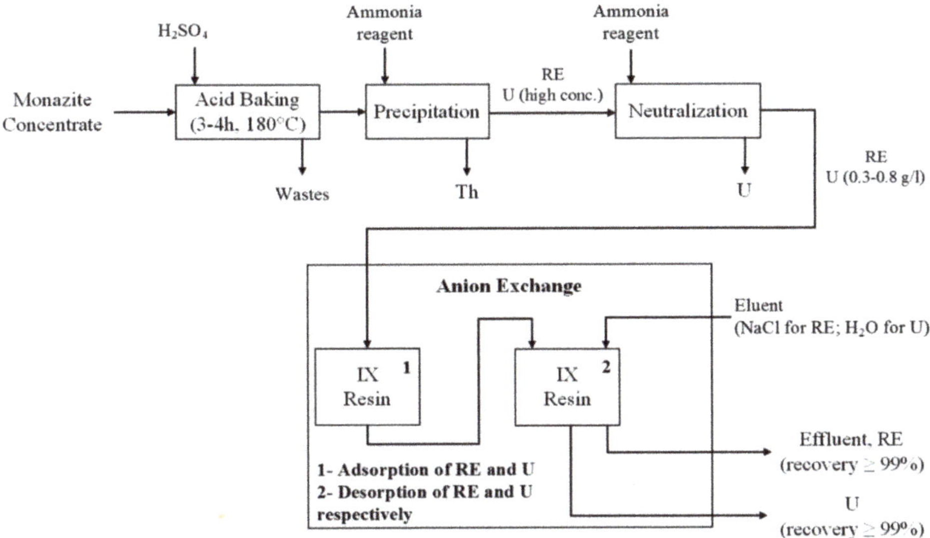

Figure 8. Flow diagram of uranium removal from REE.

6. Separation by Membranes

Membranes have emerged as a new method for recovering thorium and uranium from the REE. At first glance, membranes were used due to their selectivity to individually recover thorium and uranium from other metal ions in liquid solutions [109–111], e.g., recovery of thorium by either graphene oxide (GO) or a silica membrane [112,113]. Such a recovery occurred by the formation of a complex between the element of interest and the membrane. In general, membranes have proved to be efficient and selective in acidic conditions at a pH of 4 to 5.5 for recovery of uranium and lower than 4 to recover thorium. However, to increase the recovery of radioactive elements, it is required to functionalize the membrane using an additive with a higher selectivity towards target elements [114–119]. Membrane functionalization is done by polymer adsorption at the membrane surface. Several functionalization methods are described in detail by Xu et al. [120]. The functionalized membrane has two main purposes, increasing the membrane resistance to the acid media, i.e., at pH lower than 4, and allowing for a better affinity of the membrane toward some ions by modifying some of the membrane parameters, e.g., surface rugosity, conductivity or hydrophobicity [120,121]. In this case, membrane functionalization is thus synonym of a increasing the selectivity of the membrane towards thorium and uranium versus REE (neodymium, europium, and samarium) at ambient conditions and pH below 2. For instance, Li et al. [109] functionalized graphene oxide with PDA (GO-PDA) and showed that, as in Figure 9 adapted from [109], it is more efficient for simultaneous recovery of thorium and uranium as compared with a graphene oxide membrane. The selectivity increase was caused by the surface modification of the functionalization that generated porous channels in the membrane that allowed the REE to pass through at low pH while being impermeable to thorium and uranium [109].

Despite the encouraging performance of the functionalized membranes for simultaneous recovery of radionuclides from the REE, further study is essential to evaluate the effect of various functionalized groups towards selectivity or the regeneration capability of the membranes, especially at high acidic conditions.

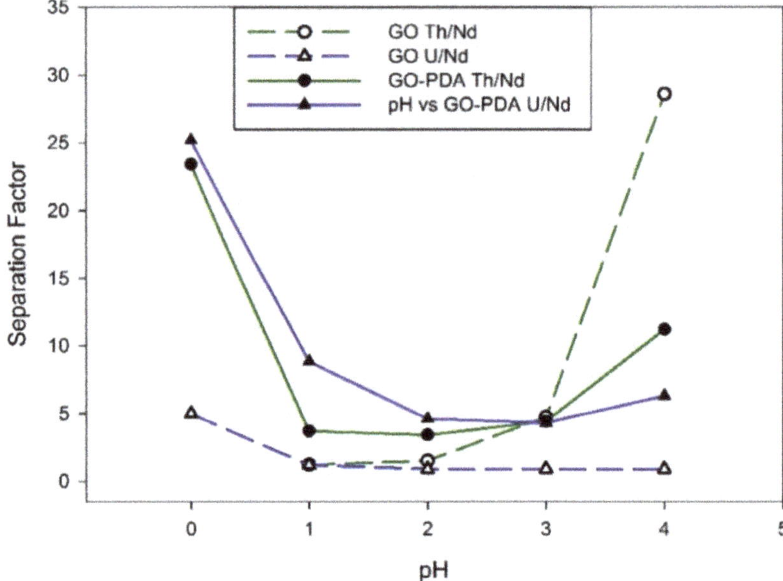

Figure 9. Separation factor of actinides/Nd by graphene oxide (GO) and functionalized graphene oxide with PDA (GO-PDA) membranes. (Reproduced with permission from ref. [109], copyright (2012), Elsevier).

7. Selection of a Separation Process and Potential Waste Management Approaches

Selection of an appropriate process from the presented methods for separating radioactive elements from REE requires various considerations such as the composition of the fresh ore, the applied upstream processes, operating conditions of the reagent, temperature and pH, scale of the process, purification range, and economy of the process, as well as advantages and limitations in each separation techniques. Table 7 summarizes these considerations for leaching, precipitation, solvent extraction, and ion chromatography.

As the separation of the radioactive elements from the REE is required in the REE supply chain, a new problem appears after this step since the presence of radioactive elements in tailings issues a waste management problem. Even though several waste management techniques exist such as water dilution [122], used when thorium and uranium are in the aqueous phase, or the safe storage when they are in the solid phase [123], those seem unreasonable due to high water or space occupation as well as health problem they may generate. One of the most recent propositions was to use thorium as a co-product of the REE industry as it would reduce the waste management problems, and could be used as a feed in the new generation of nuclear reactors [124,125]. As a matter of the fact, thorium is already a co-product of the titanium industry, and its recovery from the REE industry would represent the third most important thorium resource after titanium and uranium [124,126].

Table 7. Advantages and disadvantages of different methods for separating radioactive elements from REE.

Table	Advantages	Limitations	Recovery (%)	Scalability [4]
Leaching	• Simultaneous recovery of Th & U from REE; • Use of cost-effective reagents;	• Must have a particle size of ore/cake similar or smaller than liberation-equivalent size to avoid higher loss of REE [1]; • Feed (REE-radioactive mixture) must be in solid phase;	Th: >98 [3a] U: 65–95 [3a]	+++
Precipitation	• Simultaneous recovery of Th & U from REE; • Being pH dependent, possible individual precipitation of Th or U from REE; • Use of cost-effective reagents;	• Highly dependent on reagent, pH and sometimes temperature; • Possible co-precipitation of REE with Th & U if right pH is not maintained [2]; • Difficult to separate both Th & U simultaneously from REE with high recovery in one single step2 although it can be possible;	Th ≥ 98 U > 80 [3a] Th > 98 U > 90 [3b]	++
Solvent Extraction	• High selectivity towards Th & U; • High recovery of Th & U in individual separation steps;	• Low recovery of Th & U in simultaneous separation; • Multiple separation steps required for individual separation of Th & U; • Use of chemicals: cost of reagents and environmental impact;	Th ≥ 70 U ≈ 55 [3a] Th > 95 U > 95 [3b]	+++
Ion chromatography	• Extraction of Th & U in one step; • High recovery of both radioactive elements;	• Low flow rates; • Selection of the scale-up technology and configuration (batch, column, expanded/fluidized-bed or suspended bed); • Anion exchange only extracts U (with the presence of ppm);	Th: 90–99 U: 90–99	++

[1] Liberation-equivalent size is the optimum size to separate the desired REE minerals off the ore grain, normally found in physical beneficiation steps. [2] Normally pH should be below 6.5 to avoid co-precipitation of REE with radioactive elements, but in fact, pH is dependent on the liquor and reagent. [3a] Simultaneous separation of Th & U from REE; [3b] Individual separation of Th & U from REE; [4] Scalability factors: +++: easy scale-up (process continuous, well-known or developed in other industries); ++: scale-up with some difficulties (process semi-continuous or discontinuous or difficulty to separate two phases). References: ([127,128]); Leaching ([129,130]); Precipitation ([131,132]); SX ([130,133–135]); Ion chromatography ([136–138]).

8. Conclusions

Radioactive elements (thorium and uranium) are commonly associated with REE-bearing minerals, where the concentration depends on the mineral, the formation of the rocks, and the geographical position of the deposit. The presence of radioactive elements causes various problems in the environment and waste management. The extraction of thorium and uranium is necessary to ensure having a low radioactive product, while the loss of REE is minimized. The utilization of conventional hydrometallurgical processes such as selective precipitation, leaching, and solvent extraction for the extraction of radioactive elements is often conducted using complex industrial processes. There is a new trend of investigating new separation processes, such as the ion-exchange chromatography and membrane separation. These processes are yet at lab-scale development stage, but they seem to result in a more selective separation of Th and U from REE. Further research and development activities are required to maturate such processes and to evolve new technologies for economically viable applications at an industrial scale.

The most critical parameters to be controlled in these methods are the operating conditions (pH, and temperature), reagent type, and upstream processes. Depending on the process requirements and limitations, either one- or multi-steps processing would be applied to efficiently separate radioactive elements from REE. Considering the efficiency and the cost of the process, a specific process can be selected with regard to the advantages and limitations in each process.

Author Contributions: A.C.G. reviewed literature, wrote the first draft, and revised the manuscript. M.L. scientifically reviewed and edited the manuscript. A.A. scientifically reviewed and edited the manuscript. J.C. was the supervisor and also scientifically reviewed the manuscript. All authors have read and agreed to the published version of the manuscript.

Funding: This work is part of a NSERC/CRD funding supplied by Natural Science and Engineering Research Council of Canada and Niobec company in Province of Quebec of Canada.

Acknowledgments: Authors would like to sincerely acknowledge the support of NSERC and Niobec.

Conflicts of Interest: The authors declare no conflict of interest.

References

1. Alami, D. Environmental Applications of Rare-Earth Manganites as Catalysts: A Comparative Study. *Environ. Eng. Res.* **2013**, *18*, 211–219. [CrossRef]
2. Bouzigues, C.; Gacoin, T.; Alexandrou, A. Biological applications of rare-earth based nanoparticles. *ACS Nano* **2011**, *5*, 8488–8505. [CrossRef] [PubMed]
3. Chen, L.; Si, Z.C.; Wu, X.D.; Weng, D.; Ran, R.; Yu, J. Rare earth containing catalysts for selective catalytic reduction of NOx with ammonia: A Review. *J. Rare Earths* **2014**, *32*, 907–917. [CrossRef]
4. Gupta, C.K.; Krishnamurthy, N. *Extractive Metallurgy of Rare Earths*; CRC Press: Boca Raton, FL, USA, 2004.
5. Hu, T.; Deng, R.D.; Yang, X.F. Applications and challenges of rare earth resources in China. In Proceedings of the 2013 3rd International Conference on Information Science, Automation and Material System, ISAM 2013, Guangzhou, China, 13–14 April 2013; pp. 159–162.
6. Jordens, A.; Cheng, Y.P.; Waters, K.E. A review of the beneficiation of rare earth element bearing minerals. *Miner. Eng.* **2013**, *41*, 97–114. [CrossRef]
7. Bhargava, S.K.; Ram, R.; Pownceby, M.; Grocott, S.; Ring, B.; Tardio, J.; Jones, L. A review of acid leaching of uraninite. *Hydrometallurgy* **2015**, *151*, 10–24. [CrossRef]
8. Jaireth, S.; Hoatson, D.M.; Miezitis, Y. Geological setting and resources of the major rare-earth-element deposits in Australia. *Ore Geol. Rev.* **2014**, *62*, 72–128. [CrossRef]
9. Clow, G.; Salmon, B.; Lavigne, M.; McDonough, B.; Pelletier, P.; Vallières, D. *Technical Report on Expansion Options at the Niobec Mine, Québec, Canada*; IAMGOLD Corporation: Toronto, ON, Canada, 2011.
10. Bosserman, P.J. *Recovery of Cerium*; Google Patents: Mountain View, CA, USA, 1995.
11. Zou, D.; Chen, J.; Li, D.Q. Separation chemistry and clean technique of cerium(IV): A review. *J. Rare Earths* **2014**, *32*, 681–685. [CrossRef]

12. Simandl, G.J. Geology and economic sifnificance of current and future rare earth element sources. In Proceedings of the 51st Annual Conference of Metallurgists, Niagara Falls, ON, Canada, 30 September–3 October 2012.

13. Mowafy, A.M. Biological leaching of rare earth elements. *World J. Microbiol. Biotechnol.* **2020**, *36*, 61. [CrossRef]

14. Driscoll, M.O. An Overview of Rare Earth Minerals Supply and Applications. In *Materials Science Forum*; Trans Tech Publications Ltd.: Zurich, Switzerland, 1991; pp. 409–420.

15. Jones, A.P.; Wall, F.; Williams, C.T. *Rare Earth Minerals: Chemistry, Origin and Deposits*; Springer: London, UK, 1996.

16. Zhang, J.; Edwards, C. A Review of Rare Earth Mineral Processing Technology. In Proceedings of the 44th Annual Canadian Mineral Processors Operators Conference, Ottawa, ON, Canada, 19–21 January 2010.

17. Chambers, D.B.; Lowe, L.M.; Feasby, D.G. Radiological aspects of naturally occuring radioactive materials (NORM) in the ore processing and production of rare earth element concentrates. In Proceedings of the 51st Annual Conference of Metallurgists, COM, Niagara Falls, ON, Canada, 30 September–3 October 2012.

18. Feasby, D.G.; Chambers, D.B.; Lowe, L.M. Assesment and management of radioactivity in rare earth elements production. In Proceedings of the Rare Earth Elements (COM 2013), West Westmount, QC, Canada, 27–31 October 2013.

19. Park, B.T. Management of thorium and uranium in mining and processing of rare earth minerals. In Proceedings of the 51st Annual Conference of Metallurgists, Niagara, ON, Canada, 30 September–3 October 2012; pp. 171–184.

20. Cheng, J.; Hou, Y.; Che, L. Flotation separation on rare earth minerals and gangues. *J. Rare Earths* **2007**, *25*, 62–66.

21. Fang, J.; Zhao, D. Separation of rare earth from tails of magnetite separation in Bao Steel's concentrator. *Met. Mine* **2003**, *321*, 47–49.

22. Goode, J.R. Thorium and rare earth recovery in Canada: The first 30 years. *Can. Metall. Q.* **2013**, *52*, 234–242. [CrossRef]

23. Kogel, J.E.; Trivedi, N.C.; Barker, J.M.; Krukowsky, S.T. *Industrial Minerals & Rocks: Commodities, Markets and Uses*, 7th ed.; Society for Mining, Metallurgy, and Exploration, Inc.: Littleton, CO, USA, 2006.

24. Dev, S.; Sachan, A.; Dehghani, F.; Ghosh, T.; Briggs, B.R.; Aggarwal, S. Mechanisms of biological recovery of rare-earth elements from industrial and electronic wastes: A review. *Chem. Eng. J.* **2020**, *397*, 124596. [CrossRef]

25. Deshmane, V.G.; Islam, S.Z.; Bhave, R.R. Selective Recovery of Rare Earth Elements from a Wide Range of E-Waste and Process Scalability of Membrane Solvent Extraction. *Environ. Sci. Technol.* **2020**, *54*, 550–558. [CrossRef] [PubMed]

26. Rivera, R.M.; Ulenaers, B.; Ounoughene, G.; Binnemans, K.; Van Gerven, T. Extraction of rare earths from bauxite residue (red mud) by dry digestion followed by water leaching. *Miner. Eng.* **2018**, *119*, 82–92. [CrossRef]

27. Reid, S.; Tam, J.; Yang, M.; Azimi, G. Technospheric Mining of Rare Earth Elements from Bauxite Residue (Red Mud): Process Optimization, Kinetic Investigation, and Microwave Pretreatment. *Sci. Rep.* **2017**, *7*, 15252. [CrossRef]

28. Borra, C.R.; Pontikes, Y.; Binnemans, K.; Van Gerven, T. Leaching of rare earths from bauxite residue (red mud). *Miner. Eng.* **2015**, *76*, 20–27. [CrossRef]

29. Davris, P.; Balomenos, E.; Panias, D.; Paspaliaris, I. Selective leaching of rare earth elements from bauxite residue (red mud), using a functionalized hydrophobic ionic liquid. *Hydrometallurgy* **2016**, *164*, 125–135. [CrossRef]

30. Tuan, L.Q.; Thenepalli, T.; Chilakala, R.; Vu, H.H.; Ahn, J.W.; Kim, J. Leaching Characteristics of Low Concentration Rare Earth Elements in Korean (Samcheok) CFBC Bottom Ash Samples. *Sustainability* **2019**, *11*, 2562. [CrossRef]

31. Abdel-Rehim, A.M. An innovative method for processing Egyptian monazite. *Hydrometallurgy* **2002**, *67*, 9–17. [CrossRef]

32. El-Nadi, Y.A.; Daoud, J.A.; Aly, H.F. Modified leaching and extraction of uranium from hydrous oxide cake of Egyptian monazite. *Int. J. Miner. Process.* **2005**, *76*, 101–110. [CrossRef]

33. Lapidus, G.T.; Doyle, F.M. Selective thorium and uranium extraction from monazite: I. Single-stage oxalate leaching. *Hydrometallurgy* **2015**, *154*, 102–110. [CrossRef]

34. Lapidus, G.T.; Doyle, F.M. Selective thorium and uranium extraction from monazite: II. Approaches to enhance the removal of radioactive contaminants. *Hydrometallurgy* **2015**, *155*, 161–167. [CrossRef]

35. Alex, P.; Hubli, R.C.; Suri, A.K. Processing of rare earth concentrates. *Rare Met.* **2005**, *24*, 210–215.

36. Eyal, Y.; Olander, D.R. Leaching of uranium and thorium from monazite: I. Initial leaching. *Geochim. Et Cosmochim. Acta* **1990**, *54*, 1867–1877. [CrossRef]

37. Panda, R.; Kumari, A.; Jha, M.K.; Hait, J.; Kumar, V.; Kumar, J.R.; Lee, J.Y. Leaching of rare earth metals (REMs) from Korean monazite concentrate. *J. Ind. Eng. Chem.* **2014**, *20*, 2035–2042. [CrossRef]

38. Shaw, K.G. *A Process for Separating Thorium Compounds from Monazite Sands*; Iowa State University: Ames, IA, USA, 1953.

39. Amer, T.E.; Abdella, W.M.; Wahab, G.M.A.; El-Sheikh, E.M. A suggested alternative procedure for processing of monazite mineral concentrate. *Int. J. Miner. Process.* **2013**, *125*, 106–111. [CrossRef]

40. Chi, R.; Xu, Z. A solution chemistry approach to the study of rare earth element precipitation by oxalic acid. *Met. Mater Trans B* **1999**, *30*, 189–195. [CrossRef]

41. Kul, M.; Topkaya, Y.; Karakaya, I. Rare earth double sulfates from pre-concentrated bastnasite. *Hydrometallurgy* **2008**, *93*, 129–135. [CrossRef]

42. Fourest, B.; Lagarde, G.; Perrone, J.; Brandel, V.; Dacheux, N.; Genet, M. Solubility of thorium phosphate-diphosphate. *New J. Chem.* **1999**, *23*, 645–649. [CrossRef]

43. Borai, E.H.; Abd El-Ghany, M.S.; Ahmed, I.M.; Hamed, M.M.; Shahr El-Din, A.M.; Aly, H.F. Modi fi ed acidic leaching for selective separation of thorium, phosphate and rare earth concentrates from Egyptian crude monazite. *Int. J. Miner. Process.* **2016**, *149*. [CrossRef]

44. Krebs, D.G.I.; Furfaro, D. The Kvanefjeld process. In Proceedings of the Alta 2013 Uranium-REE Conference, Perth, Australia, 25 May–1 June 2013.

45. Pawlik, C. Recovery of rare earth elements from complex and low grade deposits. In Proceedings of the ALTA 2013 Uranium-REE Conference, Perth, Australia, 25 May–1 June 2013.

46. Vijayalakshmi, R.; Mishra, S.L.; Singh, H.; Gupta, C.K. Processing of xenotime concentrate by sulphuric acid digestion and selective thorium precipitation for separation of rare earths. *Hydrometallurgy* **2001**, *61*, 75–80. [CrossRef]

47. Bearse, A.E.; Calkins, G.D.; Clegg, J.W.; Filbert, J.R.B. Thorium and rare earths from monazite. *Chem. Eng. Prog.* **1954**, *50*, 235–239.

48. Mackowski, S.J.; Raiter, R.; Soldenhoff, K.H.; Ho, E.M. Recovery of Rare Earth Elements. U.S. Patent 7,993,612 B2, 9 August 2011.

49. Yu, B.; Verbaan, N.; Pearse, G.; Britt, S. Beneficiation and extraction of REE from GEOMEGA resources' Montviel project. In Proceedings of the Rare Earth Elements (COM 2013), West Westmount, QC, Canada, 30 September–3 October 2013.

50. Grimaldi, F.S. The analytical chemistry of uranium and thorium. In Proceedings of the United Nations International Conference on Peaceful Uses of Atomic Energy, Geneva, Switzerland, 8 August 1995; pp. 605–617.

51. Tomazic, B.; Branica, M. Separation of uranium(VI) from rare earths(III) by hydrolytic precipitation. *Inorg. Nucl. Chem. Lett.* **1968**, *4*, 377–380. [CrossRef]

52. Kang, M.J.; Han, B.E.; Hahn, P.S. Precipitation and adsorption of uranium (VI) under various aqueous conditions. *Environ. Eng. Res.* **2002**, *7*, 149–157.

53. Abreu, R.D.; Morais, C.A. Purification of rare earth elements from monazite sulphuric acid leach liquor and the production of high-purity ceric oxide. *Miner. Eng.* **2010**, *23*, 536–540. [CrossRef]

54. Carter, G.; Everest, D.A.; Wells, R.A. Selective oxalate precipitation of thorium from sulfate leach solutions derived from monazite sands. *J. Appl. Chem.* **1960**, *10*, 149–155. [CrossRef]

55. Sozanski, A. Separation of Trace Quantities of Thorium from Lanthanide Solutions by the Method of Coprecipitation. *Przem. Chem.* **1978**, *57*, 360–362.

56. Chi, R.; Xu, Z.; Zhang, Z.; He, Z.; Ruan, Y. Review on separation and treatment of thorium resources. In Proceedings of the Conference of Metallurgists Proceedings, Vancouver, BC, Canada, 28 September–1 October 2014.

57. Crouse, D.J.; Brown, K.B. The Amex Process for Extracting Thorium Ores with Alkyl Amines. *Ind. Eng. Chem.* **1959**, *51*, 1461–1464. [CrossRef]

58. Amaral, J.C.B.S.; Morais, C.A. Thorium and uranium extraction from rare earth elements in monazite sulfuric acid liquor through solvent extraction. *Miner. Eng.* **2010**, *23*, 498–503. [CrossRef]

59. Borai, E.H.; Shahr El-Din, A.M.; El-Sofany, E.A.; Sakr, A.A.; El-Sayed, G.O. Extraction and Separation of Some Naturally Occurring Radionuclides from Rare Earth Elements by Different Amines. *Arab J. Nucl. Sci. Appl.* **2014**, *47*, 48–60.

60. Li, D.Q.; Zuo, Y.; Meng, S.L. Separation of thorium(IV) and extracting rare earths from sulfuric and phosphoric acid solutions by solvent extraction method. *J. Alloy. Compd.* **2004**, *374*, 431–433. [CrossRef]

61. Crouse, D.J.; Brown, K.B. *Recovery of Thorium, Uranium and Rare Earths from Monazite Sulfate Liquors by the Amine Extraction (AMEX) Process*; Oak Ridge National Laboratory: Oak Ridge, TN, USA, 1959.

62. Cox, J.J.; Ciuculescu, T.; Altman, K.; Hwozdyk, L. *Technical Report on the Eco Ridge Mine Project, Elliot Lake Area, Ontario, Canada*; Roscoe Postle Associates Inc.: Toronto, ON, Canada, 2012.

63. Chunhua, Y.; Jiangtao, J.; Chunsheng, L.; Sheng, W.; Guangxia, X. Rare Earth Separation in China. *Tsingshua Sci. Technol.* **2006**, *11*, 241–247.

64. Wang, Y.L.; Li, Y.L.; Li, D.Q.; Liao, W.P. Kinetics of thorium extraction with di-(2-ethylhexyl) 2-ethylhexyl phosphonate from nitric acid medium. *Hydrometallurgy* **2013**, *140*, 66–70. [CrossRef]

65. Zhao, J.M.; Zuo, Y.; Li, D.Q.; Liu, S.Z. Extraction and separation of cerium(IV) from nitric acid solutions containing thorium(IV) and rare earths(III) by DEHEHP. *J. Alloy. Compd.* **2004**, *374*, 438–441. [CrossRef]

66. Liu, J.J.; Wang, W.W.; Li, D.Q. Interfacial behavior of primary amine N1923 and the kinetics of thorium(IV) extraction in sulfate media. *Colloid Surf. A* **2007**, *311*, 124–130. [CrossRef]

67. Cianetti, C.; Danesi, P.R. Kinectics and mechanism in metal solvent extraction by some organophosphorous extractants. In Proceedings of the International Solvent Extraction Conference, ISEC, Denver, CO, USA, 26 August–2 September 1983.

68. Biswas, S.; Pathak, P.N.; Singh, D.K.; Roy, S.B. Comparative Evaluation of Tri-n-butyl Phosphate (TBP) and Tris(2-ethylhexyl) Phosphate (TEHP) for the Recovery of Uranium from Monazite Leach Solution. *Sep. Sci. Technol.* **2013**, *48*, 2013–2019. [CrossRef]

69. Nasab, M.E.; Sam, A.; Milani, S.A. Determination of optimum process conditions for the separation of thorium and rare earth elements by solvent extraction. *Hydrometallurgy* **2011**, *106*, 141–147. [CrossRef]

70. Gupta, C.K.; Krishnamurthy, N. Extractive Metallurgy of Rare-Earths. *Int. Mater. Rev.* **1992**, *37*, 197–248. [CrossRef]

71. Lu, Y.; Bi, Y.; Bai, Y.; Liao, W. Extraction and separation of thorium and rare earths from nitrate medium withp-phosphorylated calixarene. *J. Chem. Technol. Biotechnol.* **2013**, *88*, 1836–1840. [CrossRef]

72. Rabie, K.A.; Abdel-Wahaab, S.M.; Mahmoud, K.F.; Hussein, A.E.M.; Abd El-Fatah, A.I. Monazite- Uranium Separation and Purification Applying Oxalic- Nitrate-TBP extraction. *Arab J. Nucl. Sci. Appl.* **2013**, *46*, 30–42.

73. Zhang, Y.Q.; Xu, Y.; Huang, X.W.; Long, Z.Q.; Cui, D.L.; Hu, F. Study on thorium recovery from bastnaesite treatment process. *J. Rare Earths* **2012**, *30*, 374–377. [CrossRef]

74. Rakesh, K.B.; Suresh, A.; Rao, P.R.V. Separation of U(VI) and Th(IV) from Nd(III) by Cross-Current Solvent Extraction Mode Using Tri-iso-amyl Phosphate as the Extractant. *Solvent Extr. Ion Exch.* **2015**, *33*, 448–461. [CrossRef]

75. Li, Y.L.; Lu, Y.C.; Bai, Y.; Liao, W.P. Extraction and separation of thorium and rare earths with 5,11,17,23-tetra (diethoxyphosphoryl)-25,26,27,28-tetraacetoxycalix[4]arene. *J. Rare Earths* **2012**, *30*, 1142–1145. [CrossRef]

76. Ali, A.M.I.; El-Nadi, Y.A.; Daoud, J.A.; Aly, H.F. Recovery of thorium (IV) from leached monazite solutions using counter-current extraction. *Int. J. Miner. Process.* **2007**, *81*, 217–223. [CrossRef]

77. He, L.T.; Jiang, Q.; Jia, Y.M.; Fang, Y.Y.; Zou, S.L.; Yang, Y.Y.; Liao, J.L.; Liu, N.; Feng, W.; Luo, S.Z.; et al. Solvent extraction of thorium(IV) and rare earth elements with novel polyaramide extractant containing preorganized chelating groups. *J. Chem. Technol. Biot.* **2013**, *88*, 1930–1936. [CrossRef]

78. Sato, T. The extraction of uranium (VI) from sulphuric acid solutions by di-(2-ethyl hexyl)-phosphoric acid. *J. Inorg. Nucl. Chem.* **1962**, *24*, 699–706. [CrossRef]

79. Sato, T. The Extraction of Thorium from Hydrochloric Acid Solutions by di-(2-ethylhexyl)-phosphoric acid. *Z. Für Anorg. Und Allg. Chem.* **1968**, *358*, 296–304. [CrossRef]

80. Sato, T. Liquid-Liquid Extraction of Rare-Earth Elements from Aqueous Acid Solutions by Acid Organophosphorus Compounds. *Hydrometallurgy* **1989**, *22*, 121–140. [CrossRef]

81. Gupta, B.; Malik, P.; Deep, A. Extraction of uranium, thorium and lanthanides using Cyanex-923: Their separations and recovery from monazite. *J. Radioanal. Nucl. Chem.* **2002**, *251*, 451–456. [CrossRef]

82. Karve, M.; Gaur, C. Extraction of U(VI) with Cyanex 302. *J. Radioanal. Nucl. Chem.* **2007**, *273*, 405–409. [CrossRef]

83. Nasab, M.E.; Milani, S.A.; Sam, A. Extractive separation of Th(IV), U(VI), Ti(IV), La(III) and Fe(III) from Zarigan ore. *J. Radioanal. Nucl. Chem.* **2011**, *288*, 677–683. [CrossRef]

84. Belova, V.V.; Egorova, N.S.; Voshkin, A.A.; Khol'kin, A.I. Extraction of rare earth metals, uranium, and thorium from nitrate solutions by binary extractants. *Theor. Found. Chem. Eng.* **2015**, *49*, 545–549. [CrossRef]

85. Singh, H.; Mishra, S.L.; Vijayalakshmi, R. Uranium recovery from phosphoric acid by solvent extraction using a synergistic mixture of di-nonyl phenyl phosphoric acid and tri-n-butyl phosphate. *Hydrometallurgy* **2004**, *73*, 63–70. [CrossRef]

86. Singh, S.K.; Dhami, P.S.; Tripathi, S.C.; Dakshinamoorthy, A. Studies on the recovery of uranium from phosphoric acid medium using synergistic mixture of (2-Ethyl hexyl) Phosphonic acid, mono (2-ethyl hexyl) ester (PC88A) and Tri-n-butyl phosphate (TBP). *Hydrometallurgy* **2009**, *95*, 170–174. [CrossRef]

87. Sreenivasulu, B.; Suresh, A.; Sivaraman, N.; Vasudeva Rao, P.R. Solvent extraction studies with some fission product elements from nitric acid media employing tri-iso-amyl phosphate and tri-n-butyl phosphate as extractants. *J. Radioanal. Nucl. Chem.* **2014**, *303*, 2165–2172. [CrossRef]

88. Jain, V.K.; Pandya, R.A.; Pillai, S.G.; Shrivastav, P.S. Simultaneous preconcentration of uranium(VI) and thorium(IV) from aqueous solutions using a chelating calix[4]arene anchored chloromethylated polystyrene solid phase. *Talanta* **2006**, *70*, 257–266. [CrossRef]

89. Patil, N.N.; Shinde, V.M. Extraction study of uranium(VI) and thorium(IV) salicylates with triphenylarsine oxide. *J. Radioanal. Nucl. Chem.* **1997**, *222*, 21–24. [CrossRef]

90. Singh, H.; Gupta, C.K. Solvent Extraction in Production and Processing of Uranium and Thorium. *Miner. Process. Extr. Metall. Rev.* **2000**, *21*, 307–349. [CrossRef]

91. Borai, E.H.; Mady, A.S. Separation and quantification of 238U, 232Th and rare earths in monazite samples by ion chromatography coupled with on-line flow scintillation detector. *Appl. Radiat. Isot. Incl. Datainstrumentation Methods Use Agric. Ind. Med.* **2002**, *57*, 463–469. [CrossRef]

92. Jeyakumar, S.; Mishra, V.G.; Das, M.K.; Raut, V.V.; Sawant, R.M.; Ramakumar, K.L. Separation behavior of U(VI) and Th(IV) on a cation exchange column using 2,6-pyridine dicarboxylic acid as a complexing agent and its application for the rapid separation and determination of U and Th by ion chromatography. *J. Sep. Sci.* **2011**, *34*, 609–616. [CrossRef]

93. Pin, C.; Zalduegui, J.F.S. Sequential separation of light rare-earth elements, thorium and uranium by miniaturized extraction chromatography: Application to isotopic analyses of silicate rocks. *Anal. Chem. Acta* **1997**, *339*, 79–89. [CrossRef]

94. Soran, M.L.; Curtui, M.; Marutoiu, C. Separation of U(VI) and Th(IV) from some rare earths by thin layer chromatography with di-(2-ethylhexyl)-dithiophosphoric acid on silica gel. *J. Liq. Chromatogr. Relat. Technol.* **2005**, *28*, 2515–2524. [CrossRef]

95. Sivaraman, N.; Kumar, R.; Subramaniam, S.; Rao, P.R.V. Separation of lanthanides using ion-interaction chromatography with HDEHP coated columns. *J. Radioanal. Nucl. Chem.* **2002**, *252*, 491–495. [CrossRef]

96. Ostapenko, V.; Vasiliev, A.; Lapshina, E.; Ermolaev, S.; Aliev, R.; Totskiy, Y.; Zhuikov, B.; Kalmykov, S. Extraction chromatographic behavior of actinium and REE on DGA, Ln and TRU resins in nitric acid solutions. *J. Radioanal. Nucl. Chem.* **2015**, *306*, 707–711. [CrossRef]

97. Ling, L.; Wang, N.H. Ligand-assisted elution chromatography for separation of lanthanides. *J. Chromatogr. A* **2015**, *1389*, 28–38. [CrossRef]

98. Soran, M.-L.; Hodişan, T.; Curtui, M.; Casoni, D. TLC separation of rare earths using di(2-ethylhexyl)dithiophosphoric acid as complexing reagent. *J. Planar Chromatogr. Mod. Tlc* **2005**, *18*, 160–163. [CrossRef]

99. Korkisch, J.; Hazan, I. Anion-exchange behaviour of uranium and other elements in the presence of aliphatic di- and tricarboxylic acids. *Talanta* **1964**, *11*, 523–530. [CrossRef]

100. Dev, K.; Pathak, R.; Rao, G.N. Sorption behaviour of lanthanum(III), neodymium(III), terbium(III), thorium(IV) and uranium(VI) on Amberlite XAD-4 resin functionalized with bicine ligands. *Talanta* **1999**, *48*, 579–584. [CrossRef]

101. Karve, M.; Rajgor, R.V. Amberlite XAD-2 impregnated organophosphinic acid extractant for separation of uranium(VI) from rare earth elements. *Desalination* **2008**, *232*, 191–197. [CrossRef]

102. Merdivan, M.; Duz, M.Z.; Hamamci, C. Sorption behaviour of uranium(VI) with N,N-dibutyl-N'-benzoylthiourea Impregnated in Amberlite XAD-16. *Talanta* **2001**, *55*, 639–645. [CrossRef]

103. Metilda, P.; Sanghamitra, K.; Mary Gladis, J.; Naidu, G.R.; Prasada Rao, T. Amberlite XAD-4 functionalized with succinic acid for the solid phase extractive preconcentration and separation of uranium(VI). *Talanta* **2005**, *65*, 192–200. [CrossRef] [PubMed]

104. Metwally, E.; Saleh, A.S.; El-Naggar, H.A. Extraction and Separation of Uranium (VI) and Thorium (IV) Using Tri-n-dodecylamine impregnated resins. *J. Nucl. Radiochem. Sci.* **2005**, *6*, 119–126. [CrossRef]

105. Prabhakaran, D.; Subramanian, M.S. Extraction of U(VI), Th(IV), and La(III) from acidic streams and geological samples using AXAD-16-POPDE polymer. *Anal. Bioanal. Chem.* **2004**, *380*, 578–585. [CrossRef] [PubMed]

106. Rabie, K.A.; AbdElMoneam, Y.K.; Abdelfattah, A.I.; Demerdahs, M.; Salem, A.R. Adaptation of anion exchange process to decontaminate monazite rare earth group from its uranium content. *Int. J. Res. Eng. Technol.* **2014**, *3*, 374–382.

107. Singh, B.N.; Maiti, B. Separation and preconcentration of U(VI) on XAD-4 modified with 8-hydroxy quinoline. *Talanta* **2006**, *69*, 393–396. [CrossRef]

108. Ang, K.L.; Li, D.; Nikoloski, A.N. The effectiveness of ion exchange resins in separating uranium and thorium from rare earth elements in acidic aqueous sulfate media. Part 1. Anionic and cationic resins. *Hydrometallurgy* **2017**, *174*, 147–155. [CrossRef]

109. Li, Z.; Chen, F.; Yuan, L.; Liu, Y.; Zhao, Y.; Chai, Z.; Shi, W. Uranium(VI) adsorption on graphene oxide nanosheets from aqueous solutions. *Chem. Eng. J.* **2012**, *210*, 539–546. [CrossRef]

110. Li, F.; Yang, Z.; Weng, H.; Chen, G.; Lin, M.; Zhao, C. High efficient separation of U(VI) and Th(IV) from rare earth elements in strong acidic solution by selective sorption on phenanthroline diamide functionalized graphene oxide. *Chem. Eng. J.* **2018**, *332*, 340–350. [CrossRef]

111. Song, W.; Wang, X.; Wang, Q.; Shao, D.; Wang, X. Plasma-induced grafting of polyacrylamide on graphene oxide nanosheets for simultaneous removal of radionuclides. *Phys. Chem. Chem. Phys.* **2015**, *17*, 398–406. [CrossRef]

112. Sadeghi, S.; Sheikhzadeh, E. Solid phase extraction using silica gel modified with murexide for preconcentration of uranium (VI) ions from water samples. *J. Hazard Mater.* **2009**, *163*, 861–868. [CrossRef]

113. Xu, H.; Li, G.; Li, J.; Chen, C.; Ren, X. Interaction of Th(IV) with graphene oxides: Batch experiments, XPS investigation, and modeling. *J. Mol. Liq.* **2016**, *213*, 58–68. [CrossRef]

114. Cheng, W.; Wang, M.; Yang, Z.; Sun, Y.; Ding, C. The efficient enrichment of U(vi) by graphene oxide-supported chitosan. *Rsc Adv.* **2014**, *4*, 61919–61926. [CrossRef]

115. Li, Y.; Wang, C.L.; Liu, C.L. Synthesis and Th(IV) sorption characteristics of functionalised graphene oxide. *J. Radioanal. Nucl. Chem.* **2014**, *302*, 489–496. [CrossRef]

116. Liu, X.; Li, J.; Wang, X.; Chen, C.; Wang, X. High performance of phosphate-functionalized graphene oxide for the selective adsorption of U(VI) from acidic solution. *J. Nucl. Mater.* **2015**, *466*, 56–64. [CrossRef]

117. Pan, N.; Li, L.; Ding, J.; Li, S.; Wang, R.; Jin, Y.; Wang, X.; Xia, C. Preparation of graphene oxide-manganese dioxide for highly efficient adsorption and separation of Th(IV)/U(VI). *J. Hazard Mater.* **2016**, *309*, 107–115. [CrossRef] [PubMed]

118. Shao, D.; Hou, G.; Li, J.; Wen, T.; Ren, X.; Wang, X. PANI/GO as a super adsorbent for the selective adsorption of uranium(VI). *Chem. Eng. J.* **2014**, *255*, 604–612. [CrossRef]

119. Sun, Y.; Shao, D.; Chen, C.; Yang, S.; Wang, X. Highly efficient enrichment of radionuclides on graphene oxide-supported polyaniline. *Environ. Sci. Technol.* **2013**, *47*, 9904–9910. [CrossRef]

120. Xu, Z.; Wan, L.; Huang, X. Functionalization Methods for Membrane Surfaces. *Surf. Eng. Polym. Membr.* **2009**, 64–79. [CrossRef]

121. Li, M.-P.; Zhang, X.; Zhang, H.; Liu, W.-L.; Huang, Z.-H.; Xie, F.; Ma, X.-H.; Xu, Z.-L. Hydrophilic yolk-shell ZIF-8 modified polyamide thin-film nanocomposite membrane with improved permeability and selectivity. *Sep. Purif. Technol.* **2020**, *247*. [CrossRef]

122. Srinivasan, M.; Subba Rao, K.; Dingankar, M.V. The "Actinide Waste" Problem in Perspective. In Proceedings of the Indo-Japan Seminar on Thorium utilization, Bombay, India, 10–13 December 1990.

123. Haridasan, P.P.; Pillai, P.B.M.; Tripathi, R.M.; Puranik, V.D. Operational radiation protection associated with thorium processing in India. In Proceedings of the Naturally Occurring Radioactive Material (NORM VI), Marrakesh, Morocco, 22–26 March 2010.

124. Ault, T.; Krahn, S.; Croff, A. Assessment of the Potential of By-Product Recovery of Thorium to Satisfy Demands of a Future Thorium Fuel Cycle. *Nucl. Technol.* **2015**, *189*, 152–162. [CrossRef]

125. Dekusar, V.M.; Kolesnikova, M.S.; Nikolaev, A.I.; Maiorov, V.G.; Zilberman, B.Y. Mineral reserves of naturally radioactive thorium-bearing raw materials. *At. Energy* **2012**, *111*, 185–194. [CrossRef]

126. Humphries, M. *Rare Earth Elements: The Global Supply Chain*; Diane Publishing: Darby, PA, USA, 2010.

127. Donati, G.; Paludetto, R. Scale up of chemical reactors. *Catal. Today* **1997**, *34*, 483–533. [CrossRef]

128. Dudukovic, M.P. Challenges and Innovations in Reaction Engineering. *Chem. Eng. Commun.* **2008**, *196*, 252–266. [CrossRef]

129. Forbes, L.K. The design of a full-scale industrial mineral leaching process. *Appl. Math. Model.* **2001**, *25*, 233–256. [CrossRef]

130. Arroyo, F.; Fernández-Pereira, C.; Bermejo, P. Demonstration Plant Equipment Design and Scale-Up from Pilot Plant of a Leaching and Solvent Extraction Process. *Minerals* **2015**, *5*, 298–313. [CrossRef]

131. Bałdyga, J. Mixing and Fluid Dynamics Effects in Particle Precipitation Processes. *Kona Powder Part. J.* **2016**, *33*, 127–149. [CrossRef]

132. Marchisio, D.L.; Rivautella, L.; Barresi, A.A. Design and scale-up of chemical reactors for nanoparticle precipitation. *Aiche J.* **2006**, *52*, 1877–1887. [CrossRef]

133. De Santana, A.O.; Dantas, C.C. Scale up of the mixer of a mixer-settler model used in a uranium solvent extraction process. *J. Radioanal. Nucl. Chem.* **1995**, *189*, 257–268. [CrossRef]

134. Geeting, M.W.; Brass, E.A.; Brown, S.J.; Campbell, S.G. Scale-up of Caustic-Side Solvent Extraction Process for Removal of Cesium at Savannah River Site. *Sep. Sci. Technol.* **2008**, *43*, 2786–2796. [CrossRef]

135. Kurniawansyah, F.; Mammucari, R.; Tandya, A.; Foster, N.R. Scale—Up and economic evaluation of the atomized rapid injection solvent extraction process. *J. Supercrit. Fluids* **2017**, *127*, 208–216. [CrossRef]

136. Carta, G. Scale-Up and Optimization in Preparative Chromatography: Principles and Biopharmaceutical Applications. Chromatographic Science Series, Volume 88 Edited by Anurag, S. Rathore (Pharmacia Corporation, Chesterfield, MO) and Ajoy Velayudhan (Oregon State University). Marcel Dekker, Inc.: New York, Basel. 2002. xvi + 342 pp. $150.00. ISBN 0-8247-0826-1. *J. Am. Chem. Soc.* **2003**, *125*, 3398–3399. [CrossRef]

137. Heuer, C.; Hugo, P.; Mann, G.; Seidel-Morgenstern, A. Scale up in preparative chromatography. *J. Chromatogr. A* **1996**, *752*, 19–29. [CrossRef]

138. Rathore, A.S.; Velayudhan, A. Guidelines for optimization and scale up in preventive chromatography. *Biopharm Int.* **2003**, *16*, 1.

Publisher's Note: MDPI stays neutral with regard to jurisdictional claims in published maps and institutional affiliations.

Article

Evaluation of the Use of Electric Arc Furnace Slag and Ladle Furnace Slag in Stone Mastic Asphalt Mixes with Discarded Cellulose Fibers from the Papermaking Industry

Juan María Terrones-Saeta *, Jorge Suárez-Macías, Francisco Javier Iglesias-Godino and Francisco Antonio Corpas-Iglesias

Department of Chemical, Environmental, and Materials Engineering, Higher Polytechnic School of Linares, University of Jaen, Scientific and Technological Campus of Linares, Linares, 23700 Jaen, Spain; jsuarez@ujaen.es (J.S.-M.); figodino@ujaen.es (F.J.I.-G.); facorpas@ujaen.es (F.A.C.-I.)
* Correspondence: terrones@ujaen.es; Tel.: +34-675-201-939

Received: 1 November 2020; Accepted: 19 November 2020; Published: 21 November 2020

Abstract: The construction sector is one of the most demanding of raw materials that exist at present. In turn, the greenhouse gas emissions that it produces are important. Therefore, at present there are several lines of research in which industrial by-products are incorporated for the manufacture of bituminous mixtures and the reduction of CO_2 emissions, framed inside the circular economy. On the base of the aforementioned, in this research, bituminous mixtures of the Stone Mastic Asphalt type were developed with electric arc furnace slag, ladle furnace slag and discarded cellulose fibers from the papermaking industry. To this end, the waste is first characterized physically and chemically, and its properties evaluated for use in bituminous mixtures. Later, different groups of samples are conformed with conventional materials and with the waste in order to be able to compare the physical and mechanical properties of the obtained bituminous mixtures. The physical tests carried out were bulk density, maximum density and void index, as well as the Marshall test for the evaluation of the strength and plastic deformations of all the bituminous mixtures manufactured. The study and evaluation of the results showed that the incorporation of slag makes it possible to absorb a greater percentage of bitumen and obtain better mechanical properties, while maintaining a similar deformation and void content. Therefore, it is feasible to use the mentioned slags to create sustainable, resistant and suitable pavements for important traffic.

Keywords: pavement; bituminous mixtures; electric arc furnace slag; ladle furnace slag; cellulose fibers; stone mastic asphalt; sustainability; steel; circular economy

---

1. Introduction

Road construction is an essential activity for the economic development of a nation and the enhancement of social welfare. Moreover, road transport accounts in different countries for a high percentage of total goods transport, being essential for short and medium distance communication. Therefore, the construction of higher quality and with greater safety roads for vehicles is an unquestionable fact [1,2]. However, this type of infrastructure affects the environment throughout its life cycle assessment [3].

The environmental impact produced by the construction of roads begins with their laying out, altering the landscape. Subsequently, for construction a series of materials are required in significant quantities which are mainly extracted from nearby quarries. In turn, during the manufacture of bituminous mixtures creates CO_2 emissions and fossil fuels are consumed. Transport equipment, extension of the bituminous mixture and compaction also represent an important source of greenhouse gas emissions. Once the infrastructure has been executed, the conservation and maintenance work [4], as well as the continuous flow of vehicles, implies a significant effect on the environment during their working life. At the end of their working life [5], the aged materials are removed and dumped in landfills in most cases, without taking advantage of the usefulness they still offer [6].

According to the scheme detailed above, corresponding to the so-called Linear Economy, significant greenhouse gas emissions are produced throughout the life cycle assessment of the road. Consequently, and in line with the new Circular Economy [7], these emissions must be reduced with different methods [8]. Among these different forms of reducing environmental impact is the use of industrial by-products as raw materials [9]. In this way, the extraction of natural materials is reduced, with the consequent decrease in gas emissions, and the deposition of industrial waste in landfills is avoided [10]. Furthermore, the use of the techniques of manufacturing bituminous mixtures more sustainable with the environment and with a much more optimized processes it also offers a significant reduction in environmental impact. In turn, the development of sustainable materials with industrial by-products, with a longer working life and with a higher quality, also creates the reduction of greenhouse gas emissions [11,12]. Finally, the use of aged materials for the manufacture of new materials avoids the dumping of waste in landfills and reduces the extraction of new raw materials. In this manner, the environmental impact is significantly reduced and the flow of materials is closed [13].

In line with the comments above, various investigations have been carried out in which waste has been incorporated for the manufacture of bituminous mixtures and as a substitute for traditional aggregates. Among these wastes are recycled concrete waste [14], copper slag [15], ceramic and brick dust [16], polymer waste [17], recycled glass [18], recovered asphalt pavement [19] and crumb tire rubber [20], among others.

The use of waste is therefore a good option within the Circular Economy that tries to obtain final products of similar quality. However, in this research, Stone Mastic Asphalt (SMA) type bituminous mixtures are developed with electric arc furnace slags and ladle furnace slags in order to improve the properties of the final mixture with respect to those made with virgin materials [21]. This is made possible by optimizing the strength characteristics of the electric arc furnace slag and the cementitious qualities of the ladle furnace slag. Furthermore, the use of industrial by-products derived from the steel of the siderurgical industry allows it to be considered a sustainable material.

Stone Mastic Asphalt (SMA) bituminous mixtures have a discontinuous grading. This discontinuous grading gives them greater resistance to plastic strains, a better surface texture, greater friction of the tire with the pavement [22], greater permeability to evacuate rainwater, and even greater absorption of noise caused by the contact of the tire with the pavement. At the same time, the incorporation of a higher percentage of bitumen compared to other types of discontinuous grading bituminous mixtures, gives it greater resistance to repetitive traction loads and consequently a longer working life [23,24]. This higher percentage of bitumen is achieved by the addition of fibers. These fibers absorb the excess bitumen and prevent it from bleeding out during the working life of the pavement. Therefore, Stone Mastic Asphalt has a high quality and the resistance suitable for use on roads with important traffic during their working lives.

However, the discontinuous grading of the detailed mixture, as well as the required quality, make the use of high-strength aggregates necessary. Aggregates of higher quality and mechanical resistance mainly correspond to siliceous rocks that are difficult to extract and process, producing important CO_2 gas emissions in their extraction and continuous wear of the equipment during processing [25]. Therefore, the use of high resistance electric arc furnace slags, with excellent shape and reduced price [26], means an important reduction of the environmental impact [27]. In addition, the coating of the electric arc furnace slag with bitumen, reduces in most cases, the possible leaching of contaminating elements that it may contain.

The electric arc furnace slag has been used in road infrastructures as an aggregate in concrete pavements [28,29], demonstrating good mechanical behavior of the resulting material. They have also been used as substitutes for natural aggregate in different percentages in hot mix asphalts, showing excellent results in terms of workability, rigidity and fatigue resistance [30,31]. At the same time, warm mix asphalt has been developed with electric arc furnace slag [32,33], reflecting the improvement in the mechanical properties of the bituminous mixes manufactured [34]. Stone mastic asphalt mixtures have even been made with partial replacement of the aggregate with electric arc furnace slag [35], demonstrating that bituminous mixtures with slag were more resistant to cracking at low temperatures than those that incorporated natural aggregate.

In turn, siliceous aggregates have less adhesion with the bitumen than calcareous aggregates, mainly due to their chemical composition and compatibility between materials. Therefore, to execute a correct mastic that coats the aggregates, that supports the traction loads during the working life and avoids the bleeding of bitumen, calcareous filler or cement is usually used. Cement is one of the materials which provides greater resistance to mixing; however, its manufacturing is a process with a significant environmental impact, as it is a high source of greenhouse gas emissions. To solve this fact, in this research, ladle furnace slag was used as a filler. Ladle furnace slags have been studied in different investigations as additives to cement [36–38] or even for soil stabilization [39], showing very interesting cementitious properties [40,41]. Nevertheless, very few investigations have been carried out in which ladle furnace slag is used as a filler in bituminous mixtures [39] and even fewer in mixtures of such high quality as Stone Mastic Asphalt.

On the other hand, for bituminous mixture containing a higher percentage of bitumen to have adequate resistance to repeated traction loads and that no bitumen bleeding occurs, cellulose fibers must be incorporated. These cellulose fibers, introduced in a low percentage into the bituminous mix, are capable of retaining the bitumen in the mix and forming a quality mastic in conjunction with the bitumen and filler. Specially treated commercial fibers are usually used for this purpose; however, in this research and with the aim of making a sustainable mix, cellulose fibers that have been discarded by the papermaking industry were incorporated. These cellulose fibers discarded by the papermaking industry have no current use, so in most cases they are deposited in landfills.

In conclusion, this research develops a quality hot mix asphalt, Stone Mastic Asphalt type, for roads with important traffic with electric arc furnace slag as a coarse and fine aggregate, with ladle furnace slag as a filler and with discarded cellulose fibers from the papermaking industry as an additive. For this purpose, the waste was initially characterized and its properties compared with conventional materials. Subsequently, different families of samples were conformed by increasing percentages of bitumen and the physical properties and Marshall Stability of the mixtures obtained were evaluated. Finally, an optimal material combination was obtained for the asphalt mixtures developed, and the advantages of using the waste over virgin materials were compared.

The tests carried out, as well as their quality limits, will be governed by Spanish regulations, which in turn coincide with European regulations. This Spanish regulation corresponds to the Circular Order OC 3/2019 [42] and was selected because of the profusion of these techniques that have reached the Spanish territory, there existing an infinity of success cases. However, the comparison of the results obtained in bituminous mixtures with waste and bituminous mixtures with traditional materials, objectively reflects the quality of the incorporation of the by-products, and the results can easily be extrapolated to other international regulations.

The results showed that the incorporation of electric arc furnace slag, ladle furnace slag and cellulose fibers created an SMA mix with a higher percentage of bitumen and better mechanical performance, compared to the use of traditional aggregates and fillers.

2. Materials and Methods

This section describes the materials used for the development of the research, as well as the scientific methodology followed to reach the final conclusions. The final objective is the study of the benefits of incorporating the waste mentioned for the manufacture of SMA-type bituminous mixtures.

2.1. Materials

The materials used in this project are mainly waste and commercial materials. These materials are detailed in Section 2.1, defining their origin, production process and particular characteristics, making possible the reproduction of the present tests.

It should be noted that the industrial waste from this research (electric arc furnace slag (EAFS), ladle furnace slag (LFS) and cellulose fibers) was supplied by the producing company in an unaltered form. The process that has been carried out on these wastes is detailed in the following sections.

In turn, it should also be mentioned that the tests carried out have been executed for different production batches of the waste. In this way it has been confirmed that the physical and chemical properties of the waste are maintained over time. This fact is essential, since if the characteristics of the waste were to be modified to a large extent it would make their use in the construction of road infrastructure unfeasible, since large quantities of materials are consumed and could lead to changes in the final characteristics of the bituminous mixtures. It can therefore be stated that the waste studied maintains its physical and chemical properties over time, unlike other wastes such as sewage sludge, cutting sludge, etc.

Finally, it should be mentioned that the ladle furnace slag, electric arc furnace slag, cellulose fibers, as well as the hornfels aggregates and calcareous filler, were dried at a temperature of 105 ± 2 °C for 24 h in order to eliminate the humidity in them. The elimination of the humidity from the materials is intended to avoid introducing more variables into the methodology and to provide objective results. In the subsequent manufacturing process in industry, this humidity of the materials should simply be taken into account in order to take the appropriate corrections, if it was necessary.

2.1.1. Electric Arc Furnace Slag (EAFS)

The electric arc furnace slag used comes from the siderurgical industry located in the region of Andalucía, Spain. These slags have a continuous grading with different particle sizes up to a maximum of 22 mm. The existence of particles smaller than 0.063 mm is negligible, and there are mainly coarse and fine aggregates, in smaller quantities. An irregular shape of the particles can be observed by the processes of their formation.

It may be pointed out that electric arc furnace slag is formed in the metallurgical industry in the first stage called melting and in the electric arc furnace. These furnaces are fed with soft iron or steel scrap. In this melting stage, a series of phases are carried out such as oxidation, to remove manganese and silicon impurities, dephosphorization and the formation of foaming slag. All the impurities are accumulated in this foaming slag. The slag is extracted, forming the electric arc furnace slag after cooling and watering with water.

The production company then crushes the material and performs an economical particle size classification for filler of embankments. These slags are used in the present investigation.

In turn, the mission of the electric arc furnace slag is to replace the traditionally used coarse and fine siliceous aggregate. Therefore, it provides the necessary mineral skeleton of the bituminous mix, and it must be sufficiently resistant to support the repeated compressive loads of the traffic, as well as the roughness to provide good friction between the tire and the pavement. The slag from the electric arc furnace was washed and sieved by different sieves, obtaining the grading fractions necessary for the conformation of the grading curve.

2.1.2. Ladle Furnace Slag (LFS)

The ladle furnace slag comes, like the electric arc furnace slag, from the area of Andalucía, Spain. These slags have a very fine particle size derived directly from their formation process.

Ladle furnace slag is produced in the refining stage, after the melting stage in which the electric arc furnace slag is produced. The refining stage includes a series of phases such as deoxidation, allowing the removal of metal oxides from the furnace, desulphurization and decarburization of the steel. For this phase to take place, the liquid from the electric arc furnace is transferred to the ladle furnace, being covered with slag, and continuously stirred by blowing inert gas, usually argon. Finally, this slag of much smaller particle size is removed and deposited in the vicinity for cooling.

The ladle furnace slag was taken directly from the producing industry as an undisturbed sample and will serve as a filler for the bituminous mixtures conformed. These ladle furnace slags provide the desired cementitious characteristics, which have been confirmed by various authors. For this purpose, they were sieved after drying at 105 ± 2 °C for 24 h by the 0.063 mm sieve.

2.1.3. Cellulose Fiber from the Papermaking Industry

Cellulose fibers are currently an unused waste produced from the cardboard manufacturing industry.

These fibers are formed in the process of producing packaging paper from recycled paper. The recycled paper is grinded with water to put the fibers in suspension, and then submitted to a physical separation with different sieves. Finally, a cyclonic separation is carried out. The waste from this cyclonic separation is transferred to a press to remove some of the water contained in the waste. This waste, after being pressed, is the one used in this research and is called cellulose fiber discarded by the paper industry.

Detailed cellulose fibers are the additive that was incorporated into the bituminous mix for the retention of a higher percentage of bitumen in the mix. These fibers have been taken from the production industry and have undergone a process of adaptation for use in bituminous mixtures. This process consists of a washing with a 30% sodium hydroxide solution. This pre-treatment is carried out with a double objective; on the one hand, the organic reactions that could be produced are paralyzed; on the other hand, any natural waxes that could be adhered to the fibers and that would prevent the correct adhesion with the bitumen of the bituminous mix are removed. Once this pre-treatment has been carried out, they are ground to achieve the smallest possible fiber size, making it possible to homogenize them during the mixing process in the bituminous mixture.

2.1.4. Bitumen

The bitumen used is a 50/70 bitumen as defined by European regulations, both numbers being the penetration rate at which it oscillates. This hard penetration bitumen is usually used in the Spanish regions due to the existing hot climates. It is a commercial bitumen without additives. Its technical data can be seen in Table 1.

Table 1. Technical specifications of the bitumen used.

Characteristics	Unit	Standard	Min	Max
Fresh binder				
Penetration (25 °C)	0.1 mm	UNE-EN 1426 [43]	50	70
Penetration index	-	UNE-EN 12591 [44]	−1.5	0.7
Softening point (R & B)	°C	UNE-EN 1427 [45]	46	54
Fraass point	°C	UNE-EN 12593 [46]	-	−8
Solubility in xylene	%	UNE-EN 12592 [47]	99.0	-
Flash point	°C	UNE-EN ISO 2592 [48]	230	-
Resistance to Hardening 163 °C (UNE-EN 12607-1) [49]				
Mass loss	%	UNE-EN 12607-1 [49]	-	0.5
Retained penetration	%	UNE-EN 1426 [43]	50	-
Increase in softening point (R & B)	°C	UNE-EN 1427 [45]	-	11

2.1.5. Hornfels Aggregate

Hornfels aggregate is a commonly used aggregate on important traffic roads mainly due to its excellent characteristics. This aggregate comes from the area of Andalucía, Spain, just like the other materials.

Hornfels rocks are a type of contact metamorphic rocks, very hard and with great resistance to the cycles of freezing and thawing. It contains a high proportion of quartz, graphite, biotite, iron oxide or feldspars, so it can be considered a quality siliceous rock.

The extraction of this material in quarries, being a hard rock, consumes a great quantity of explosives, since its resistance to fragmentation is high. In addition, its siliceous composition means that treatment and processing equipment often wear out, compared to limestone stone.

It is therefore an aggregate of excellent quality, in which significant greenhouse gas emissions are emitted during its extraction and with which electric arc furnace slag is to be compared. Therefore, its function within the bituminous mixtures created is that of a coarse and fine aggregate. To this end, as with the slag, the aggregate is received from the quarry and washed, to be subsequently sieved by different sieves that can form the selected grading curve.

2.1.6. Calcareous Filler

The problems derived from the lack of adhesion between the siliceous aggregates and the bitumen make the use of filler of limestone type common. Calcareous aggregates have much lower resistance than siliceous ones, as well as a lower resistance to the abrasion caused by the tire. Therefore, its use in important traffic roads is not usual or recommended.

However, its use as an inert filler makes possible the formation of a mastic of acceptable quality that coats the siliceous aggregates and forms a structure capable of withstanding the loads of traffic. Therefore, the function of the calcareous filler in the present investigation is the comparison with the properties of the bituminous mixtures conformed with it, with those of the mixtures conformed with ladle furnace slag.

The calcareous filler supplied by the producing company had a very fine particle size and did not need to be sieved, unlike the ladle furnace slag. This filler was dried at 105 ± 2 °C for 24 h to avoid the existence of water during the conformation of the bituminous mixtures.

2.2. Methodology

The methodology followed in the present investigation is composed of a series of logically ordered tests to obtain objective results on the quality of the execution of Stone Mastic Asphalt type bituminous mixtures with electric arc furnace slag, ladle furnace slag and cellulose fibers from the papermaking industry. To this end, bituminous mixtures manufactured were compared with bituminous mixtures conformed with commercial materials.

Based on this, the wastes were analyzed to determine their physical properties and chemical composition. In this way, the suitability of the materials for forming SMA mixtures for roads with high vehicle traffic was evaluated.

Subsequently, mixtures were conformed with traditional aggregates and with waste, as well as increasing percentages of bitumen. The groups of samples conformed were analyzed to obtain the physical and resistant properties, through the Marshall test.

Finally, and after evaluating the properties of the different mixtures, the optimum combination of materials was obtained for each family of samples studying the advantages of using electric arc furnace slags, ladle furnace slags and cellulose fibers.

The following sub-sections describe each of the research phases in detail.

2.2.1. Characterization of Raw Materials

The waste and commercial materials were treated as detailed before in order to be able to carry out physical and chemical characterization tests, as well as for use in subsequent tests.

Firstly, the electric arc furnace slag, ladle furnace slag and cellulose fibers were analyzed by elemental analysis, determining the percentage of carbon, nitrogen, hydrogen and sulfur in the samples. In turn, the slags were subjected to the X-ray fluorescence test, as they are inorganic materials, unlike cellulose fibers from the papermaking industry.

Once their chemical composition had been determined and the presence of contaminating elements that could prejudice the final bituminous mixture analyzed, a series of physical tests were carried out on the different wastes according to the function that each one plays within the bituminous mixture.

The cellulose fibers from the papermaking industry were evaluated with a scanning electron microscope (Carl Zeiss, Oberkochen, Germany) at different magnifications and after metallization with carbon. The size of the fibers obtained after pre-treatment and the existence of agglomerations that could impair the homogeneous distribution of the fibers in the bituminous mixture were thus observed.

The ladle furnace slag was subjected to particle density tests (standard UNE-EN 1097-7) [50], to evaluate the possible volumetric corrections required; bulk density tests (standard UNE-EN 1097-3) [51], to determine whether it is a powdery material that is detrimental to its proportioning; and plasticity index tests (standards UNE 103103 and UNE 103104) [52,53], to evaluate the possible existence of clayey particles that could create expanding problems in the final mix.

The electric arc furnace slag was subjected to particle density tests (standard UNE-EN 1097-7) [50], to determine whether volumetric corrections were necessary; a sand equivalent test (standard UNE-EN 933-8) [54], to evaluate the percentage of colloidal particles that could damage the final mixture; percentage of crushed surface tests (standard UNE-EN 933-5) [55]; and flakiness index tests (standard UNE-EN 933-3) [56], for the qualification of the aggregate, since the SMA mixture resists the loads of traffic on the mineral skeleton, and therefore the particles must have certain shapes; resistance to fragmentation tests (standard UNE-EN 1097-2) [57], to qualify the hardness of the material and its suitability for high traffic; resistance to freezing and thawing cycles tests (standard UNE-EN 1367-1) [58], to evaluate the aggregate's resistance to thermal fatigue; and determination of the value of polished stone (standard UNE-EN 1097-8) [59], to quantify the effect on the aggregate of the continuous tire friction of the with the pavement and, consequently, its durability through time.

2.2.2. Conformed of Bituminous Mixtures and Tests

Once the previous tests had been carried out, specific for each material and in accordance with the role that each material plays in the mixture, we proceeded to make the bituminous mixtures reflected in Table 2 with the materials detailed.

Table 2. Families of bituminous mixtures conformed with electric arc furnace slag, Hornfels aggregate, ladle furnace slag, calcareous filler and fibers from the papermaking industry.

Samples Groups	ACFC	ASFC	ASFS
Coarse aggregate	Hornfels aggregate	EAFS	EAFS
Fine Aggregate	Hornfels aggregate	EAFS	EAFS
Filler	Calcareous	Calcareous	LFS
Additives	Papermaking waste	Papermaking waste	Papermaking waste

As shown in Table 2, there are families of bituminous mixtures conformed with virgin materials and families conformed with waste. In this way, the qaulity of the incorporation of waste is easily comparable.

The materials that perform the function of aggregate, whether waste or natural aggregates, were dried and sieved by different sieves to obtain the desired grading curve. The grading curve used corresponds to the intermediate grading curve established by the grading envelope detailed in the regulations of Circular Order OC 3/2019 [42]. The selection of this grading curve is motivated by an essential reason: to compare the difference between bituminous mixtures made with natural aggregates and those made with slag. To do so, they must have the same grading, thus avoiding secondary variables that could mask the final conclusions. In addition, cellulose fibers discarded from the papermaking industry were incorporated into all bituminous mixtures in a percentage of 0.5% in mass and regarding conventional aggregate (conventional aggregate density 2.65 t/m^3), as indicated by various studies on this type of mixture. The grading curve for the three families of samples is shown in Figure 1.

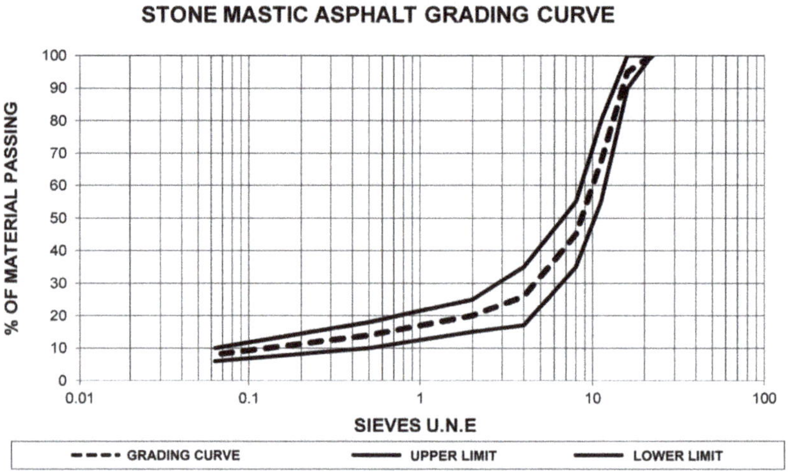

Figure 1. Grading curve of the different families of bituminous mixtures (ACFC, ASFC and ASFC) type SMA.

Once the grading curve was defined, different groups of samples were conformed of the three types of bituminous mixtures detailed in Table 2. In order to be able to compare the results faithfully, and given that the electric arc furnace slag has a higher density than the hornfels aggregates, the percentage of bitumen by volume was proportioned. This proportioning by volume allows an objective evaluation of the bitumen absorption capacity of bituminous mixtures with slag, since if mass proportioning was done, the optimum bitumen percentage would not be comparable due to the high density of the slag.

Based on the comments above, the families of mixtures were manufactured with percentages of bitumen in volume and regarding aggregate from 15% to 18% in 0.5% increments. To this end, the aggregates (natural or waste) were heated in an oven to a temperature of 180 ± 5 °C for 1 h, as was the bitumen and cellulose fibers, and then mixed in an automatic planetary mixer (MECÁNICA CIENTÍFICA S.A., Madrid, Spain) for 10 ± 1 min. The resulting mixture was extracted and compacted by a Marshall compactor (MECÁNICA CIENTÍFICA S.A., Madrid, Spain) with 50 blows per side to each specimen (standard UNE-EN 12697-30) [60]. The conformed specimens were left at ambient temperature for 24 h for subsequent mechanical stripping. A total of 8 Marshall-type samples were made for each percentage of bitumen in each family.

Once the groups of samples with increasing percentages of bitumen from each family of samples had been obtained, the physical properties were characterized. The tests carried out were on the maximum density of the bituminous mixture (standard UNE-EN 12697-5) [61] and bulk density (standard UNE-EN 12697-6) [62]. In turn, the void characteristics of the bituminous mixtures obtained were calculated (standard UNE-EN 12697-8) [63].

The Marshall test was carried out to evaluate the mechanical resistance of the families of bituminous mixtures conformed (standard UNE-EN 12697-14) [64]. With this test, the plastic deformations that occur in each bituminous mix can be evaluated, this being an essential characteristic due to the high percentage of bitumen that SMA bituminous mixes have.

2.2.3. Determination of Optimal Material Combinations and Comparison of the Results

Once the mechanical and physical properties of the three families of samples with different bitumen percentages had obtained, the optimum combination of materials was then obtained. This optimum combination of materials was calculated graphically, taking Marshall stability as the main property. In other words, the percentage of volume of bitumen which provided the highest Marshall stability of each family (ACFC, ASFC and ASFS) was calculated, provided that permissible values were obtained for the physical properties and deformation of the bituminous mixtures.

With the optimum combinations of materials for the three families of samples (ACFC, ASFC and ASFS), samples were again made to evaluate the physical and mechanical properties obtained graphically, thus corroborating the quality of the material selection. In turn, binder drainage tests UNE-EN 12697-18 [65] were carried out, to evaluate that the fibers fulfilled their function within the conformed bituminous mixtures and that there were no bleeding of bitumen due to their high percentage; wheel-tracking tests [66] were also conducted, to evaluate the durability of the mixture before the continuous passage of vehicles.

The results obtained from the different sample families for their optimal material combination were compared. In this manner, the influence of the use of electric arc furnace slag and ladle furnace slag in the manufacture of Stone Mastic Asphalt mixtures with cellulose fibers from the papermaking industry can be objectively evaluated.

3. Results and Discussion

This section describes the results of the trials mentioned in the methodology, as well as the discussion about them. The series of trials logically ordered will condition the final conclusions, there being at all times a continuous process of feedback.

3.1. Characterization of Raw Materials

A significant percentage of waste is used in this research. These wastes are electric arc furnace slag, ladle furnace slag and discarded cellulose fibers from the paper industry. The use of waste has a number of environmental advantages as discussed above; however, this waste must be physically and chemically characterized in order not to induce problems in the final material.

Firstly, the discarded cellulose fibers from the paper industry were analyzed. These fibers were analyzed, after the treatment described in the methodology, in an elemental analyzer (TruSpec Micro, LECO, St. Joseph, MI, USA) to detect the percentage of carbon, nitrogen, hydrogen and sulfur in the sample. This test is essential for the material under study since, unlike slag, it is organic in nature. The results of the elemental analysis of the cellulose fiber waste from the papermaking industry are detailed in Table 3.

Table 3. Elemental analysis of cellulose fibers discarded by the papermaking industry.

Sample	Nitrogen, %	Carbon, %	Hydrogen, %	Sulfur, %
Cellulose fibers	0.447 ± 0.008	44.489 ± 0.325	5.884 ± 0.178	0.000 ± 0.000

As can be seen, the percentages of carbon and hydrogen are high, as they correspond to an organic material. On another point, the percentage of nitrogen contained in the sample is low, a fact that should be taken into account as it could damage the final bituminous mix. In turn, the percentage of sulfur is null. The low proportion of elements such as nitrogen and sulfur therefore ensures that these fibers are well incorporated into the bituminous mixtures, since otherwise they could leach contaminant elements and even affect the characteristics of the bituminous mixtures. It is important to note that the sum of the elements analyzed does not correspond to 100% of the chemical composition, so there must be other inorganic chemical elements in the fibers analyzed. These inorganic chemical elements could correspond to the treatment carried out on the cellulose fibers before their use for conforming bituminous mixtures, this main element being sodium, since the fibers are treated with sodium hydroxide.

At the same time, and in order to characterize the cellulose fibers of the papermaking industry in a complete way, the scanning electron microscope test was carried out. This test aims to identify the shape of the fibers with high magnification, focusing mainly on their size and the existence of agglomerations. Both of these detailed factors have a significant influence on the correct mixing of the fibers with the aggregates and the bitumen. The scanning electron microscope therefore provided sufficient physical information to evaluate the suitability of the fibers for homogenization within the bituminous mix and, consequently, the increased retention of bitumen and the elimination of bleeding from the bituminous mix. Figure 2 shows the image of the cellulose fibers obtained with a scanning electron microscope.

Figure 2. Image of the cellulose fibers of the paper industry obtained with the scanning electron microscope in the secondary option.

As can be seen in Figure 2, the cellulose fibers analyzed have millimetric dimensions, and there are no agglomerations of these fibers that could damage their homogeneous distribution in the bituminous mix. Therefore, they are considered to be suitable for use.

Once the cellulose fibers were analyzed, they were characterized the ladle furnace slag. The ladle furnace slag was used as a filler. Therefore, these slags must form, together with the bitumen and cellulose fibers, a mastic of adequate quality to resist the continuous traction loads of the pavement. For this reason, it is essential to characterize them chemically, determining the existence of chemical cementitious compounds or polluting chemical elements that must be controlled in subsequent processes. Table 4 shows the results of the elemental analysis of the ladle furnace slag.

Table 4. Elemental analysis of the ladle furnace slag.

Sample	Nitrogen, %	Carbon, %	Hydrogen, %	Sulfur, %
LFS	0.007 ± 0.001	3.405 ± 0.068	1.386 ± 0.026	0.000 ± 0.000

Elemental analysis of the ladle furnace slag shows that it does indeed correspond to an inorganic material. The low percentage of nitrogen and sulfur, being the latter, of which is very harmful to the final bituminous mix, should be noted. If significant percentages are available of sulfur in the slag, a leachate test should be carried out later to confirm the retention of this element in the bituminous mixture. On the other hand, and due to the ladle furnace slag production process, the existing percentages of carbon and hydrogen come directly from the carbonate compounds and hydration of the oxides present in the slag, as reflected by X-ray fluorescence. This process is natural in this type of material and is mainly due to the open-air exposure of the waste after its extraction.

The X-ray fluorescence test provided sufficient information about the other chemical elements; this test is detailed in Table 5.

Table 5. Results of the X-ray fluorescence of ladle furnace slag.

Compound	Wt, %	Est. Error
CaO	40.19	0.25
MgO	19.38	0.20
SiO_2	12.49	0.17
Al_2O_3	7.29	0.13
Fe_2O_3	2.38	0.08
MnO	0.936	0.047
S	0.548	0.027
TiO_2	0.486	0.024
BaO	0.240	0.012
Na_2O	0.118	0.042
Cr_2O_3	0.1100	0.0055
Cl	0.0833	0.0042
SrO	0.0733	0.0037
ZnO	0.0681	0.0034
K_2O	0.0506	0.0025
ZrO_2	0.0425	0.0021
V_2O_5	0.0179	0.0017
P	0.0138	0.0012
CuO	0.0117	0.0010
NiO	0.0082	0.0011
PbO	0.0048	0.0010
Nb_2O_5	0.0046	0.0006
MoO_3	0.0028	0.0009
Co_3O_4	0.0021	0.0009
SeO_2	0.0012	0.0005

The X-ray fluorescence (Thermo Fisher Scientific, Waltham, MA, USA) test shows a chemical composition of the ladle furnace slag that is logical and derived from its production process. The existence of calcium oxides, magnesium oxides and silicon oxides in a higher proportion is mainly due to the material added to the ladle furnace for steel purification. The incorporation of lime or dolomites in the ladle furnace creates this composition of the steel oxides. In addition, the function of ladle furnace slag is the deoxidation and desulphurization of steel, so it is logical to find metal oxides and sulfur in its composition. However, there are no chemical elements that could directly damage the mechanical characteristics of the bituminous mixtures, nor are there any polluting elements in large proportion that could be leached out later and cause environmental pollution.

On the other hand, the physical properties of the ladle furnace slag were quantified. The main tests to determine these properties, for a material that plays the role of filler in the bituminous mix, are detailed in Table 6.

Table 6. Density and plasticity tests for the fine portion of ladle furnace slag.

Test	Standard	Value/Unit
Particle density	UNE-EN 1097-7 [50]	2.71 ± 0.07 t/m^3
Bulk density	UNE-EN 1097-3 [51]	0.75 ± 0.01 t/m^3
Plasticity index	UNE 103103/UNE 103104 [52,53]	No plasticity

It can be seen how the particle density of ladle furnace slag is slightly higher than that of a commercial calcareous filler. At the same time, the bulk density in kerosene of the slag reflects the behavior of a powdery material, which without producing proportioning problems in the factory if it has a reduced particle size is capable of adhering correctly with the bitumen and forming a quality mastic. The non-existence of plasticity avoids subsequent problems of expansiveness due to the existence of clayey particles. This lack of plasticity is due to the chemical composition of the ladle furnace slag, as it is mainly composed of calcium and magnesium oxides.

On the other hand, electric arc furnace slag plays the role of a coarse and fine aggregate in the bituminous mix. The tests carried out must therefore check the suitability of the slag for this purpose. For chemical characterization, the elemental analysis test was carried out; this test is reflected in Table 7.

Table 7. Elemental analysis of the electric arc furnace slag.

Sample	Nitrogen, %	Carbon, %	Hydrogen, %	Sulfur, %
EAFS	0.005 ± 0.000	0.164 ± 0.003	0.044 ± 0.001	0.000 ± 0.000

Elemental analysis of electric arc furnace slag mainly shows its inorganic composition. The low percentages of carbon and hydrogen reflect that these slags are a more stable material than ladle furnace slags, as no carbonated or hydrated processes of the chemical compounds take place. The null values of sulfur and nitrogen should be highlighted, so there will be no leaching of these elements in the final bituminous mixtures. The remaining chemical elements present in the sample of electric arc furnace slag were determined with the X-ray fluorescence test. This test is shown in Table 8.

The chemical composition of EAFS derives directly from its formation process. A high percentage of iron is to be expected, since it comes from steel, as well as a high percentage of calcium oxide due to its addition to obtain the final material. The silicon and aluminum oxides are common in the scrap that is used for the manufacture of new steels. Magnesium, manganese and chrome are also common in the composition of steel. The other elements are found in such small percentages that they cannot be extrapolated. The very low percentage of sulfur ensures that the EAFS leachate does not pose an environmental problem, as is the case with other pollutants. Otherwise, we would have to study the leaching of these chemical pollutants and compare them with the limit values established by the regulations. It should be noted that the existence of oxides, mainly calcium oxide, in the unaltered sample of electric arc furnace slag does not cause any subsequent problem of expansion in contact with water. This fact is derived from the industrial process of slag formation: after extracting the residue, the mixture is watered. This produces a carbonate of the oxides and therefore stability in its physical structure.

Table 8. X-ray fluorescence of electric arc furnace slag.

Compound	Wt, %	Est. Error
CaO	31.75	0.23
Fe_2O_3	21.96	0.21
SiO_2	17.52	0.19
Al_2O_3	12.26	0.16
MnO	6.15	0.12
MgO	5.05	0.11
Cr_2O_3	2.73	0.08
TiO_2	0.955	0.047
BaO	0.658	0.033
P_2O_5	0.319	0.016
SrO	0.186	0.0093
V_2O_5	0.159	0.0079
Nb_2O_5	0.0659	0.0033
S	0.0645	0.0032
ZrO_2	0.0551	0.0028
K_2O	0.0289	0.0016
CuO	0.0254	0.0017
ZnO	0.0245	0.0016
Co_3O_4	0.0147	0.0016
Eu_2O_3	0.0137	0.0065
WO_3	0.0104	0.0031
Y_2O_3	0.0018	0.0005

It should be noted that European or American regulations governing the leaching of chemicals elements of the aggregates for roads show restrictions on heavy metals, chlorides, fluorides or sulphates. These chemical elements are in a proportion of less than 1%, and even much less, in ladle furnace slags and electric arc furnace slags, not existing in their composition in some cases. Therefore, the leaching of these elements is minimal, as the electric arc furnace slag is mainly composed of calcium oxide, iron oxide, silicon oxide and aluminum oxide, as well as the ladle furnace slag of silicon oxide, calcium oxide and magnesium oxide. None of these chemical compounds mentioned are limited in their concentration in the leachate, as they do not produce environmental pollution. In addition, the coating of the slag with bitumen of the bituminous mixture quantitatively reduces the leaching of any element, therefore, the compliance of the quality standards is assured.

Once the chemical composition of the electric arc furnace slag had been analyzed and the absence of chemical elements that could damage the final bituminous mix during its manufacture or its working life had been assessed, the physical and resistance properties of the slag were determined. Table 9 shows the physical tests carried out on electric arc furnace slag.

Table 9. Density and plasticity tests for the fine portion of ladle furnace slag.

Test	Standard	Value/Unit
Particle density (coarse aggregate)	UNE-EN 1097-7 [50]	3.13 ± 0.05 t/m^3
Particle density (fine aggregate)	UNE-EN 1097-7 [50]	3.34 ± 0.07 t/m^3
Sand Equivalent test	UNE-EN 933-8 [54]	$77 \pm 2\%$
Broken surfaces (coarse aggregate)	UNE-EN 933-5 [55]	$100 \pm 1\%$
Flakiness index	UNE-EN 933-3 [56]	$0 \pm 1\%$

As can be seen in Table 9, the particle density of electric arc furnace slag is higher than that of a traditional aggregate (approximately 2.65 t/m^3). This higher density is due to its chemical composition composed of metallic elements, mainly iron. However, a higher density does not affect the process of conforming bituminous mixtures or their final characteristics, it should only be taken into account for the correct proportioning of the bitumen and the additives, as well as their comparison. Therefore, since the density of the ladle furnace slag is higher than that of the Hornfels aggregate, the bitumen was proportioned by volume so that the results were comparable and it was possible to evaluate which material is capable of absorbing a higher percentage of the bitumen. On the other hand, the sand equivalent test reflects the low proportion of colloidal particles that exist in the electric arc furnace slag, so there are no subsequent problems of expansiveness in the bituminous mix due to clayey materials. The excellent results obtained from the tests on the percentage of broken surfaces and the flakiness index should be highlighted. Both tests reflect the aptitude of the slag for the conformation of Stone Mastic Asphalt type mixtures, since this type of bituminous mixture has a discontinuous grading, the compression loads of the traffic are supported by the friction of the coarse aggregate. Therefore, the coarse aggregate is required to have optimum results from these tests so that no subsequent compaction of the bituminous mix occurs due to the continuous passage of vehicles. This excellent shape of the particles in the electric arc furnace slag is due to the production process, since the continuous oxygenation of the furnace causes the irregular shapes and edges that the particles have.

As mentioned, in Stone Mastic Asphalt mixes the repeated traction loads caused by traffic are borne by the mix mastic formed by the filler, bitumen and fibers. In turn, the compression loads are mainly supported by the coarse aggregate due to friction between particles and due to the discontinuous grading. Therefore, if designing bituminous mixtures for high traffic is intended, the coarse aggregate must have an adequate resistance to avoid its fracture. This resistance has been evaluated by the tests of resistance to fragmentation (standard UNE-EN 1097-2) [57] and resistance to freezing and thawing cycles (standard UNE-EN 1367-1) [58] reflecting values of 13 ± 1% and 0.551 ± 0.016%, respectively. These results show the excellent resistance of electric arc furnace slag to both fracture and thermal fatigue and are therefore only comparable with excellent quality and very expensive aggregates, both economically and environmentally.

At the same time, the continuous friction of the tire with the pavement creates a polishing of the aggregate of the bituminous mixture, with the consequent decrease in safety for the driver. Therefore, it is essential to perform the determination test of the polished stone (standard UNE-EN 1097-8) [59]. This test reflected a value of 58 ± 1. This result ensures an adequate resistance of the slag to the continuous passage of vehicles and a durability in time of the surface roughness.

In short, and based on the results of the waste characterization tests, it can be stated that both electric arc furnace slag and ladle furnace slag and cellulose fibers from the papermaking industry have suitable characteristics for use in bituminous mixtures. However, if it is true that special precautions must be taken to achieve successful incorporation into the bituminous mix.

3.2. Conforming of Bituminous Mixtures and Tests

Once the physical and chemical characteristics of the waste, bituminous mixtures of the families detailed in Table 2 were conformed, with the grading curve defined in Figure 1 and with percentages of bitumen in volume of aggregate of 15% to 18%.

All the samples conformed were analyzed to evaluate their physical properties and resistance. The first of the tests carried out on the bituminous mixtures was the bulk density test. This test is shown graphically in Figure 3.

As can be seen, the bulk density of bituminous mixtures with electric arc furnace slag is higher than the bulk density of mixtures with hornfels aggregate. This fact is fundamentally due to the higher density of the electric arc furnace slag, which does not negatively influence the subsequent results but if is a factor to be taken into account. The difference in density between mixtures made with calcareous filler or ladle furnace slag filler is very small, since the density of both materials is similar and as well the percentage of filler incorporation being lower. In turn, the maximum density of bituminous mixtures conformed is detailed in Figure 4.

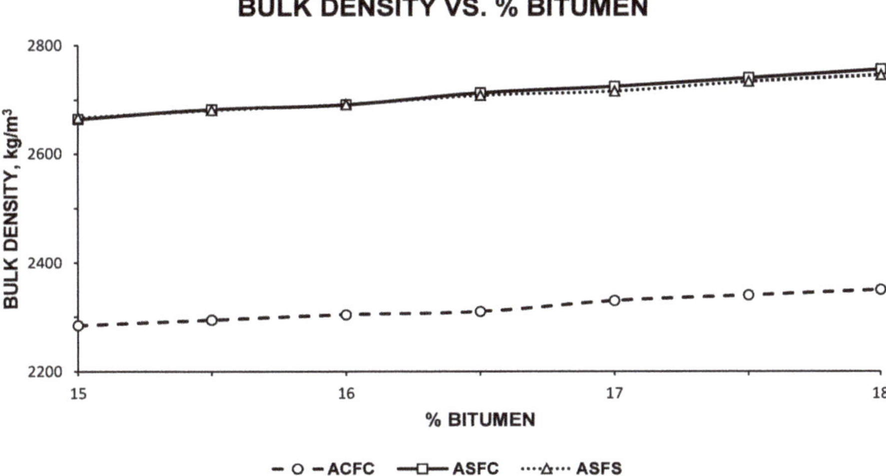

Figure 3. Bulk density of the families of bituminous mixtures ACFC, ASFC and ASFS with different percentages of bitumen in volume of aggregate.

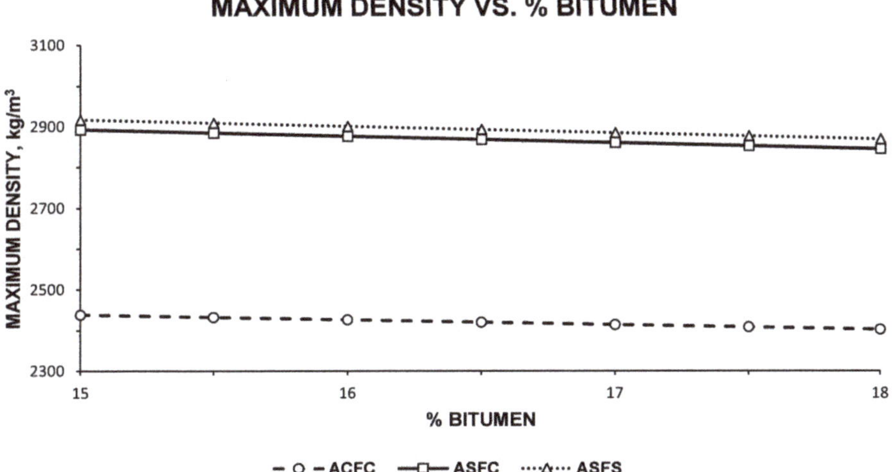

Figure 4. Maximum density of the families of bituminous mixtures ACFC, ASFC and ASFS with different percentages of bitumen in volume of aggregate.

Similarly to the previous case, the maximum density of mixtures that contain electric arc furnace slag is higher than those that incorporate hornfels aggregate, due to the higher density of this material. In turn, mixtures containing ladle furnace slag have a slightly higher density than mixtures conformed with calcareous filler. The results of this test, as well as that of bulk density, directly condition the percentage of voids in bituminous mixtures. The percentage of voids is essential to determine the behavior of the bituminous mixture, being limited by the regulations. The voids content of the different families of mixes is shown in Figure 5.

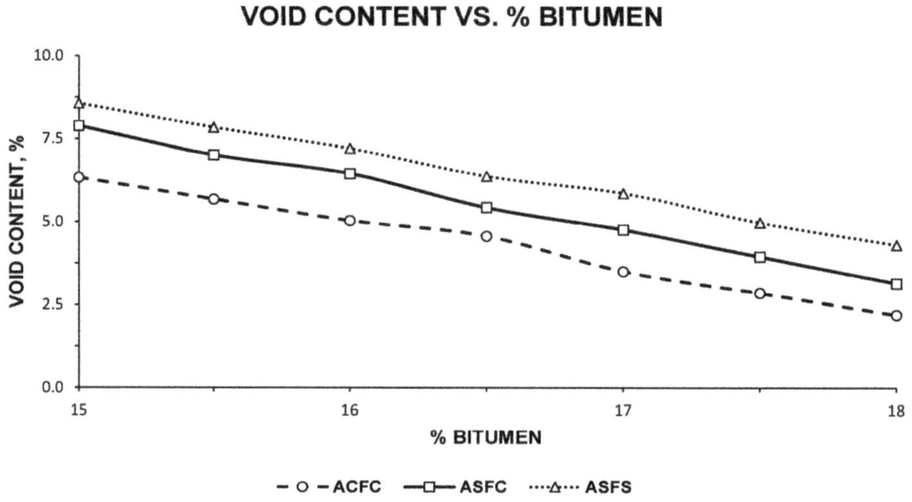

VOID CONTENT VS. % BITUMEN

Figure 5. Void content of the families of bituminous mixtures ACFC, ASFC and ASFS with different percentages of bitumen in volume of aggregate.

The percentage of voids is an essential characteristic to avoid the formation of plastic deformations, to drain rainwater from the surface, to achieve greater friction between tire and pavement, and even to reduce noise caused by vehicle traffic. Therefore, Spanish regulations limit the percentage of voids to between 4% and 7% for this type of bituminous mixture. Based on the above, it can be seen that the mixture with hornfels aggregate has a lower void content than mixtures conformed to electric arc furnace slag. This fact indicates a higher absorption of bitumen by electric arc furnace slags and ladle furnace slags.

Depending on the detailed limitations, ACFC bituminous mixtures are valid up to 17% bitumen by volume and of aggregate. Higher percentages of bitumen would develop an unacceptable void content. On the other hand, ACFC bituminous mixtures have acceptable voids percentages from 15.5% to 17.5%, according to the limitations set by the regulations. In turn, ASFS bituminous mixtures show acceptable percentages of bitumen according to the same limitations from 16% to 18%.

The Marshall test will be in responsible for showing the mechanical resistance of the bituminous mixture and, in short, within the range detailed above by the voids content, the optimum combination of materials for each family of samples. Figure 6 shows the Marshall stability of the different families of samples.

The Marshall test reflects the superior mechanical resistance of bituminous mixtures conformed to electric arc furnace slag as an aggregate and ladle furnace slag as a filler. In turn, the mixture with electric arc furnace slag and calcareous filler presents a lower resistance than the previous one but slightly higher than the resistance of the bituminous mixture formed with hornfels aggregate and calcareous filler. Two conclusions can be drawn from this fact. On the one hand, the ladle furnace slag has a significant influence on the mechanical resistance of the bituminous mixes conformed, thanks to its cementitious characteristics; on the other hand, the use of electric arc furnace slag makes it possible to absorb a higher percentage of bitumen than hornfels aggregate and obtaining better mechanical resistance. This higher percentage of bitumen, together with the filler and fibers, obtains a quality mastic to withstand repeated traffic loads and, consequently, a longer durability of the bituminous mix over time. It should be noted that if no volume proportioning had been carried out, the higher density of the electric arc furnace slag would have masked the results and the detailed conclusions could not have been obtained.

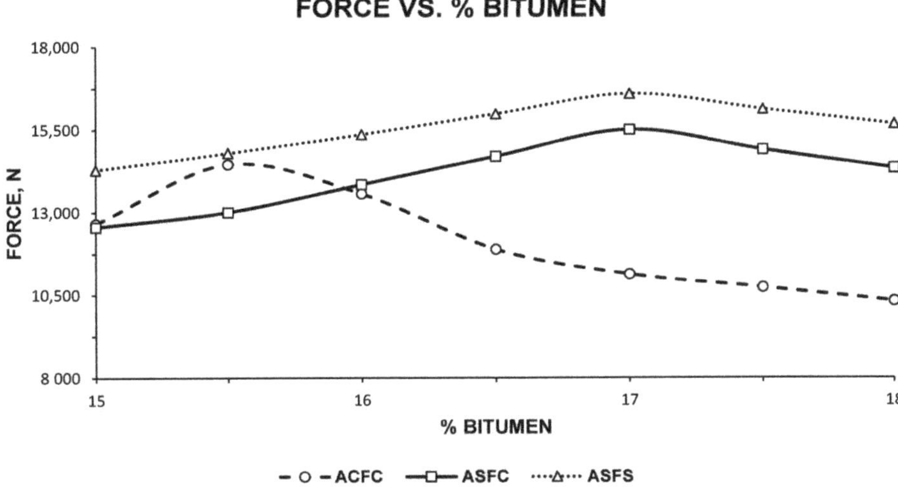

Figure 6. Marshall stability of the families of bituminous mixtures ACFC, ASFC and ASFS with different percentages of bitumen in volume of aggregate.

In addition, the Marshall test reflects the possibility of assessing the plastic deformations that may occur in the pavement. Therefore, it is essential to represent and evaluate the deformation of each family of bituminous mixtures. Marshall deformation is shown in Figure 7.

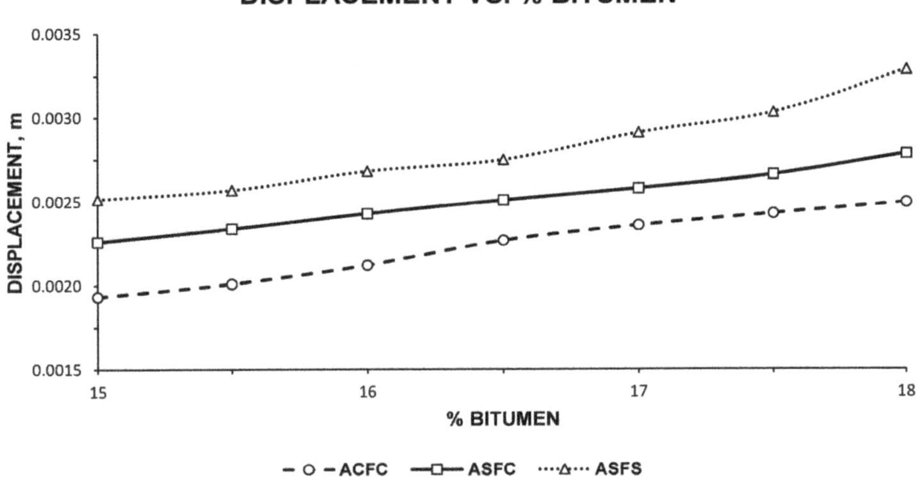

Figure 7. Marshall deformation of the families of bituminous mixtures ACFC, ASFC and ASFS with different percentages of bitumen in volume of aggregate.

Marshall deformation, or displacement during the test, is limited by the Spanish regulations for this type of bituminous mixture. The acceptable range of Marshall deformation results is 2 to 3 mm. Therefore, the ACFC family of bituminous mixtures has valid deformation percentages from 15.5% onwards. The ASFC family has acceptable bitumen percentages throughout its range to obtain adequate Marshall deformations. Finally, the ASFS family has adequate deformations, according to the regulations, in the percentages of bitumen from 15% to 17.5%. It is worth noting the greater deformation of the bituminous mixtures conformed with electric furnace slag and, in particular, of the mixtures that incorporate ladle furnace slag filler. This greater deformation is mainly due to the higher percentage of bitumen; however, the variations between the different families are small.

3.3. Determination of Optimal Material Combinations and Comparison of Results

Once the families of bituminous mixtures (ACFC, ASFC and ASFS) had been physically and mechanically analyzed, the optimum combination of materials was obtained. For this purpose, the Marshall test was taken as the reference test, since the aim is to obtain a resistant bituminous mixture without problems of plastic deformation. Therefore, the percentage of bitumen that provided the greatest Marshall stability was selected, provided that the other physical properties were acceptable according to the regulations.

With this optimal combination of materials for each family, the previous tests were again carried out to corroborate the quality of the mixtures, as well as the binder drainage test (standard UNE-EN 12697-18) [65] and the wheel-tracking tests (standard UNE-EN 12697-22) [66]. The results of all the tests carried out for the optimum combination of materials in each family are detailed in Table 10.

Table 10. Test results for the optimal combination of materials of the different families of bituminous mixtures ACFC, ASFC and ASFS.

Test	Standard	ACFC	ASFC	ASFS
Optimal percentage of bitumen (in volume and of aggregate)	-	15.50%	17%	17%
Bulk density	UNE-EN 12697-6 [62]	2291 ± 46 t/m^3	2716 ± 53 t/m^3	2721 ± 54 t/m^3
Maximum density	UNE-EN 12697-5 [61]	2436 ± 49 t/m^3	2857 ± 57 t/m^3	2887 ± 58 t/m^3
Void content	UNE-EN 12697-8 [63]	$5.9 \pm 0.1\%$	$4.9 \pm 0.1\%$	$5.8 \pm 0.1\%$
Stability Marshall	UNE-EN 12697-14 [64]	14586 ± 287 N	15524 ± 307 N	16658 ± 331 N
Marshall Deformation	UNE-EN 12697-14 [64]	0.0020 ± 0.0001 mm	0.0026 ± 0.0001 mm	0.0028 ± 0.0001 mm
Binder drainage	UNE-EN 12697-18 [65]	$0 \pm 0\%$	$0 \pm 0\%$	$0 \pm 0\%$
Wheel tracking (10,000 cycles; 60 °C)	UNE-EN 12697-22 [66]	0.07 ± 0.01 mm	0.05 ± 0.01 mm	$0.04 \pm 0.01\%$

As can be seen in Table 10, the results reflected from the previous tests for the optimum combination of materials from the three families of bituminous mixtures are acceptable according to the Spanish standard ORDER OC 3/2019. In addition, the three detailed mixtures reflect excellent results in comparison with other types of bituminous mixtures and can all be used for important traffic roads.

Nevertheless, and with the aim of evaluating the influence that the incorporation of waste has on the bituminous mixture, the best results obtained by the ASFS family, the conform made up of electric arc furnace slag and ladle furnace slag, should be highlighted. With similar but slightly lower results, there is the ASFC bituminous mixture and, finally, with important differences in terms of resistance and bitumen percentages, there is the ACFC mixture.

More specifically, the densities of the three families differ because of the higher density of the slags used. However, the percentage of voids obtained is similar even if a higher percentage of bitumen is used in the mixtures with electric arc furnace slag. This higher percentage of bitumen, without producing bleeding problems as confirmed by the binder drainage test, implies a higher resistance of the mix to withstand the repeated traction loads of traffic. This fact is reflected by the wheel tracking test, with the best values being achieved in the ASFS and ASFC mixes. The Marshall stability of the bituminous mixtures reflects the excellent behavior of the use of electric arc furnace slag and ladle furnace slag, being the family with the highest resistance. Finally, we should comment on the correct functioning of cellulose fibers from the paper industry as an additive in the mixtures, since it has allowed the incorporation of higher percentages of bitumen in all the families without the production of bleeding.

4. Conclusions

The research methodology followed and the results of the tests, as well as the discussions held, reflect a series of partial conclusions that will lead to the final conclusion. In turn, it should be mentioned that the initial hypothesis was the aptitude of electric arc furnace slag, ladle furnace slag and cellulose fibers from the papermaking industry for the conformation of Stone Mastic Asphalt type bituminous mixtures. The partial conclusions obtained are described below:

- The cellulose fibers discarded from the papermaking industry have a mainly organic composition, without the presence of chemical elements such as sulfur in large proportions that could cause environmental pollution problems. The size of the fibers is millimetric, and there are no agglomerations of the same after their treatment.
- Ladle furnace slag has an inorganic composition, being composed mainly of calcium, magnesium and silicon oxides. These oxides are mainly responsible for the cementitious characteristics. At the same time, it has a particle density that is slightly higher than that of a conventional aggregate, as well as a reduced particle size as demonstrated by its apparent density. However, the ladle furnace slag does not possess any plasticity.
- Electric arc furnace slag has an inorganic composition consisting mainly of metal element oxides. The density of these slags is high, not reflecting the existence of a high percentage of colloidal particles. In turn, the shape of the particles makes them suitable for use in high traffic bituminous mixtures, since, together with their mechanical resistance, they are capable of withstanding the compressive loads of traffic without problems of fractures or deformations. The resistance to abrasion, by contact with the tire, is another of the particularities that it possesses.
- Bituminous mixtures conformed with cellulose fibers from the paper industry are suitable for retention of a higher percentage of bitumen without causing bleeding problems. This fact has been corroborated by the binder drainage test of the mixtures made with the optimum combination of materials.
- Bituminous mixtures conformed with electric arc furnace slag provide greater Marshall stability, with similar deformation values. In addition, they are capable of absorbing a higher percentage of bitumen than a conventional aggregate, forming a mastic of suitable quality for the repeated traction loads of traffic. This fact has been corroborated with the wheel tracking test.
- The ladle furnace slag gives the mix greater Marshall stability, even being used in a low proportion and as a filler in the bituminous mixes.
- The optimum combination of materials in the ACFC, ASFC and ASFS mixtures reflects a similar percentage of void index. However, the maximum and bulk density of the slag mixtures is higher than that of hornfels aggregate.

Based on the partial conclusions derived from the methodology presented in this investigation and according to the tests carried out, it can be stated that the incorporation of electric arc furnace slag, ladle furnace slag and cellulose fibers from the paper industry, creates Stone Mastic Asphalt type bituminous mixtures with a higher percentage of bitumen, greater Marshall stability, better behavior towards permanents deformations and with a similar content of voids or deformations. The quality of these materials is therefore assured for use in bituminous mixtures dedicated to roads for important traffic. It is important to highlight the importance of using unused waste to conformed materials, not only of similar but superior quality, as it reduces their disposal in landfills and avoids the extraction of new raw materials.

Author Contributions: Conceptualization, F.A.C.-I., F.J.I.-G., J.M.T.-S. and J.S.-M.; methodology, F.A.C.-I., F.J.I.-G., J.M.T.-S. and J.S.-M.; software, J.M.T.-S. and J.S.-M.; validation, F.A.C.-I. and F.J.I.-G.; formal analysis, F.A.C.-I. and F.J.I.-G.; investigation, J.M.T.-S. and J.S.-M.; resources, F.A.C.-I.; data curation, F.J.I.-G.; writing—original draft preparation, J.S.-M.; writing—review and editing, J.M.T.-S.; visualization, J.M.T.-S.; supervision, F.A.C.-I.; project administration, J.S.-M.; funding acquisition, F.A.C.-I. All authors have read and agreed to the published version of the manuscript.

Funding: This research received no external funding.

Acknowledgments: Technical and human support provided by CICT of Universidad de Jaén (UJA, MINECO, Junta de Andalucía, FEDER) is gratefully acknowledged.

Conflicts of Interest: The authors declare no conflict of interest.

References

1. Shi, X.; Mukhopadhyay, A.; Zollinger, D.; Grasley, Z. Economic input-output life cycle assessment of concrete pavement containing recycled concrete aggregate. *J. Clean. Prod.* **2019**, *225*, 414–425. [CrossRef]
2. Arabani, M.; Azarhoosh, A.R. The effect of recycled concrete aggregate and steel slag on the dynamic properties of asphalt mixtures. *Constr. Build. Mater.* **2012**, *35*, 1–7. [CrossRef]
3. Plati, C. Sustainability factors in pavement materials, design, and preservation strategies: A literature review. *Constr. Build. Mater.* **2019**, *211*, 539–555. [CrossRef]
4. Amin, M.; Khan, K.; Saleem, M.; Khurram, N.; Niazi, M. Influence of Mechanically Activated Electric Arc Furnace Slag on Compressive Strength of Mortars Incorporating Curing Moisture and Temperature Effects. *Sustainability* **2017**, *9*, 1178. [CrossRef]
5. Pérez-Martnez, P.J. Energy consumption and emissions from the road transport in Spain: A conceptual approach. *Transport* **2012**, *27*, 383–396. [CrossRef]
6. Turk, J.; Mauko Pranjić, A.; Mladenovič, A.; Cotič, Z.; Jurjavčič, P. Environmental comparison of two alternative road pavement rehabilitation techniques: Cold-in-place-recycling versus traditional reconstruction. *J. Clean. Prod.* **2016**, *121*, 45–55. [CrossRef]
7. Morseletto, P. Targets for a circular economy. *Resour. Conserv. Recycl.* **2020**, *153*, 104553. [CrossRef]
8. Menaria, Y.; Sankhla, R. Use of Waste Plastic in Flexible Pavements-Green Roads. *Open J. Civ. Eng.* **2015**, *5*, 299–311. [CrossRef]
9. Al-Busaltan, S.; Al Nageim, H.; Atherton, W.; Sharples, G. Green Bituminous Asphalt relevant for highway and airfield pavement. *Constr. Build. Mater.* **2012**, *31*, 243–250. [CrossRef]
10. Jin, R.; Li, B.; Zhou, T.; Wanatowski, D.; Piroozfar, P. An empirical study of perceptions towards construction and demolition waste recycling and reuse in China. *Resour. Conserv. Recycl.* **2017**, *126*, 86–98. [CrossRef]
11. Demirbas, A. Waste management, waste resource facilities and waste conversion processes. *Energy Convers. Manag.* **2011**, *52*, 1280–1287. [CrossRef]
12. Anthonissen, J.; Van den bergh, W.; Braet, J. Review and environmental impact assessment of green technologies for base courses in bituminous pavements. *Environ. Impact Assess. Rev.* **2016**, *60*, 139–147. [CrossRef]
13. Zheng, X.; Easa, S.M.; Ji, T.; Jiang, Z. Incorporating uncertainty into life-cycle sustainability assessment of pavement alternatives. *J. Clean. Prod.* **2020**, *264*, 121466. [CrossRef]
14. Sangiorgi, C.; Lantieri, C.; Dondi, G. Construction and demolition waste recycling: An application for road construction. *Int. J. Pavement Eng.* **2015**, *16*, 530–537. [CrossRef]
15. Lori, A.R.; Hassani, A.; Sedghi, R. Investigating the mechanical and hydraulic characteristics of pervious concrete containing copper slag as coarse aggregate. *Constr. Build. Mater.* **2019**, *197*, 130–142. [CrossRef]
16. Lokesh, Y. Study On the Effect of Stone, Dust, Ceramic Dust and Brick Dust as Fillers on the Strength, Physical and Durability Properties of Bituminous Concrete (BC-II) Mix. *Int. J. Appl. Eng. Res.* **2018**, *13*, 203–208.
17. Kalantar, Z.N.; Karim, M.R.; Mahrez, A. A review of using waste and virgin polymer in pavement. *Constr. Build. Mater.* **2012**, *33*, 55–62. [CrossRef]
18. Zakaria, N.M.; Hassan, M.K.; Ibrahim, A.N.H.; Rosyidi, S.A.P.; Yusoff, N.I.M.; Mohamed, A.A.; Hassan, N. The use of mixed waste recycled plastic and glass as an aggregate replacement in asphalt mixtures. *J. Teknol.* **2018**, *80*, 79–88. [CrossRef]
19. Song, W.; Huang, B.; Shu, X. Influence of warm-mix asphalt technology and rejuvenator on performance of asphalt mixtures containing 50% reclaimed asphalt pavement. *J. Clean. Prod.* **2018**, *192*, 191–198. [CrossRef]
20. Fakhri, M.; Javadi, S.; Sedghi, R.; Arzjani, D.; Zarrinpour, Y. Effects of deicing agents on moisture susceptibility of the WMA containing recycled crumb rubber. *Constr. Build. Mater.* **2019**, *227*, 116581. [CrossRef]
21. Escorias de Acería de Horno de Arco Eléctrico|CEDEX. Available online: http://www.cedexmateriales.es/catalogo-de-residuos/25/escorias-de-aceria-de-horno-de-arco-electrico/ (accessed on 29 April 2020).

22. Pranjić, I.; Deluka-Tibljaš, A.; Cuculić, M.; Šurdonja, S. Influence of Pavement Surface Macrotexture on Pavement Skid Resistance. *Transp. Res. Procedia* **2020**, *45*, 747–754. [CrossRef]
23. Mao, X.; Wang, J.; Yuan, C.; Yu, W.; Gan, J. A Dynamic Traffic Assignment Model for the Sustainability of Pavement Performance. *Sustainability* **2018**, *11*, 170. [CrossRef]
24. Zhang, H.; Hu, Z.; Hou, S.; Xu, T. Aging behaviors of bitumen degraded by the microbial consortium on bituminous pavement. *Constr. Build. Mater.* **2020**, *254*, 119333. [CrossRef]
25. Alkins, A.E.; Lane, B.; Kazmierowski, T. Sustainable Pavements. *Transp. Res. Rec. J. Transp. Res. Board* **2008**, *2084*, 100–103. [CrossRef]
26. Ameri, M.; Hesami, S.; Goli, H. Laboratory evaluation of warm mix asphalt mixtures containing electric arc furnace (EAF) steel slag. *Constr. Build. Mater.* **2013**, *49*, 611–617. [CrossRef]
27. Esther, L.A.; Pedro, L.G.; Irune, I.V.; Gerardo, F. Comprehensive analysis of the environmental impact of electric arc furnace steel slag on asphalt mixtures. *J. Clean. Prod.* **2020**, *275*, 123121. [CrossRef]
28. Lam, M.N.T.; Le, D.H.; Jaritngam, S. Compressive strength and durability properties of roller-compacted concrete pavement containing electric arc furnace slag aggregate and fly ash. *Constr. Build. Mater.* **2018**, *191*, 912–922. [CrossRef]
29. Lam, M.N.T.; Jaritngam, S.; Le, D.H. Roller-compacted concrete pavement made of Electric Arc Furnace slag aggregate: Mix design and mechanical properties. *Constr. Build. Mater.* **2017**, *154*, 482–495. [CrossRef]
30. Kavussi, A.; Qazizadeh, M.J. Fatigue characterization of asphalt mixes containing electric arc furnace (EAF) steel slag subjected to long term aging. *Constr. Build. Mater.* **2014**, *72*, 158–166. [CrossRef]
31. Pasetto, M.; Baldo, N. Experimental evaluation of high performance base course and road base asphalt concrete with electric arc furnace steel slags. *J. Hazard. Mater.* **2010**, *181*, 938–948. [CrossRef]
32. Ziaee, S.A.; Behnia, K. Evaluating the effect of electric arc furnace steel slag on dynamic and static mechanical behavior of warm mix asphalt mixtures. *J. Clean. Prod.* **2020**, *274*, 123092. [CrossRef]
33. Motevalizadeh, S.M.; Sedghi, R.; Rooholamini, H. Fracture properties of asphalt mixtures containing electric arc furnace slag at low and intermediate temperatures. *Constr. Build. Mater.* **2020**, *240*, 117965. [CrossRef]
34. Skaf, M.; Manso, J.M.; Aragón, Á.; Fuente-Alonso, J.A.; Ortega-López, V. EAF slag in asphalt mixes: A brief review of its possible re-use. *Resour. Conserv. Recycl.* **2017**, *120*, 176–185. [CrossRef]
35. Wu, S.; Xue, Y.; Ye, Q.; Chen, Y. Utilization of steel slag as aggregates for stone mastic asphalt (SMA) mixtures. *Build. Environ.* **2007**, *42*, 2580–2585. [CrossRef]
36. Sideris, K.K.; Tassos, C.; Chatzopoulos, A.; Manita, P. Mechanical characteristics and durability of self compacting concretes produced with ladle furnace slag. *Constr. Build. Mater.* **2018**, *170*, 660–667. [CrossRef]
37. Herrero, T.; Vegas, I.J.; Santamaría, A.; San-José, J.T.; Skaf, M. Effect of high-alumina ladle furnace slag as cement substitution in masonry mortars. *Constr. Build. Mater.* **2016**, *123*, 404–413. [CrossRef]
38. Sáez-De-Guinoa Vilaplana, A.; Ferreira, V.J.; López-Sabirón, A.M.; Aranda-Usón, A.; Lausín-González, C.; Berganza-Conde, C.; Ferreira, G. Utilization of Ladle Furnace slag from a steelwork for laboratory scale production of Portland cement. *Constr. Build. Mater.* **2015**, *94*, 837–843. [CrossRef]
39. Montenegro-Cooper, J.M.; Celemín-Matachana, M.; Cañizal, J.; González, J.J. Study of the expansive behavior of ladle furnace slag and its mixture with low quality natural soils. *Constr. Build. Mater.* **2019**, *203*, 201–209. [CrossRef]
40. Adolfsson, D.; Engström, F.; Robinson, R.; Björkman, B. Cementitious Phases in Ladle Slag. *Steel Res. Int.* **2011**, *82*, 398–403. [CrossRef]
41. Shi, C. Characteristics and cementitious properties of ladle slag fines from steel production. *Cem. Concr. Res.* **2002**, *32*, 459–462. [CrossRef]
42. Orden Circular OC 3/2019 Sobre Mezclas Bituminosa Tipo SMA. Available online: https://www.mitma.gob.es/recursos_mfom/comodin/recursos/oc3_2019.pdf (accessed on 29 October 2020).
43. UNE-EN 1426:2015 Bitumen and Bituminous Binders—Determination of Needle Penetration. Available online: https://www.une.org/encuentra-tu-norma/busca-tu-norma/norma/?c=N0055820 (accessed on 29 September 2020).
44. UNE-EN 12591:2009 Bitumen and Bituminous Binders—Specifications for Paving Grade Bitumens. Available online: https://www.une.org/encuentra-tu-norma/busca-tu-norma/norma?c=N0044322 (accessed on 29 October 2020).

45. UNE-EN 1427:2015 Bitumen and Bituminous Binders—Determination of the Softening Point—Ring and Ball Method. Available online: https://www.une.org/encuentra-tu-norma/busca-tu-norma/norma?c=N0055821 (accessed on 29 September 2020).

46. UNE-EN 12593:2015 Bitumen and Bituminous Binders—Determination of the Fraass Breaking Point. Available online: https://www.une.org/encuentra-tu-norma/busca-tu-norma/norma?c=N0055822 (accessed on 29 October 2020).

47. UNE-EN 12592:2015 Bitumen and Bituminous Binders—Determination of Solubility. Available online: https://www.une.org/encuentra-tu-norma/busca-tu-norma/norma?c=N0054574 (accessed on 29 October 2020).

48. UNE-EN ISO 2592:2018 Petroleum and Related Products—Determination of Flash and Fire Points—Cleveland Open Cup Method. Available online: https://www.une.org/encuentra-tu-norma/busca-tu-norma/norma?c=N0060082 (accessed on 30 October 2020).

49. UNE-EN 12607-1:2015 Bitumen and Bituminous Binders—Determination of the Resistance to Hardening under Influence of Heat and Air—Part 1: RTFOT Method. Available online: https://www.une.org/encuentra-tu-norma/busca-tu-norma/norma?c=N0054566 (accessed on 30 October 2020).

50. UNE-EN 1097-7:2009 Tests for Mechanical and Physical Properties of Aggregates—Part 3: Determination of Loose Bulk Density and Voids. Available online: https://www.une.org/encuentra-tu-norma/busca-tu-norma/norma?c=N0042553 (accessed on 16 September 2020).

51. UNE-EN 1097-3:1999 Tests for Mechanical and Physical Properties of Aggregates—Part 3: Determination of Loose Bulk Density and Voids. Available online: https://www.une.org/encuentra-tu-norma/busca-tu-norma/norma/?c=N0009465 (accessed on 16 September 2020).

52. UNE 103103:1994 Determination of the Liquid Limit of a Soil by the Casagrande Apparatus Method. Available online: https://www.une.org/encuentra-tu-norma/busca-tu-norma/norma/?c=N0007830 (accessed on 29 September 2020).

53. UNE 103104:1993 Test for Plastic Limit of a Soil. Available online: https://www.une.org/encuentra-tu-norma/busca-tu-norma/norma?c=N0007831 (accessed on 29 September 2020).

54. UNE-EN 933-8:2012+A1:2015/1M:2016 Tests for Geometrical Properties of Aggregates—Part 8: Assessment of Fines—Sand Equivalent Test. Available online: https://www.une.org/encuentra-tu-norma/busca-tu-norma/norma?c=N0056257 (accessed on 16 September 2020).

55. UNE-EN 933-5:1999/A1:2005 Tests for Geometrical Properties of Aggregates—Part 5: Determination of Percentage of Crushed and Broken Surfaces in Coarse Aggregate Particles. Available online: https://www.une.org/encuentra-tu-norma/busca-tu-norma/norma/?c=N0034842 (accessed on 16 September 2020).

56. UNE-EN 933-3:2012 Tests for Geometrical Properties of Aggregates—Part 3: Determination of Particle Shape—Flakiness Index. Available online: https://www.une.org/encuentra-tu-norma/busca-tu-norma/norma?c=N0049063 (accessed on 16 September 2020).

57. UNE-EN 1097-2:2010 Tests for Mechanical and Physical Properties of Aggregates—Part 2: Methods for the Determination of Resistance to Fragmentation. Available online: https://www.une.org/encuentra-tu-norma/busca-tu-norma/norma/?c=N0046026 (accessed on 16 September 2020).

58. UNE-EN 1367-1:2008 Tests for Thermal and Weathering Properties of Aggregates—Part 1: Determination of Resistance to Freezing and Thawing. Available online: https://www.une.org/encuentra-tu-norma/busca-tu-norma/norma/?c=N0040756 (accessed on 16 September 2020).

59. UNE-EN 1097-8:2010 Tests for Mechanical and Physical Properties of Aggregates—Part 8: Determination of the Polished Stone Value. Available online: https://www.une.org/encuentra-tu-norma/busca-tu-norma/norma?c=N0044542 (accessed on 20 September 2020).

60. UNE-EN 12697-30:2019 Bituminous Mixtures—Test Methods—Part 30: Specimen Preparation by Impact Compactor. Available online: https://www.une.org/encuentra-tu-norma/busca-tu-norma/norma?c=N0062608 (accessed on 30 October 2020).

61. UNE-EN 12697-5:2020 Test Methods—Part 5: Determination of the Maximum Density. Available online: https://www.une.org/encuentra-tu-norma/busca-tu-norma/norma?c=N0063145 (accessed on 29 September 2020).

62. UNE-EN 12697-6:2012 Bituminous Mixtures—Test Methods for Hot Mix Asphalt—Part 6: Determination of Bulk Density of Bituminous Specimens. Available online: https://www.une.org/encuentra-tu-norma/busca-tu-norma/norma/?c=N0049868 (accessed on 29 September 2020).

63. UNE-EN 12697-8:2020 Bituminous Mixtures—Test Methods—Part 8: Determination of Void Characteristics of Bituminous Specimens. Available online: https://www.une.org/encuentra-tu-norma/busca-tu-norma/norma/?c=N0063146 (accessed on 29 September 2020).
64. UNE-EN 12697-14:2001 Bituminous Mixtures—Test Methods for Hot Mix Asphalt—Part 14: Water Content. Available online: https://www.une.org/encuentra-tu-norma/busca-tu-norma/norma?c=N0025364 (accessed on 30 October 2020).
65. UNE-EN 12697-18:2018 Bituminous Mixtures—Test Methods—Part 18: Binder Drainage. Available online: https://www.une.org/encuentra-tu-norma/busca-tu-norma/norma?c=N0061210 (accessed on 31 October 2020).
66. UNE-EN 12697-22:2008+A1:2008 Bituminous Mixtures—Test Methods for Hot Mix Asphalt—Part 22: Wheel Tracking. Available online: https://www.une.org/encuentra-tu-norma/busca-tu-norma/norma?c=N0040736 (accessed on 31 October 2020).

Publisher's Note: MDPI stays neutral with regard to jurisdictional claims in published maps and institutional affiliations.

 *metals*

Article

Recovery of Cobalt from the Residues of an Industrial Zinc Refinery

Laurence Boisvert [1], Keven Turgeon [1], Jean-François Boulanger [2], Claude Bazin [1,*] and Georges Houlachi [3]

[1] Department of Mining, Metallurgical and Materials Engineering, Université Laval,
    Québec, QC G1V 0A6, Canada; laurence.boisvert.1@ulaval.ca (L.B.); keven.turgeon.1@ulaval.ca (K.T.)
[2] Institut de Recherche sur les Mines et l'Environnement,
    Université du Québec en Abitibi Témiscamingue (UQAT), Rouyn-Noranda, QC J9X 5E4, Canada;
    Jean-Francois.Boulanger@uqat.ca
[3] Institut de Recherche, Hydro-Québec, Varennes, QC J3X 1S1, Canada; houlachi.georges@hydroquebec.com
* Correspondence: claude.bazin@gmn.ulaval.ca; Tel.: +1-418-656-5914

Received: 4 November 2020; Accepted: 19 November 2020; Published: 22 November 2020

Abstract: The electrolytic production of metallic zinc from processing zinc sulfide concentrates generates a residue containing cadmium, copper, and cobalt that need to be removed from the electrolytic zinc solution because they are harmful to the zinc electro-winning process. This residue is commonly sent to other parties that partly recover the contained elements. These elements can generate revenues if recovered at the zinc plant site. A series of laboratory tests were conducted to evaluate a method to process a zinc plant residue with the objective of recovering cobalt into a salable product. The proposed process comprises washing, selective leaching, purifying and precipitation of cobalt following its oxidation. The process allows the production of a cobalt rich hydroxide precipitate assaying 45 ± 4% Co, 0.8 ± 0.2% Zn, 4.4 ± 0.7% Cu, and 0.120 ± 0.004% Cd at a 61 ± 14% Co recovery. Replicating the whole process with different feed samples allowed the identification of the critical steps in the production of the cobalt product; one of these critical steps being the control of the oxidation conditions for the selective precipitation step.

Keywords: zinc residue; cobalt hydroxide; cementation; leaching; oxidative precipitation

---

1. Introduction

The conventional roast-leach zinc extraction process yields a solid residue consisting of a mixture of zinc, copper, cadmium, and cobalt. Some zinc smelting plants process that residue [1–3] to recover the contained valuable metals. The high value of cobalt makes it an excellent candidate for a first step in the development of a process to recover the metals contained in that residue [1]. Indeed, cobalt is currently considered as a critical material [4] as it is used in the making of Li-ion batteries [1,5], increasingly strategic for the shift toward green energy or more precisely toward a 100% electric vehicle market [4]. The «critical» status of cobalt is related to uncertainties about the supply of the metal. In fact, 60% of the world's cobalt is mined in the Congo and 80% of its production is processed in China [4]. In the case of a cobalt supply disruption due to a natural disaster, a change of government or a boycott [6], a zinc residue that contains more than 2% Co [1] becomes an interesting alternative to primary cobalt.

The extraction of cobalt from zinc plant residues is not discussed in many papers except in a recent review [1]. Few papers [2,3,7,8] were found to describe processes to recover cobalt from zinc residues. Wang and Zhou described a process [2] to treat a zinc residue containing active carbon and organic compounds used to capture the cobalt and the manganese from the zinc solution prior to the Zn electro-winning step. The process developed for that residue consists of a washing stage followed by two roasting steps at different temperatures, leaching, and precipitation of iron and manganese followed by anion exchange and solvent extraction of cobalt using extractant P507. Fattahi et al. described a process [3] to extract cobalt from the zinc residues of Iranian Zn smelters that add permanganate to the zinc solution in order to oxidize Co(II) to Co(III) which is precipitated by increasing the pH. The precipitated zinc residue contains about 2% Co and more than 10% Mn. The authors [3] propose a reductive leaching followed by a precipitation of cobalt sulfide (CoS) using Na_2S to selectively recover the cobalt from the manganese. Although the conventional approach for the purification of zinc solution is the cementation of Cu, Cd, and Co onto zinc dust [1,9], it was not possible to find in the Western literature a process dedicated to the treatment of this type of residue, although a qualitative description of a possible process can be found in [1]. Li et al. [10] studied the rate of leaching of a zinc residue but did not indicate if the residue is obtained by cementation on zinc dust and the authors did not attempt to process the leach solution to obtain a salable cobalt product.

This paper describes a processing scheme to produce a cobalt rich compound from a zinc plant residue produced by the cementation process. The objective of the test work presented in this paper is not to optimize an existing flowsheet but to propose a process to obtain a salable cobalt compound from a zinc residue produced by cementation on zinc dust. As requested by the industrial partner, the proposed process should use only open tank reactors for leaching and precipitation without resorting to solvent extraction nor ion exchange to obtain the cobalt product. This last constraint complicates the process to be developed as selective extractants are available for cobalt [1,7]. However, if the economics of the process are not favorable, the use of SX will be investigated but it is unlikely that the addition of a SX plant could move the process economics toward more profitable conditions.

2. Materials and Methods

2.1. Instrumentation and Reagents

The solid and liquid samples are assayed using a MP-AES 4200 (microwave-plasma atomic emission spectrometer) from Agilent Technology (Santa Clara, CA, USA). For solid samples, 0.5 g sub samples are taken and digested in aqua regia for analysis using the MP-AES. The pH and Eh of the solution are, respectively, measured using a Fisher Scientific accumet XL600 pH-meter (Waltham, MA, USA), an Orion pH probe, and an Orion Oxidation Reduction Potential (ORP) probe from Thermo scientific (Waltham, MA, USA). Table 1 presents the reagents used for the experimentation.

Table 1. Reagents used for the experimentation.

Reagent	Composition	Brand and Purity
Sulfuric acid	H_2SO_4	Fisher, 98% purity
Sodium hydroxide	NaOH	Fisher, 98.8% purity
Ammonium persulfate	$(NH_4)_2S_2O_8$	Alfa Aesar, 98% purity

2.2. Provenance of the Zinc Residue

The zinc residue used for the test work is provided by the CEZinc refinery [11] in Valleyfield, QC, Canada. Figure 1 shows the zinc extraction process and identifies the origin of the Co-bearing zinc residue. The plant processes zinc sulfide concentrates assaying more than 50% Zn, 0–5% Pb, less than 2% Cu, 0.5% Cd and from 50–200 g/t Co. The main impurities are iron (>8%) and sulfur (>30%). The extraction process of zinc begins by a roasting of the zinc sulfide concentrate to oxidize sulfur into SO_2 that is subsequently converted into sulfuric acid (H_2SO_4). The roasting also transforms the zinc sulfide into zinc oxide which is soluble in weakly acidic solutions. The roasted product is leached in sequence with weak and strong sulfuric acid to solubilize the zinc oxide. The metallic impurities (Fe, Cu, Cd, Co) follow the zinc into the solution. The solution then undergoes a neutralization, during which the solubilized iron is precipitated as jarosite [8]. The purification of the iron-free solution from the remaining metallic impurities is done using cementation on zinc dust [9] where copper, cadmium, and cobalt displace the zinc of the zinc powder through the reaction:

$$M_{aq}^{2+} + Zn_s \rightarrow Zn_{aq}^{2+} + M_s \tag{1}$$

where M stands for copper, cadmium, or cobalt. The reduced copper, cadmium, and cobalt are plated onto the surface of the zinc powder. The cemented powder is separated from the solution by a leaf press filter. The recovered solid is the «Zinc Plant Residue» (ZPR) considered in the following study. This residue is currently transferred to another plant for further processing. Zinc is finally electro-won from the purified solution (See Figure 1). The spent electrolyte is recycled back to the leaching step.

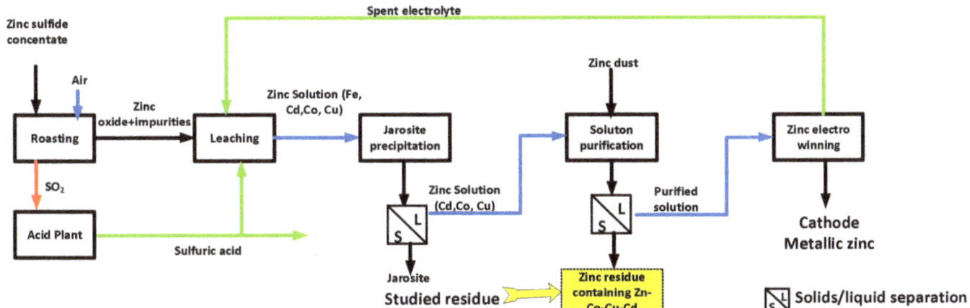

Figure 1. Simplified flowsheet of the zinc extraction process.

2.3. Characterization of the Zinc Plant Residue (ZPR)

A 20 kg sample of ZPR was collected by the CEZinc personnel at the discharge of the press filter. The sample was shipped wet to the laboratory for the test work. The received sample was split wet into ten parts after spreading the material onto a plastic sheet. One 2 kg sample was put in an oven for drying overnight and the remaining material was kept wet for the subsequent test work. A portion of the dried sample is shown in Figure 2a. The material is strongly agglomerated with practically unbreakable lumps. Sulfuric acid and sulfates are likely responsible for the observed particles binding. It rapidly appeared that this material could not be characterized (chemical composition, size distribution, etc.) as is and it was decided to wash the residue in water to remove any excess of sulfuric acid before drying. Washing was carried out by mixing 500 g of wet residue in 2700 mL of water (15% solids in mass) in a beaker for 90 min at room temperature (25 °C). The washed residue was separated from the solution by vacuum filtration and dried overnight in an oven. The dried washed residue is shown in Figure 2b and is found to be more amenable to characterization than the raw residue.

(**a**)   (**b**)

Figure 2. ZPR as is and after washing with water. (**a**) Raw residue after drying; (**b**) Residue after washing and drying.

2.3.1. Specific Gravity

The measured specific gravity (Gas pycnometer, HumiPyc model 2 from Instruquest, Coconut Creek, FL, USA) of the washed ZPR is 3.95 g/cm^3. Since the densities of zinc and of the cemented metals are all above 5 g/cm^3 this result indicates that the residue is not made exclusively of pure metals.

2.3.2. Size Distribution

Figure 3 shows the particle size distribution obtained by sieving the washed ZPR on a Tyler screen series from 1.2 mm down to 0.038 mm. The ZPR is coarse with a D_{80} of about 900 μm. The material coarseness will pose a problem for the sampling for assaying of the residue. Indeed, since assaying of the sample implies collecting a 0.5–1.0 g sample for the digestion prior to analysis using the emission spectrometer (Section 2.1), one can expect a significant variability in the assays of the residue due to the fundamental error of sampling [12]. For ore type material, this error is reduced by pulverizing the sample prior to sampling. However, the metallic and ductile nature of the ZPR makes impossible the pulverization of the sample, and thus one should expect a variability in the assays that data reconciliation [13–16] as applied here should be able to attenuate.

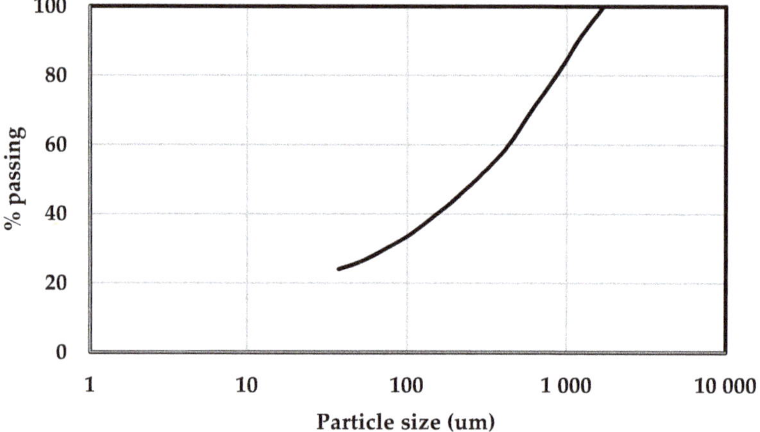

Figure 3. Size distribution of the residue.

2.3.3. Chemical Composition

Table 2 gives the chemical composition of the ZPR sample. The measured assays of the sample are compared (see Table 2) to the typical composition of the ZPR provided by the CEZinc plant. The general proximity of the sample composition with the typical composition of the residue provided by CEZinc confirms that the received sample is representative of the residue usually released by the plant. The difference in the composition is attributed to the washing as discussed later. The Co content is consistent with typical ZPR from other zinc smelters [1–3].

Table 2. Composition of the ZPR.

Elements	CEZ Typical Composition %	Sample Composition * %		
Zn	20–25	16.0	±	0.5
Cd	2–6	3.5	±	0.2
S	-	5.7	±	0.8
Cu	15–25	11.9	±	0.4
Ca	-	0.76	±	0.04
Co	2–4	1.7	±	0.1
Ni	0–2	0.44	±	0.04
Fe	0–1	0.34	±	0.02
Mn	-	0.22	±	0.01
Pb	9–12	4.0	±	0.1

*: Average ± standard deviation of three samples.

2.3.4. X-ray Diffraction (XRD)

Figure 4 shows the XRD pattern (Instrument: Aeris, Malvern Panalytical (Malvern, UK) obtained for the residue. The XRD shows the presence of metallic zinc, copper, and possibly metallic lead. The peaks for Cd and Co are not visible due to the low contents of these elements. The presence of Pb is confirmed by the assays (see Table 2) and is reported under the form of PbO_2 for the residue of an Iranian zinc smelter [3] and under the form of $PbSO_4$ [10] for a Chinese zinc residue. The XRD results of Figure 4 also show the presence of copper, zinc, and calcium sulfates likely responsible for the observed low specific gravity of the residue (Section 2.3.1). These sulfates, except gypsum, could be completely removed by a longer or slightly acidic or hot water wash of the residue as reported in [2].

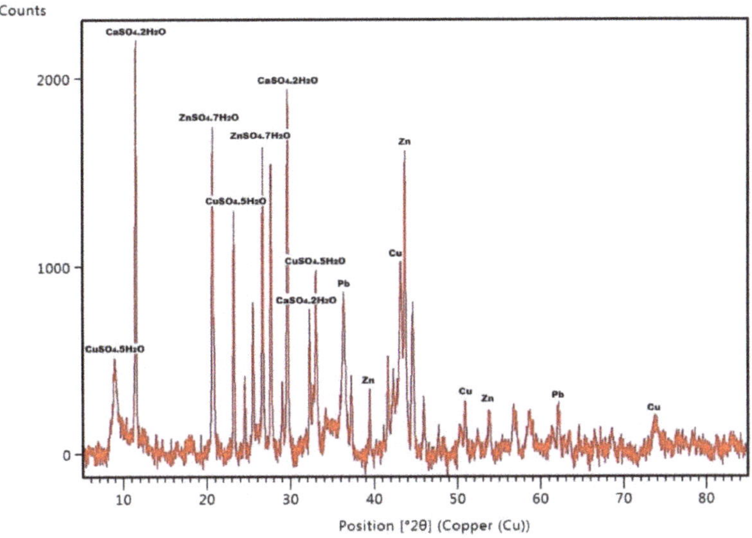

Figure 4. XRD powder pattern of the CEZinc ZPR.

3. Results

3.1. Washing of the ZPR

The first step of the proposed process is the washing of the ZPR. Initially done to eliminate the sulfuric acid from the ZPR, the washing step provides an economic way to pre-concentrate the cobalt by allowing a partial elimination of the soluble zinc and cadmium sulfates contained in the

ZPR. The results of a washing test are shown in Figure 5 and Table 3. Washing is carried out at room temperature using water at 15% solids in mass. Figure 5 shows the measured concentrations of Zn, Cd, Ca, Co, and Cu during washing. The solution pH falls from 7.2 to 6.3 during the washing step due to the release of the residual sulfuric acid. Table 3 gives the proportions of metals removed from the ZPR calculated using:

$$D_m = 100 \frac{V x_m}{V x_m + W_{ZPR} y_{ZPR}} \tag{2}$$

where D_m is the dissolved proportion of metal m, V is the volume (L) of the wash solution, x_m the metal content (g/L) in the solution. W_{ZPR} and y_{ZPR} are, respectively, the weight (g) and the fraction of metal m in the dried washed ZPR. Results presented in Table 3 show that, respectively, 20% and 43% of the zinc and cadmium contained in the raw ZPR are removed by a 30 min wash. About 4% of the cobalt contained in the ZPR is lost during the operation. Higher impurity removal can be achieved by washing the ZPR for 90 min but at increased cobalt losses as shown in Figure 5. The zinc is likely under an insoluble metallic form as metallic zinc dust is used for the cementation, while cadmium is likely present as a sulfate. The 4% Co dissolution is an indication that some cobalt is under the form of sulfate, with the remaining being metallic. The washing step can be viewed as a selective leaching operation for which the leaching conditions are adjusted to target specific metals [1,2]. The use of diluted acid in replacement of water could yield a better elimination of zinc and cadmium at the expenses of more important cobalt losses into the wash solution [10].

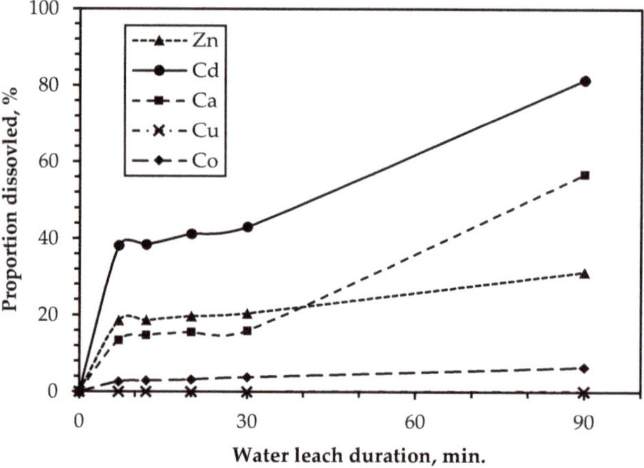

Figure 5. Dissolved proportion of metals in time (25 °C, 15% solids in mass).

Table 3. Dissolved metals after 30 min of washing with water (25 °C, 15% solids in mass).

Metal	% Dissolved *		
Zn	20	±	3
Cd	43	±	2
Cu	0.03	±	0.02
Ca	16	±	1
Co	4	±	1
Ni	2	±	1
Fe	0.4	±	0.2
Mn	84	±	3
Pb	0.1	±	0.2

*: average ± standard deviation of three tests.

3.2. Preparation of ZPR Samples for the Co Extraction Tests

The leaching tests are carried out using samples prepared by washing 75 g of raw ZPR for 30 min using the above described approach. The washed material is dried at low temperature in an oven. The use of 75 g batches was found to be a good compromise between the variability of the 75 g batch composition [17] and the mass of material to be manipulated during the experimentation. Table 4 gives the assays obtained by assaying three (3) randomly selected 75 g batches.

Table 4. Chemical assays (%) of three randomly selected feed samples (Smp = Sample; Avg = Average; Std-dev =Standard deviation RSD: = Std-Dev/Avg).

Elements	Smp #1	Smp #2	Smp #3	Avg	Std-Dev	RSD (%)
Zn	16.4	16.2	15.5	16.0	0.5	3
Cd	3.7	3.5	3.3	3.5	0.2	6
Cu	11.5	12.1	12.2	11.9	0.4	3
Ca	0.80	0.77	0.72	0.76	0.04	5
Co	1.68	1.80	1.63	1.71	0.09	5
Ni	0.47	0.47	0.40	0.44	0.04	9
Fe	0.36	0.34	0.31	0.34	0.02	6
Mn	0.23	0.21	0.23	0.22	0.01	5
Pb	3.8	4.1	4.0	4.0	0.1	3

3.3. Leaching of the Washed ZPR

The next step of the process is to leach the metals out of the ZPR into an aqueous solution from whichi it will be possible to separate the cobalt from the other metals. The leaching tests are conducted in a mechanically agitated beaker for 30 min using a weight L/S ratio of 15/1 at 80 °C. Leaching tests are carried out with sulfuric acid as it is the acid used in the zinc plant that provided the ZPR. Table 5 gives the average proportion of metals dissolved (±standard deviation) of three (3) leaching tests and the average metal contents in the liquor. About 65% of the solids in the ZPR are dissolved during the leaching process with 98% of the cobalt effectively leached off the ZPR, a performance similar to that reported in [10].

Table 5. Proportion of the metals dissolved during the leaching of the ZPR (Average (±standard deviation) of the results of three (3) leaching tests and the average composition of the liquor (Ratio L/S: 15/1; 100 g/L H_2SO_4, Eh = 90 mV, pH = 0.2, 30 min, 80 °C).

Species	% Dissolved			Liquor Concentration (mg/L)
Solids	65	±	1	
Zn	97.1	±	0.4	11,800
Cd	91	±	2	370
Cu	37	±	3	3600
Ca	88	±	3	252
Co	97	±	1	1170
Ni	97	±	1	345
Fe	98	±	2	270
Mn	91	±	16	26
Pb	0.2	±	0.3	6

3.4. Selective Precipitation of the Cobalt from the Pregnant Liquor Solution

The main impurities in the pregnant liquor solution are Zn, Cd, Ni, Fe, and Cu (Table 5). The approach used to selectively separate cobalt from these elements begins with an oxidation of iron and manganese to precipitate of these oxidized elements by increasing the pH near 3.0 while Cu, Zn, Cd, and Co(II) remain in the solution. The next step is to oxidize Co(II) to Co(III) and to precipitate Co(III) by increasing the pH above 3.0 leaving Zn, Cd, Ni and Cu in the solution.

Ammonium PerSulfate or APS ((NH$_4$)$_2$S$_2$O$_8$), as discussed in [1,8], provides the required oxidation potential for iron, manganese and subsequently cobalt. The sequence of the steps followed to precipitate the cobalt is shown in Figure 6.

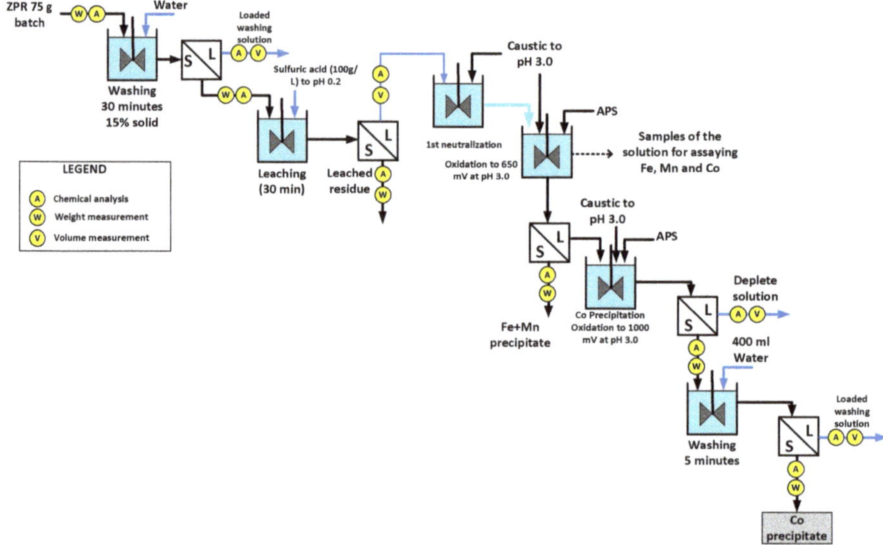

Figure 6. Process for the extraction of cobalt from the ZPR.

3.4.1. Precipitation of Fe-Mn

Sodium hydroxide is firstly added to the pregnant liquor solution to bring the pH to 3.0. APS is then added to the solution to increase the Redox potential from 90 to 650 mV. During the oxidation process, the pH of the solution is maintained at 3.0 by a regular addition of NaOH. Samples of the solution were collected at different Eh values and analyzed for Fe, Mn, and Co. As anticipated [1] Fe and Mn precipitate prior to Co a behavior confirmed by the variations of the metal contents in the solution as shown in Figure 7. According to Figure 7, the optimal Eh for the precipitation of Fe and Mn without precipitation of Co is 650 mV and the minimum Eh for the precipitation of Co is 1000 mV. These Eh values are coherent with those reported in [1]. Table 6 gives the proportions of the metals in the leached solution that are removed by the oxidation to 650 mV. Iron and manganese are almost completely removed from the solution. Ideally, cobalt should not be precipitated at this stage; however, it was observed that a close control of the Eh was not always possible as the Redox probe shows a significant measurement variability causing cobalt losses (~17%) in the Fe–Mn precipitate. The difficulty in the control of the Eh is attributed to the probe and to its sensitivity to the APS addition. Indeed, a small addition of APS can make a large local step in the Redox potential of the solution. It is likely that a process solely based on the use of an APS dosage [1,8] is not recommended to achieve a selective separation. Cleary the oxidant addition should be based on a reliable measurement of the ReDox potential of the solution and not on the dosage of the oxidant.

Once the Redox potential reaches 650 mV, the addition of APS is stopped and the iron and manganese precipitate are removed by filtration (see Figure 6). The filtration of the sludgy Fe-Mn precipitate is difficult and is believed to be one of the reasons for the loss of cobalt at this stage. Ensuring a rapid and adequate filtration of the solution after the end of the reaction seems critical to limit cobalt losses. The solution from that solid/liquid separation advances to the cobalt precipitation step (Figure 6). Results of Table 6 show that Mn and Fe are effectively removed from the solution at the expenses of some cobalt losses. The large standard deviation observed for the fraction of cobalt precipitated is attributed to Eh measurement problems as discussed above and difficulties in the filtration of the precipitate from one test to another one. Improving on the accuracy of the Eh measurement method and of the solids/liquid separation of the Fe–Mn precipitate could significantly improve the global recovery of cobalt.

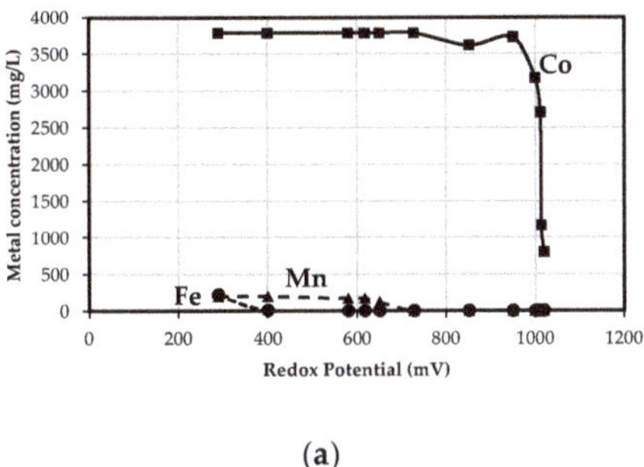

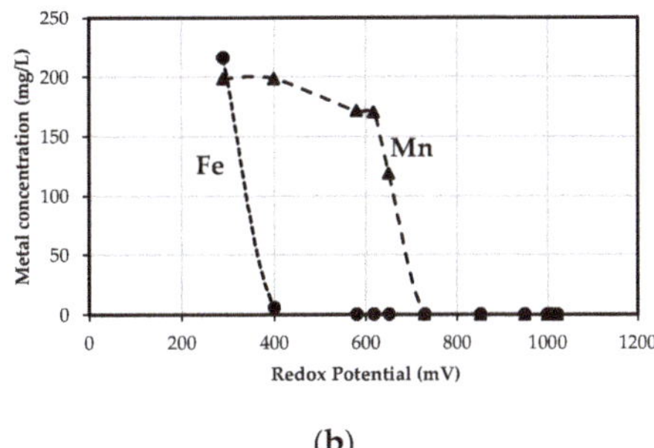

(a)   (b)

Figure 7. Precipitation of Fe and Mn with increasing the Eh while maintaining a pH of 3.0 (**a**) Fe, Mn, and Co contents in the solution during the oxidation to 1000 mV. (**b**) Fe and Mn contents in the solution during the oxidation below 1000 mV.

Table 6. Metal precipitation during the oxidation for Fe-Mn removal of the leach solution (Average (±standard deviation) results of three (3) tests (Eh = 90 to 650 mV, pH = 3.0, 80 °C, 15 min).

Species	% Precipitated		
Zn	0.3	±	0.2
Cd	0.6	±	0.2
Cu	1.1	±	0.4
Ca	1.1	±	0.6
Co	17	±	8
Ni	0.6	±	0.3
Fe	80	±	20
Mn	90	±	10

3.4.2. Precipitation of Co

The cobalt precipitation is carried out by adding APS to bring the Redox to 1000 mV and keeping the solution at 80 °C to accelerate the precipitation [1]. The precipitation of cobalt under the form of CoOOH is slow as shown in Figure 8a. This observed behavior is consistent with previously reported results [1] reproduced in Figure 8b. Table 7 gives the variation of the solution composition from the leach step to the Co precipitation one. Zn, Cd, and Cu remain in the solution while cobalt precipitates with some of the iron and manganese remaining in the solution after the first oxidation step. Table 8 gives the proportions of the different metals precipitated during the cobalt precipitation step. About 2% of the copper precipitates with the cobalt and represents a critical impurity as discussed later. The fact that the observed Co precipitation rate in Figure 8 is significantly less than that reported in [1] is an indication that there is room to improve the proposed processing scheme in order to reduce the precipitation time and subsequently the volume of the precipitation vessels for a continuous process.

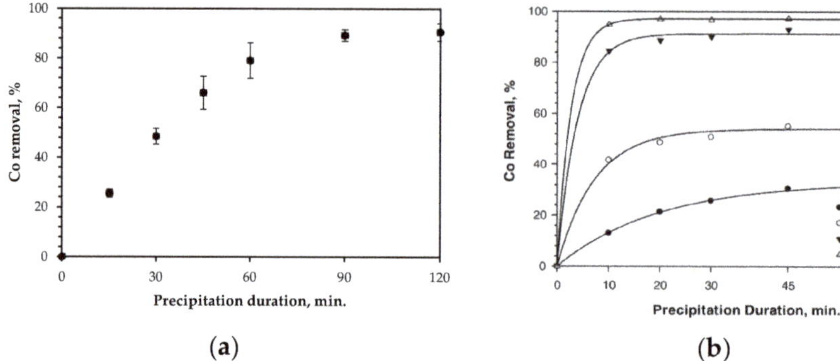

Figure 8. Cobalt precipitation rates. (**a**) Observed rate of Co precipitation (pH = 3.0, Eh = 1000 mV, 80 °C, 1 g/L Co). (**b**) Rates of Co precipitation reported in [1] B-0.

Table 7. Composition of the solutions from the ZPR leach, Fe-Mn and Co precipitation steps.

Solution	Volume (mL)	All Concentrations in mg/L					
		Zn	Cd	Cu	Co	Fe	Mn
Leach solution	841	11,800	370	3600	1170	270	26
After Fe-Mn ppt	944	10,511	327	3200	860	46	1.3
After Co ppt	975	10,111	318	3000	84	16	0.4

Table 8. Metal precipitation during the cobalt precipitation stage (Average ± standard deviation) for three (3) tests (Eh = 1000 mV, pH = 3.0, 80 °C, 120 min).

Species	% Precipitated		
Zn	0.4	±	0.1
Cd	1	±	0.3
Cu	2	±	1
Ca	1	±	0.4
Co	89	±	3

3.4.3. Washing and Composition of the Cobalt Precipitate

The cobalt precipitate is washed with water at 80 °C at 1% solids for 5 min, filtered, dried overnight, weighed, and assayed. Table 9 gives the composition of the cobalt product obtained from 3 independent tests. Table 9 compares the average composition of the cobalt product to a target cobalt hydroxide feed for a cobalt refinery [18]. Results show that some tunings (likely at the cobalt precipitation stage) of the process are still required to lower the zinc and copper contents to meet the composition of the feed material for the Co-refinery.

Table 9. Composition (%) of the produced cobalt hydroxide and specifications for the feed of a custom cobalt refinery.

Material	Co	Cd	Cu	Zn	Fe	Mn	Ni	Pb
Produced Co hydroxide *	45 ± 4	0.120 ± 0.004	4.4 ± 0.7	0.8 ± 0.2	2 ± 1	<0.1	<0.1	0.3 ± 0.1
Custom refinery Co hydroxide feed [19]	23.2	n.a.	1.61	0.19	2.39	3.27	0.39	n.a.

*: Average ± standard deviation of three (3) replicated tests.

The XRD powder pattern of the cobalt product is shown in Figure 9. Results confirm that the produced material is under the form of CoO(OH) and Co(OH)$_2$. The presence of these compounds was expected based on observations from other researchers [1].

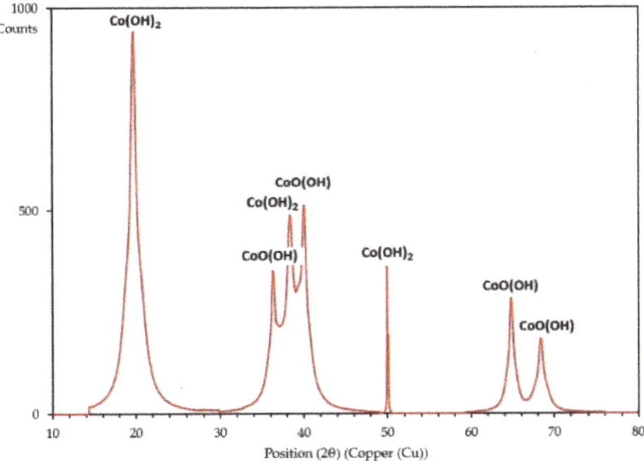

Figure 9. XRD powder pattern of the cobalt product.

3.5. Overall Process Performance and Quality of the Product

The process flow sheet for the extraction of cobalt from the ZPR is shown in Figure 10. The flowsheet has similarities with the flowsheet presented in [1]. The similarity between the two flowsheets developed independently is an indication that this processing scheme is a viable route for the recovery of cobalt from a ZPR produced by cementation on zinc dust.

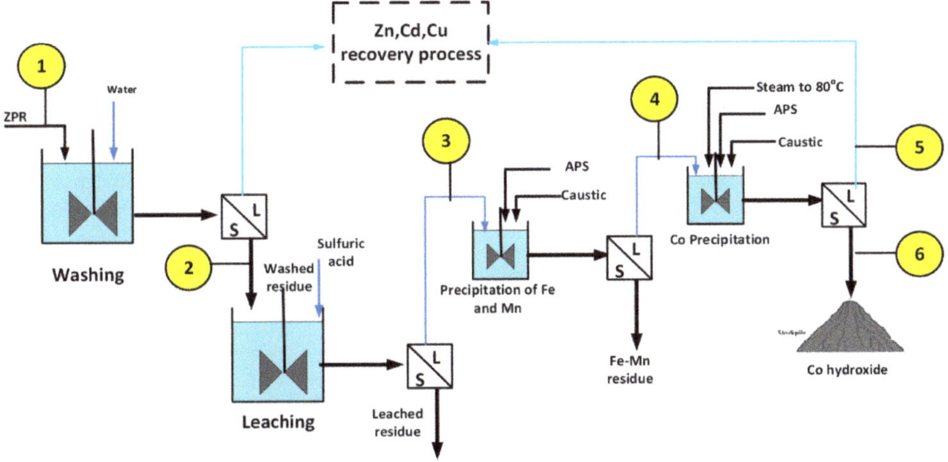

Figure 10. ZPR processing circuit to extract cobalt (See Table 10 for the composition of streams 1–6).

In order to assess the robustness of the proposed process, the whole processing sequence of Figure 10 was repeated with three ZPR feed samples. The previous sections gave the results of the repetitions obtained for the various stages of the process. Table 10 summarizes these results in terms of cobalt distribution. The repetitions show that the critical step in the proposed process is the oxidation/precipitation of the iron and manganese prior to the cobalt precipitation. The main cobalt losses occur at the Fe-Mn precipitation step. A problem or inaccuracy in the measurement of the Redox potential and the difficulties in the filtration of the Fe-Mn sludge cause some Co precipitation and an incomplete removal of impurities that will subsequently contaminate the cobalt product. Cobalt losses also occur at the final precipitation step, where recovery could be improved by using a pH above 3.0 and allowing more time for precipitation [1]. However, the use of a higher pH and of an increase precipitation time may cause more impurities to precipitate with the cobalt. This optimization problem of maximizing Co recovery under the constraint of an acceptable product purity can be adequately tackled down using a factorial Design Of Experiments (DOE) [19] if the process is to be continued with an optimization phase. The aspect of the process sensitivity to operating conditions, particularly the control of the Redox potential, addressed here is seldom discussed in the literature dealing with the recovery of cobalt from a ZPR, while it is as important as the process itself, especially if it is to go on to piloting.

Table 10. Distribution (%) of the cobalt in the process streams of the circuit of Figure 9.

Process Stream	Stream Number as in Figure 10	Avg.	Std-Dev
ZPR	1	100	100
Washed solids ZPR	2	86	3
Leach solution	3	84	4
Co ppt feed solution	4	70	10
Spent solution from Co ppt	5	7	1
Cobalt hydroxide	6	62	14

In summary, the proposed processing flow sheet yields a Co product assaying 45 ± 4% Co at a recovery of 62%. Improvements to the reproducibility of the method can be achieved by improving the control of the Redox potential and the solids/liquid separation of the precipitated Fe-Mn sludge. These results cannot be compared to those of other processes used for processing a ZPR produced by cementation on zinc dust, as it was not possible to find such data in the literature.

Ideally, to avoid paying refining charges from the selling of the cobalt hydroxide product to a custom Co refinery, one should aim at producing electrolysis-grade cobalt at the end of the ZPR-Co process. The concentration of cobalt in the solution released by the S/L separation of the Fe-Mn precipitate is low at 3 g/L, and should be increased to at least 45 g/L [20] to allow Co electro-winning. Solvent Extraction (SX) is an approach to increase the solution concentration and remove some of the accompanying impurities (See Table 9). The other option is to stock pile the produced Co hydroxide and to dissolve it under reducing conditions to convert Co(III) into Co(II) with a controlled amount of sulfuric acid to generate the solution to feed the electrolytic cells. Figure 11 shows the two options.

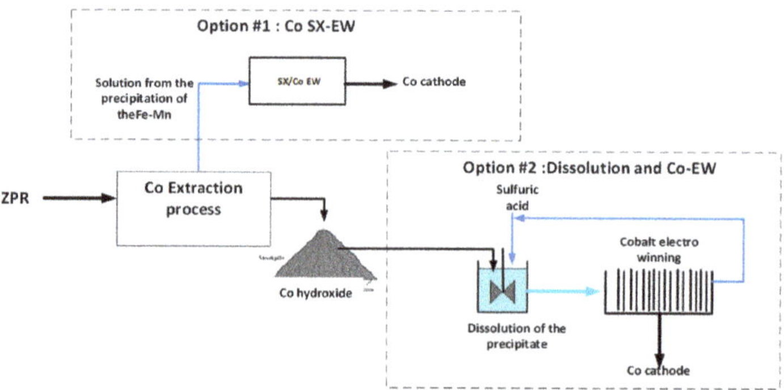

Figure 11. Processing options to produce electrolytic cobalt.

3.6. Economics of the Process

The analysis presented in this section focuses only on the operating costs of the process of Figure 10. It is a preliminary analysis and the calculated costs should be considered as Class 5 estimates (AACE International). Results are used to anticipate the economic viability of the process and decide on the continuation of the laboratory test work to optimize the proposed ZPR-Co process.

A cobalt price of 30.20 USD/kg (LME Price August 2020) is used to estimate the revenues that can be generated by processing the ZPR. The Net Smelter Return, if the cobalt hydroxide is to be sold to a refinery, is estimated by assuming that 90% of the cobalt in the hydroxide is payable and that the refining charges are 3.00 CAD/kg of Cobalt hydroxide. If the cobalt content of the produced Co hydroxide is 45% (see Table 9) then the revenues (assuming an exchange rate of 1.25 CAD/USD) generated per kg of cobalt hydroxide are:

$$NSR = 0.9 \times 0.45 \times 30.2 \times 1.25 - 3 = \frac{12.28 \; \text{CAD\$}}{\text{kg of Co hydroxide}}, \tag{3}$$

Since the test work has shown that the processing of 75 g of ZPR yields 1.2 g of Co hydroxide, the potential revenues per kg of ZPR are 0.20 CAD/kg ZPR.

The operating costs considered here are due to the reagents and exclude the energy costs for heating the solution at the precipitation stage. Table 11 gives the calculated operating costs per kg of ZPR. The reagent costs are probably over-estimated as purchasing the sulfuric acid may not be required. Indeed some spent zinc electrolyte could be used for the Co process (see Figure 1). Additionally, it is very likely that recycling some of the solution streams of the ZPR-Co process could allow a reduction in the consumption of caustic and APS. This preliminary evaluation shows that the proposed process is not viable but the process certainly deserves to be optimized to increase the overall cobalt recovery and to reduce the consumption of APS that is the main contributor to the operating costs. The option of producing metallic cobalt rather than cobalt hydroxide should also be investigated.

Table 11. Preliminary estimation of the operating costs for the ZPR-Co process of Figure 9. (The reagent prices were obtained from quotations dating of 2019 and are used here only to give an order of magnitude).

Reagent	Consumption (kg/kg of ZPR)	Price CAD\$/kg	CAD\$/kg ZPR
H_2SO_4	1.9	0.170	0.032
NaOH	0.07	0.80	0.056
APS	0.25	0.85	0.22
Total costs	-	-	0.30
Gross revenues	-	-	0.20
Benefits (losses)	-	-	(0.10)

3.7. Residues of the ZPR-Co Process

The proposed process generates two solid residues (the ZPR leached residue, the Fe–Mn precipitate) and two liquid effluents (the solutions from the ZPR washing and from the Co precipitation as indicated in Figure 12. Table 12 gives the compositions of the liquid and solid residues. The copper content of the ZPR leached solids is 36%, which is above the copper content of the ZPR (see Table 2). If the metals of the reject solutions were to be precipitated under the form of hydroxides by increasing the pH above 9.0, the precipitate would assay 13% Cu, which is still in the range of the ZPR copper content. Thus, two of the residues of the ZPR-Co process could possibly be used as copper sources and possibly be sold to a Cu smelter.

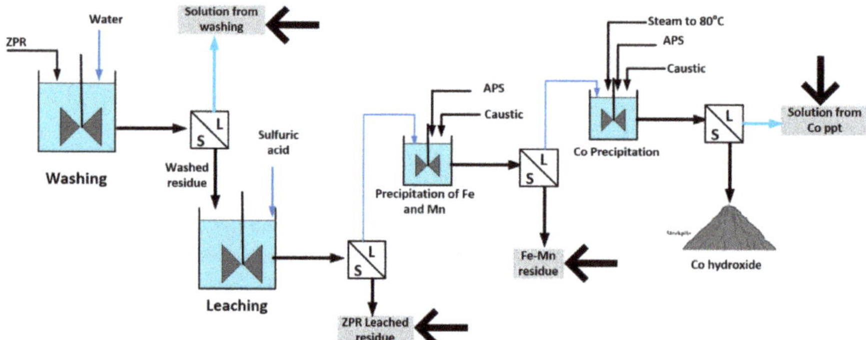

Figure 12. Liquid and solid residues generated by the ZPR-Co process.

Table 12. Metal contents in the residues of the ZPR-Co process.

Residues from the ZPR-Co Process	Content (mg/L or %)		
	Zn	Cd	Cu
Soln from washing (mg/L)	1202	3067	3.6
Soln from Co ppt (mg/L)	1111	317	3028
Leach solid residue (%)	2.06	0.22	36.4
Fe-Mn solid residue (%)	2.09	0.13	1.97

4. Conclusions

A process to obtain a cobalt rich hydroxide from the processing of a Cu–Cd–Co residue produced by cementation on zinc dust is described in this paper. The process includes a water leaching of the residue to remove soluble Zn and Cd. The washed residue is leached for 30 min with sulfuric acid (100 g/L) at a pH of 0.2. The Fe and Mn of the leach solution are precipitated by increasing the Eh to 650 mV while maintaining a pH of 3.0 with sodium hydroxide. Cobalt hydroxide is finally precipitated by increasing the Eh to 1000 mV. Repetition of the whole processing flow sheet shows that the critical steps of the process are the control of the Eh at the precipitation stage and the solids/liquid filtration of the Fe–Mn precipitate. The cobalt recovery from the zinc residues to the cobalt hydroxide is 62 ± 14%. The cobalt content of the cobalt product is 45 ± 4%. The variability in the results is mainly due to the control and measurement of the solution Redox potential

Author Contributions: Conceptualization, investigation, formal analysis methodology, data curation, writing—review and editing, L.B.; conceptualization, investigation, formal analysis methodology, data curation, supervision, writing—review and editing, K.T. and J.-F.B.; writing—original draft preparation, writing—review and editing, formal analysis, supervision, project administration, C.B.; funding acquisition, conceptualization, review, G.H. All authors have read and agreed to the published version of the manuscript.

Funding: This research was funded by Institut de recherche de Hydro-Québec.

Acknowledgments: The authors acknowledge the financial support of the Institut de recherche de Hydro-Québec and CEZinc for providing the zinc residues sample for the testwork. E. Benguerel, of CEZinc, is also acknowledged for reviewing the paper prior to publication.

Conflicts of Interest: The authors declare no conflict of interest.

References

1. Huang, Y.; Zhang, Z.; Cao, Y.; Han, G.; Peng, W.; Zhu, X.; Zhang, T.-A.; Dou, Z. Overview of cobalt resources and comprehensive analysis of cobalt recovery from zinc plant purification residue-a review. *Hydrometallurgy* **2020**, *193*, 3–12. [CrossRef]
2. Wang, Y.; Zhou, C. Hydrometallurgical process for recovery of cobalt from zinc plant residue. *Hydrometallurgy* **2002**, *63*, 225–234. [CrossRef]

3. Fattahi, A.; Rashchi, F.; Abkhoshk, E. Reductive leaching of zinc, cobalt and manganese from zinc plant residue. *Hydrometallurgy* **2016**, *161*, 185–192. [CrossRef]

4. Anonymous. *Critical Materials Rare Earths Supply Chain: A Situational White Paper*; U.S. Department of Energy, Office of Energy Efficiency & Renewable Energy: Washington, DC, USA, 2020.

5. Golmohammadzadeh, R.; Faraji, F.; Rashchi, F. Recovery of lithium and cobalt from spent lithium ion batteries (LIBs) using organic acids as leaching reagents: A review. *Resour. Conserv. Recycl.* **2018**, *136*, 418–435. [CrossRef]

6. Frankel, T.C. The Cobalt Pipeline: Tracing the Path from Deadly Hand-Dug Mines in Congo to Consumer's Phones and Laptops. *The Washington Post*, 1 October 2016. Available online: https://longform.org/posts/the-cobalt-pipeline (accessed on 3 November 2020).

7. Ashtari, P.; Pourghahramani, P. Hydrometallurgical recycling of cobalt from zinc plants residue. *J. Mater. Cycles Waste Manag.* **2018**, *20*, 155–166. [CrossRef]

8. Güler, E.; Seyrankaya, A. Precipitation of impurity ions from zinc leach solutions with high iron contents—A special emphasis on cobalt precipitation. *Hydrometallurgy* **2016**, *164*, 118–124. [CrossRef]

9. Bøckman, O.; Østvold, T. Products formed during cobalt cementation on zinc in zinc sulfate electrolytes. *Hydrometallurgy* **2000**, *54*, 65–78. [CrossRef]

10. Qian, L.; Zhang, B.; Min, X.-B.; Shen, W.-Q. Acid leaching kinetics of zinc plant purification residue. *Trans. Nonferrous Met. Soc. China* **2013**, *23*, 2786–2791.

11. CEzince Company. Available online: https://www.cezinc.com/fr/Pages/home.aspx (accessed on 5 August 2020).

12. Gy, P.M. *Sampling of Particulate Materials, Theory and Practice*; Elsevier: Amsterdam, The Netherlands, 1982.

13. Bazin, C.; Hodouin, D.; Zouadi, M. Data reconciliation and equilibrium constant estimation: Application to copper solvent extraction. *Hydrometallurgy* **2005**, *80*, 43–53. [CrossRef]

14. Hodouin, D.; Everell, M. A hierarchical procedure for adjustment and material balancing of mineral processes data. *Int. J. Miner. Process.* **1980**, *7*, 91–116. [CrossRef]

15. Hodouin, D.; Flament, F.; Bazin, C. Reliability of material balance calculations a sensitivity approach. *Miner. Eng.* **1989**, *2*, 157–169. [CrossRef]

16. Turgeon, K.; Bazin, C.; Boulanger, J.-F.; Whitty-Léveillé, L.; Larivière, D. Material balancing and estimation of equilibrium constants: Application to solvent extraction tests of rare earth elements. In Proceedings of the International Mineral Processing Congress, Quebec, QC, Canada, 11–15 September 2016.

17. Bazin, C.; Hodouin, D.; Blondin, R.M. Estimation of the variance of the fundamental error associated to the sampling of low grade ores. *Int. J. Miner. Process.* **2013**, *124*, 117–123. [CrossRef]

18. Zhan, D.; Marquez, J. First Cobalt Corporation: 2019 First Cobalt Refinery Restart. 103870-RPT-0001. Available online: https://www.firstcobalt.com/_resources/reports/20190627-Ausenco-ConceptualStudy_Rev-E.pdf (accessed on 20 November 2020).

19. Box, G.E.; Draper, N.R. *Evolutionary Operation: A Statistical Method for Process Improvement*; Wiley: New York, NY, USA, 1969; Volume 25.

20. Conard, B. History of electrolytic cobalt refining at Vale Canada (Inco). *CIM J.* **2015**, *6*, 51–58. [CrossRef]

Publisher's Note: MDPI stays neutral with regard to jurisdictional claims in published maps and institutional affiliations.

metals 

MDPI

Article

Enhanced Cementation of Co^{2+} and Ni^{2+} from Sulfate and Chloride Solutions Using Aluminum as an Electron Donor and Conductive Particles as an Electron Pathway

Sanghyeon Choi [1,*,†], Sanghee Jeon [2,†], Ilhwan Park [2], Mayumi Ito [2] and Naoki Hiroyoshi [2]

[1] Division of Sustainable Resources Engineering, Graduate School of Engineering, Hokkaido University, Sapporo 060-8628, Japan
[2] Division of Sustainable Resources Engineering, Faculty of Engineering, Hokkaido University, Sapporo 060-8628, Japan; shjun1121@eng.hokudai.ac.jp (S.J.); i-park@eng.hokudai.ac.jp (I.P.); itomayu@eng.hokudai.ac.jp (M.I.); hiroyosi@eng.hokudai.ac.jp (N.H.)
* Correspondence: cshshow351@gmail.com; Tel.: +81-11-706-6315
† These authors have contributed equally to this work and share first authorship.

Citation: Choi, S.; Jeon, S.; Park, I.; Ito, M.; Hiroyoshi, N. Enhanced Cementation of Co^{2+} and Ni^{2+} from Sulfate and Chloride Solutions Using Aluminum as an Electron Donor and Conductive Particles as an Electron Pathway. *Metals* **2021**, *11*, 248. https://doi.org/10.3390/met11020248

Academic Editor: Dariush Azizi
Received: 10 January 2021
Accepted: 30 January 2021
Published: 2 February 2021

Abstract: Cobalt and nickel have become important strategic resources because they are widely used for renewable energy technologies and rechargeable battery production. Cementation, an electrochemical deposition of noble metal ions using a less noble metal as an electron donor, is an important option to recover Co and Ni from dilute aqueous solutions of these metal ions. In this study, cementation experiments for recovering Co^{2+} and Ni^{2+} from sulfate and chloride solutions (pH = 4) were conducted at 298 K using Al powder as electron donor, and the effects of additives such as activated carbon (AC), TiO$_2$, and SiO$_2$ powders on the cementation efficiency were investigated. Without additives, cementation efficiencies of Co^{2+} and Ni^{2+} were almost zero in both sulfate and chloride solutions, mainly because of the presence of an aluminum oxide layer (Al$_2$O$_3$) on an Al surface, which inhibits electron transfer from Al to the metal ions. Addition of nonconductor (SiO$_2$) did not affect the cementation efficiencies of Co^{2+} and Ni^{2+} using Al as electron donor, while addition of (semi)conductors such as AC or TiO$_2$ enhanced the cementation efficiencies significantly. The results of surface analysis (Auger electron spectroscopy) for the cementation products when using TiO$_2$/Al mixture showed that Co and Ni were deposited on TiO$_2$ particles attached on the Al surface. This result suggests that conductors such as TiO$_2$ act as an electron pathway from Al to Co^{2+} and Ni^{2+}, even when an Al oxide layer covered on an Al surface.

Keywords: cementation; cobalt (Co); nickel (Ni); aluminum (Al); titanium dioxide (TiO$_2$); silicon dioxide (SiO$_2$)

1. Introduction

Cementation, an electrochemical deposition of noble metal ions by a less noble metal as an electron donor, is usually applied to remove/recover metal ions from dilute aqueous solutions [1–4]. The advantages of cementation are (1) recovery of metals in zero-valent form, (2) simple methods, and (3) low-energy consumption [2,5]. In this method, the overall reaction of cementation is given by Equation (1) [6–8]:

$$mN^0 + nM^{m+} \rightarrow mN^{n+} + nM^0 \tag{1}$$

The cementation reaction is divided into anodic (Equation (2)) and cathodic reactions (Equation (3)):

$$\text{Anodic } mN^0 \rightarrow mN^{n+} + nme^- \tag{2}$$

$$\text{Cathodic } nM^{m+} + nme^- \rightarrow nM^0 \tag{3}$$

The noble metal ions (M^{m+}) are deposited on the surface of a less noble elemental metal (N^0) spontaneously, and the driving force of this reaction is mainly determined by differences in the standard electrode potentials for M^{n+}/M^0 and N^{n+}/N^0 redox pairs, and it increases when the electrode potential of N^0 is low.

Aluminum (Al) can be considered as a strong reductant (electron donor) used for cementation because of its extremely low standard electrode potential (i.e., $E^0_{Al3+/Al}$ = −1.67 V vs. standard hydrogen electrode (SHE)) [7,9–11]. The practical application of Al for cementation, however, is limited due to the presence of a dense Al oxide layer (Al_2O_3) on the Al surface, which inhibits electron transfer from Al^0 to metal ions [9,12,13]. When the Al oxide layer is removed from the surface, Al can be used as an electron donor for cementation. To remove the Al oxide layer, however, high temperatures, acid/alkaline solutions, or high concentration of chloride ions are needed [2,5,9,14,15], and these extreme conditions make it difficult to use Al as an electron donor in the practical cementation processes.

Recently, the authors investigated the effects of activated carbon (AC) addition on the efficiency of cementation using Al as an electron donor for recovering gold ions from ammonium thiosulfate solution [16,17], and heavy metal ions (Co^{2+}, Ni^{2+}, Zn^{2+}, and Cd^{2+}) from acidic sulfate and chloride solutions. The results showed that cementation efficiencies of the metal ions were significantly enhanced by the addition of activated carbon (AC) even when an insulating Al oxide layer covered on the Al surface [16,17]. This "enhanced cementation using AC/Al-mixture" can be operated under mild conditions; i.e., it does not require extreme operating conditions such as high temperatures, and high concentrations of chemical reagent such as acid, base, and chloride ions. This new method may, therefore, provide a practical way to use Al, one of the strongest reductants (electron donor) for cementation to recover metal ions from dilute solutions.

Although the details of the mechanism of enhanced cementation using the AC/Al-mixture are not yet fully understood, the results of surface analysis for the cementation products have suggested that AC attached on the Al surface acted as an electron pathway from Al to noble metal ions, even in the presence of a surface Al oxide layer [17]. If this is the case and the essential role of AC is just as an electron pathway, enhanced cementation would occur even when AC is replaced by other (semi)conductors. On the other hand, as AC is a porous material and has a very large specific surface area [18], not only the electroconductivity but also large adsorption capacity of AC for metal ions may play an important role in the enhanced cementation using the AC/Al-mixture. If this is the case, replacing AC to another conductor with a low specific surface area cannot enhance the cementation using Al as an electron donor.

Cobalt (Co) and nickel (Ni) represent important strategic resources in the world market and their use is rapidly growing for renewable energy technologies and rechargeable battery productions, and the importance of the development of technologies for recovering and purifying Co and Ni is continuously increasing [19–24]. Therefore, this study aims to investigate whether the AC could be replaced with other (semi)conductors for recovery of Co and Ni from sulfate and chloride solutions. Titanium dioxide (TiO_2) was selected for a semiconductor because of its nontoxic, nonreactive, and high chemical stability, while silicon dioxide (SiO_2) was chosen for a nonconductor to clarify the mechanism(s) of the enhanced cementation using the mixture of conductor and Al [25,26].

In the present study, batch-type cementation experiments were conducted using Al as an electron donor to recover Co^{2+} and Ni^{2+} from sulfate and chloride solutions and the effects of the addition of AC, TiO_2, or SiO_2 on the recoveries of these metal ions were investigated. Surface analysis (Auger electron spectroscopy (AES)) for the cementation products were also conducted to elucidate the cementation mechanism.

2. Materials and Methods

2.1. Materials

As an electron donor, Al powder (99.99%, Wako Pure Chemical Industries, Ltd., Osaka, Japan) was used, and AC powder (99.99%, Wako Pure Chemical Industries, Ltd., Osaka,

Japan), TiO_2 powder (99.0%, rutile form, Wako Pure Chemical Industries, Ltd., Osaka, Japan), and SiO_2 powder (99.0%, Wako pure Chemical Industries, Ltd., Osaka, Japan) were used as additives. Particle size distribution of these materials, measured by laser diffraction (Microtrac® MT3300SX, Nikkiso Co. Ltd., Osaka, Japan), is shown in Figure 1. The median diameters (D_{50}) of Al, AC, TiO_2, and SiO_2 were 21.3, 38.1, 8.5, and 21.2 μm, respectively.

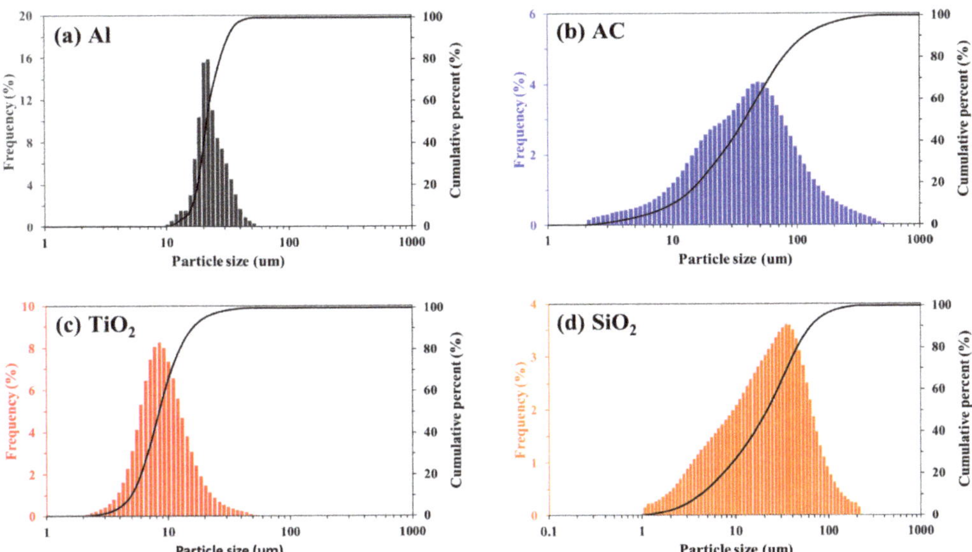

Figure 1. Particle size distribution for (**a**) aluminum (Al), (**b**) activated carbon (AC), (**c**) titanium dioxide (TiO_2), and (**d**) silicon dioxide (SiO_2) used in this study.

2.2. Recovery of Co^{2+} and Ni^{2+} from Sulfate and Chloride Solutions

2.2.1. Preparation of Co^{2+} and Ni^{2+} Solutions

The sulfate solutions containing 1 mM metal ions were prepared by dissolving $CoSO_4·7H_2O$ (99.0%, Wako Pure Chemical Industries, Ltd., Osaka, Japan) or $NiSO_4·6H_2O$ (99.0%, Wako Pure Chemical Industries, Ltd., Osaka, Japan) in deionized (DI) water (18 MΩ·cm, Mill-Q® Integral Water Purification System, Merck Millipore, Burlington, Vermont, USA). For the preparation of 1 mM metal chloride solutions, $CoCl_2·6H_2O$ (99.0%, Kishida Chemical Co., Ltd., Osaka, Japan), or $NiCl_2·6H_2O$ (98.0%, Kishida Chemical Co., Ltd., Osaka, Japan) was dissolved in DI water. The initial pH of sulfate and chloride solutions was adjusted to 4.0 using 1 M H_2SO_4 and HCl (Wako Pure Chemical Industries, Ltd., Osaka, Japan), respectively. The total concentration of SO_4^{2-} and Cl^- were fixed to 0.1 M using Na_2SO_4 (99.0%, Wako Pure Chemical Industries, Ltd., Osaka, Japan) and NaCl (99.5%, Wako Pure Chemical Industries, Ltd., Osaka, Japan) to normalize their effects on experiments.

2.2.2. Cementation Tests

The cementation tests were carried out in a 50 mL Erlenmeyer flask using a thermostat water bath shaker (Cool bath shaker, ML-10F, Taitec Corporation, Saitama, Japan) with 40 mm of shaking amplitude and 120 min^{-1} of shaking frequency at 25 °C for 24 h. (Note that these parameters were selected based on our preliminary experiments). Ten milliliters of the prepared solution were added to the flask, then ultrapure nitrogen gas (99.9%) was introduced for 15 min before experiments to maintain an oxygen-free environment. One-tenth gram of Al powder and/or a predetermined amount (0.01, 0.05, 0.1, 0.2, 0.4 g) of additive (i.e., AC, TiO$_2$, and SiO$_2$) were added to the solution. Ultrapure nitrogen gas (99.9%) was further introduced to the flask for 5 min, then the flask was tightly capped with a rubber cap and sealed with parafilm, and an experiment was conducted. After 24 h, the suspension was filtered using a syringe-driven membrane filter (pore size: 0.2 μm, LMS Co., Ltd., Tokyo, Japan); final pH of the filtrate was measured. The filtrate was diluted with 0.1 M HNO$_3$, and the concentrations of metal ions were analyzed by inductively coupled plasma atomic emission spectroscopy (ICP-AES, ICPE-9820, Shimadzu Corporation, Kyoto, Japan). The recovery efficiency of Co^{2+} and Ni^{2+} was calculated based on Equation (4):

$$\text{Recovery efficiency, } R = \frac{C_i - C_f}{C_i} \tag{4}$$

where C_i and C_f are the initial and final concentrations of metal ions, respectively.

2.2.3. Surface Analysis

The solid products obtained by filtration were washed 5 times with DI water, dried in a vacuum oven at 40 °C for 24 h, and then analyzed by Auger electron spectroscopy (AES) using JAMP-9500F (JEOL Co., Ltd., Tokyo, Japan). The dried residue was mounted on an AES holder using conductive carbon tape. The analysis was conducted under the following conditions: ultrahigh vacuum condition, ~1 × 10^{-7} Pa; probe energy, 10 kV; and probe current, 19.7 nA. The spectra were analyzed by using Spectra Investigator AES software.

3. Results and Discussion

3.1. Recovery of Co^{2+} and Ni^{2+}

3.1.1. Recovery of Co^{2+} and Ni^{2+} from Sulfate Solution

Cementation experiments for recovering Co^{2+} and Ni^{2+} from sulfate solutions (initial pH = 4) were conducted for 24 h using Al powder as an electron donor, and the effects of the dosage of additives (AC, TiO$_2$, and SiO$_2$) on the efficiency of Co and Ni recoveries were investigated. To access the adsorption of Co^{2+} and Ni^{2+} on the additives, experiments without Al were also conducted.

Figures 2a–c and 3a–c show the Co and Ni recovery efficiencies and final pH as a function of SiO$_2$, AC, and TiO$_2$ dosages, respectively. In all experiments, final pH was in the range from 5.1 to 5.6, at which Co^{2+} and Ni^{2+} do not precipitate as their hydroxide (Figures S1 and S2).

As shown in Figures 2a and 3a, without Al, the efficiencies of Co and Ni recovery were almost 0% at any dosage of SiO$_2$, suggesting that there was no adsorption of Co^{2+} and Ni^{2+} on the SiO$_2$ surface. Even with Al, the Co and Ni recovery efficiencies were also almost 0% regardless of SiO$_2$ dosage, suggesting that cementation of Co^{2+} and Ni^{2+} using Al as an electron donor did not occur. This may be due to the presence of an Al oxide layer covering the Al surface, which inhibits the electron transportation from Al to Co^{2+} and Ni^{2+} [2,27]. Because the cementation did not occur regardless of SiO$_2$ addition, the results also confirm that physical breakage of the Al oxide layer due to the collision of SiO$_2$ to Al powder in the shaking flask did not cause enhanced cementation.

As shown in Figures 2b and 3b, even without Al, the recovery efficiency of Co^{2+} and Ni^{2+} increased with increasing AC dosage, suggesting that these metal ions adsorbed on the AC surface. It has been reported that there are functional groups such as carboxyl and carbonyl groups on the surface of the activated carbon and they act as adsorption sites to metal ions through the reaction described by Equation (5) [18,28,29]. Increase in final pH indicates that not only Co^{2+} and Ni^{2+}, but also proton (H$^+$) adsorbed on AC [30,31].

$$\text{-C-COOH} + \text{M}^{2+} \rightarrow \text{-C-COOM} + 2\text{H}^+ \text{ (M = Co or Ni)} \tag{5}$$

In the range between 0.05 to 0.2 g AC dosage, recovery efficiency was much higher with Al than without Al; at 0.1 g AC dosage, the efficiency was 56% for Co and 61% for Ni with Al, while it was 31% for Co and 43% for Ni without Al. The difference of metal recovery efficiency between either with or without Al was 25% for Co and 18% for Ni, which cannot be ignored as an experimental error. This suggests that the addition of AC enhances Co and Ni cementation using Al as an electron donor (Equations (6) and (7)), even though the Al oxide layer remained on the Al surface.

$$3Co^{2+} + 2Al^{0} \rightarrow 3Co^{0} + 2Al^{3+} \tag{6}$$

$$3Ni^{2+} + 2Al^{0} \rightarrow 3Ni^{0} + 2Al^{3+} \tag{7}$$

Following these equations, it is expected that the stoichiometric amount of Al dissolves when cementation occurs; however, the dissolved Al concentration after cementation was less than 3 ppm (Tables S1 and S2), which means that most of the Al^{3+} was precipitated as Al-(oxy)hydroxide [7,32].

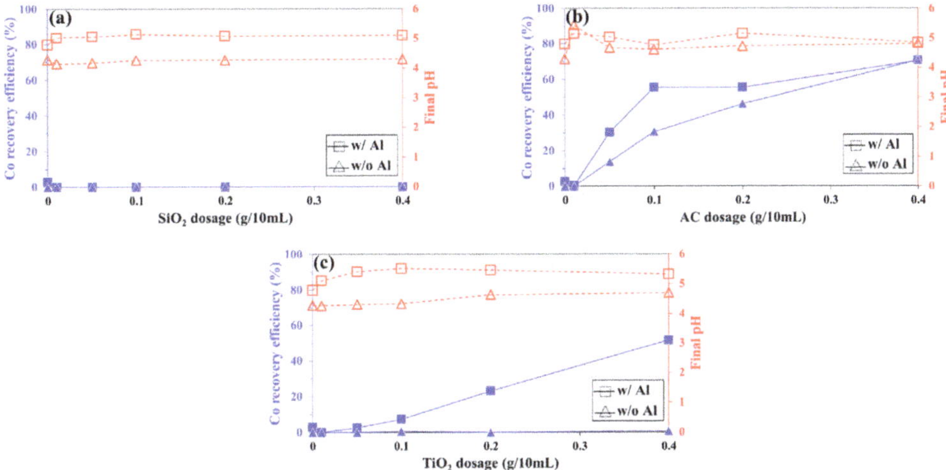

Figure 2. The effects of (**a**) SiO_2, (**b**) AC, and (**c**) TiO_2 dosages on the recovery efficiency of Co^{2+} and final pH in sulfate solutions at initial pH 4.0 for 24 h.

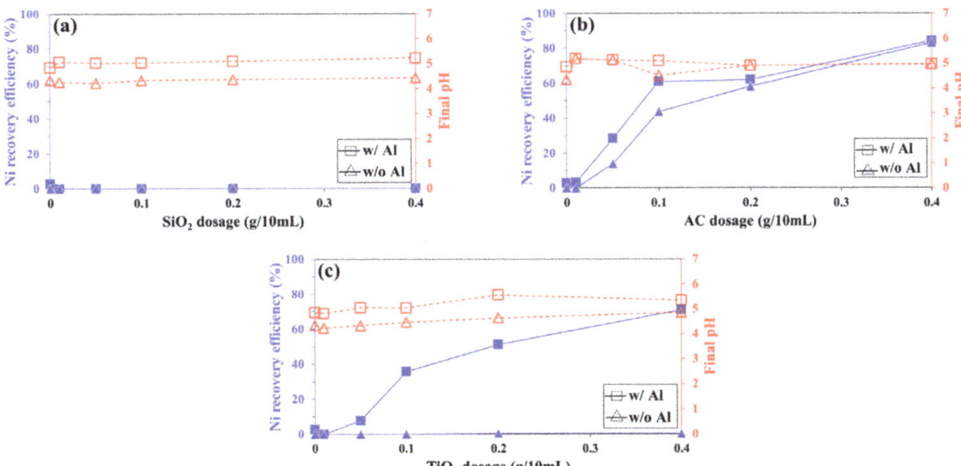

Figure 3. The effects of (**a**) SiO_2, (**b**) AC, and (**c**) TiO_2 dosages on the recovery efficiency of Ni^{2+} and final pH in sulfate solutions at initial pH 4.0 for 24 h.

As shown in Figures 2c and 3c, the recovery efficiency of Co^{2+} and Ni^{2+} without Al was almost 0% regardless of TiO_2 dosage, indicating that TiO_2 has no ability to adsorb Co^{2+} and Ni^{2+}. When 0.1 g of Al was used together with TiO_2, the recovery efficiency continuously increased with increasing TiO_2 dosage and reached the maximum value of 52% for Co and 71% for Ni with 0.4 g TiO_2. As already discussed, Co^{2+} and Ni^{2+} do not precipitate as hydroxides at the pH ranges observed in this series of experiments; the enhanced recovery of Co^{2+} and Ni^{2+} with TiO_2 and Al suggests that the addition of TiO_2 enhanced the cementation of Co^{2+} and Ni^{2+} by Al (Equations (6) and (7)). It was also confirmed that the dissolved Ti concentrations were below detection limit, indicating that TiO_2 is stable enough to be used as an agent to enhance cementation of Co^{2+} and Ni^{2+} with Al in the sulfate solution (Tables S1 and S2).

3.1.2. Recovery of Co^{2+} and Ni^{2+} from Chloride Solution

Cementation experiments for recovering Co^{2+} and Ni^{2+} from chloride solutions (initial pH = 4) were conducted for 24 h using Al powder as an electron donor, and the effects of the dosage of additives (AC, TiO_2, and SiO_2) on the efficiency of Co and Ni recovery were investigated. To access the adsorption of Co^{2+} and Ni^{2+} on the additives, experiments without Al were also conducted. Figures 4a–c and 5a–c show the Co and Ni recovery efficiencies and final pH as a function of AC, TiO_2, and SiO_2 dosages, respectively.

Similar to the sulfate system (Figures 2 and 3), final pH values of the chloride solutions (Figures 4 and 5) were less than 5.5 for Co and 6.1 for Ni (Tables S3 and S4), which means that removal of Co^{2+} and Ni^{2+} from the solutions by the formation of cobalt and nickel hydroxide precipitation does not need to be considered in this series of experiments (Figures S3 and S4).

It has been reported that in the presence of high concentrations of Cl^-, the Al oxide layer was dissolved and removed from the Al surface [13,33–35]. If the Al oxide layer is dissolved, a high concentration of dissolved Al would be detected in the solutions, but the observed results (Tables S3 and S4) showed that concentrations of Al were less than 5 ppm under all conditions. This implies that removal of the Al oxide layer did not occur under the experimental condition used here.

As shown in Figures 4a and 5a, when SiO_2 was used as an additive, the recovery efficiencies of Co^{2+} and Ni^{2+} both with and without Al were almost 0%. This indicates that in chloride solutions, Co^{2+} and Ni^{2+} were not adsorbed on SiO_2, and the cementation of Co and Ni with Al did not occur.

The results shown in Figures 4b and 5b suggest that adsorption of Co^{2+} and Ni^{2+} on AC occurred in chloride solutions, because in the absence of Al, recovery efficiencies of these ions increased with increasing AC dosage. As in sulfate solutions, in the presence of AC, enhancement of metal ion recovery by Al addition was confirmed (Figures 4b and 5b); e.g., at 0.1 g AC dosage, by adding Al, the efficiency increased from 57% to 70% for Co, and it increased from 57% to 70% for Ni. This suggests that enhanced cementation of these metal ions with AC occurred in chloride solutions.

Figures 4c and 5c show that the efficiencies of Co^{2+} and Ni^{2+} recovery in the absence of Al were almost 0% at any dosage of TiO_2, suggesting that adsorption of these ions on TiO_2 can be ignored. In the presence of Al, the efficiencies of Co^{2+} and Ni^{2+} recovery increased with increasing TiO_2 dosage; without TiO_2, the efficiencies were almost 0% for both Co and Ni while they increased to 61% for Co^{2+} and 99.9% for Ni^{2+} when 0.4 g TiO_2 was added. These results suggest clearly that addition of TiO_2 enhanced the cementation of Co and Ni by using Al as an electron donor, and indicated that AC can be replaced with TiO_2 even if its surface area is lower than AC [18,36,37].

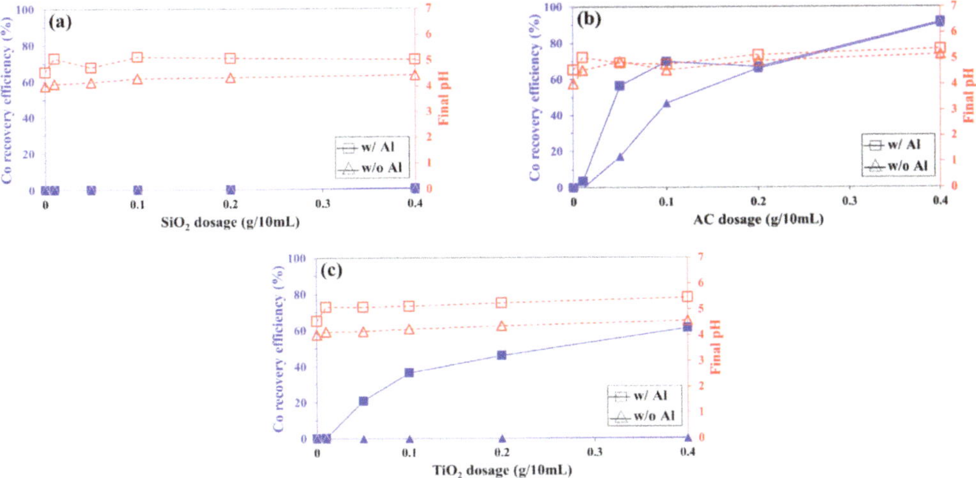

Figure 4. The effects of (**a**) SiO$_2$, (**b**) AC, and (**c**) TiO$_2$ dosages on the recovery efficiency of Co^{2+} and final pH in chloride solutions at initial pH 4.0 for 24 h.

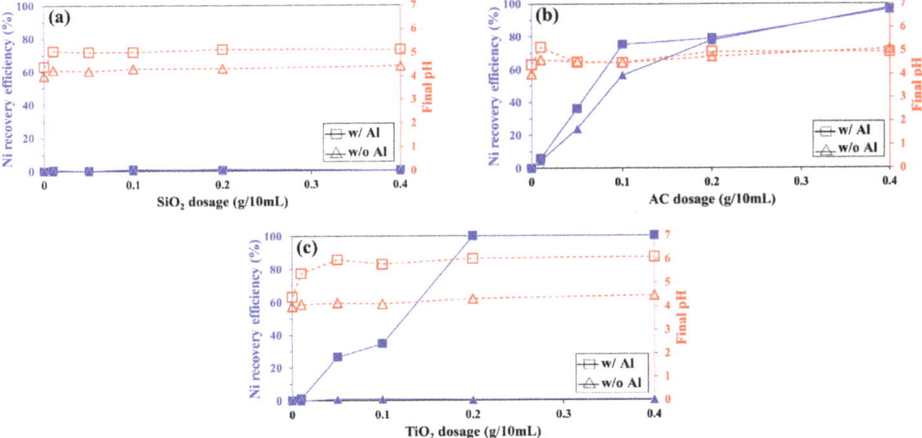

Figure 5. The effects of (**a**) SiO$_2$, (**b**) AC, and (**c**) TiO$_2$ dosages on the recovery efficiency of Ni^{2+} and final pH in chloride solutions at initial pH 4.0 for 24 h.

3.2. Surface Analysis of Deposited Co and Ni

To investigate the elemental compositions of the deposited Co and Ni, residues obtained from the Co^{2+} and Ni^{2+} recovery experiment from chloride solutions using 0.4 g of TiO$_2$ and 0.1 g of Al were analyzed by AES. Figures 6 and 7 show the AES photomicrographs (Figures 6a and 7a) and scan results of Co (Figure 6b,c) and Ni (Figure 7b,c). In both AES photomicrographs, many small gray particles and light particles are attached together onto the surface of the dark particle. The wide AES spectra of the dark particle (point 1 in Figures 6b and 7b) show strong signals of Al and O, indicating that these particles are assigned to Al powder. The small gray particles correspond to TiO$_2$ because of Ti and O signals observed at point 2 in Figures 6b and 7b. Meanwhile, light particles are observed at point 3 in Figures 6b and 7b are most likely the deposited Co and Ni, respectively.

To identify the elemental composition of the deposited Co and Ni, the narrow AES spectra in the range of 750–785 eV for Co and 830–858 eV for Ni were analyzed (Figures 6c and 7c). These spectra were fitted using reference spectra of Co, CoO, and Co_3O_4 for Co composition, and Ni and NiO for Ni composition. Fitting results indicate that the deposited Co consisted of metallic Co (93.1%) and CoO (6.9%), while the deposited Ni was composed of metallic Ni (86.2%) and NiO (13.8%). The analysis range of Auger is 0.3–5 nm, which is a near-surface analysis [38], so it is speculated that only the outermost surfaces of deposited Co and Ni were oxidized due to the oxidation of metallic Co and Ni during the dry process.

These results suggest that Co and Ni were deposited on TiO_2 particles attached to the Al surface and TiO_2 can act as an electron pathway from Al to Co^{2+} and Ni^{2+}, even if the Al oxide layer remains on the Al surface. These results showed that physical separation (i.e., ultrasonification) could be applied as the postcementation process for $Co/Ni–TiO_2$ particle and Al separation. Afterward, it is expected that only Co and Ni would be dissolved in aqueous solutions, while TiO_2 would not be dissolved because TiO_2 is more stable than Co and Ni.

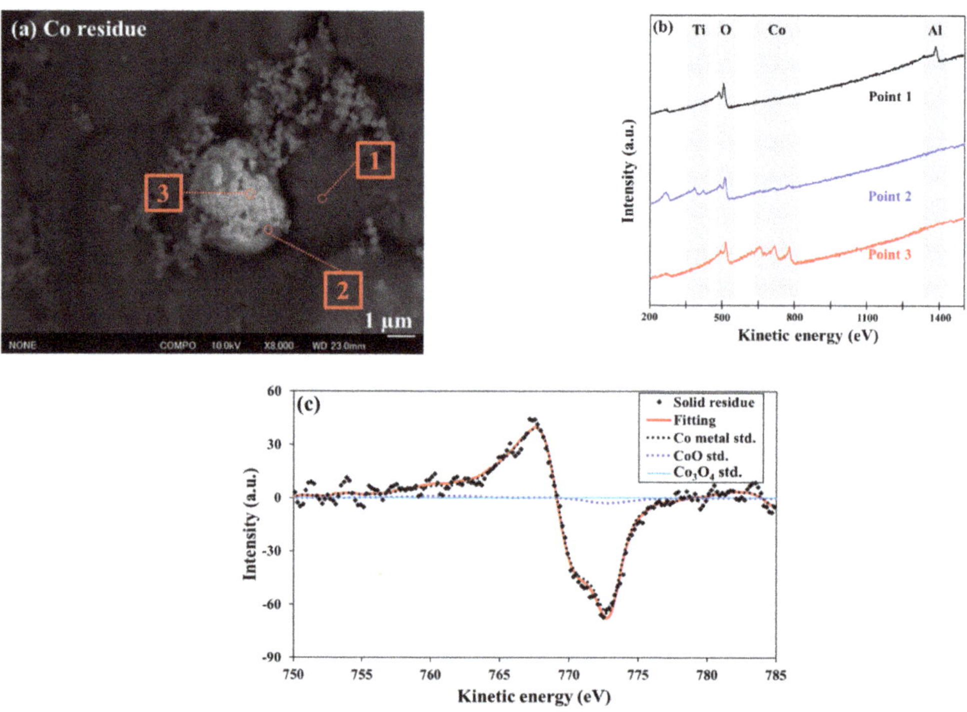

Figure 6. Auger electron spectroscopy (AES) results of the residue obtained after cementation of Co^{2+} from chloride solution using TiO_2/Al: (**a**) photomicrograph, (**b**) wide scan energy spectrum of each point, and (**c**) the narrow scan energy spectrum of the Co peak with fitting spectra of Co, CoO, and Co_3O_4.

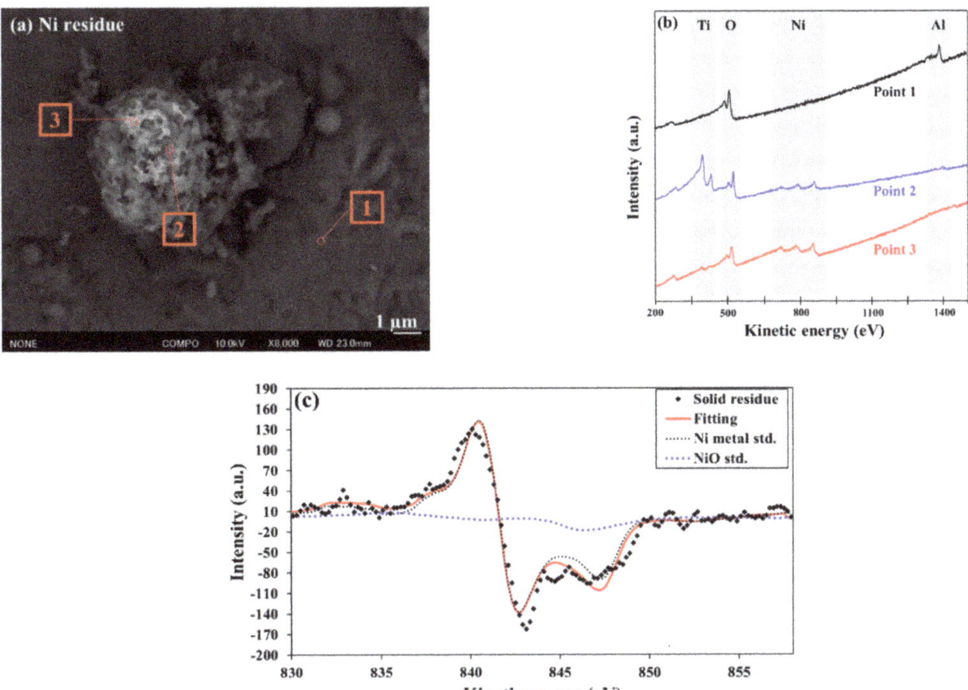

Figure 7. Auger electron spectroscopy (AES) results of the residue obtained after cementation of Ni^{2+} from chloride solution using TiO_2/Al: (**a**) photomicrograph, (**b**) wide scan energy spectrum of each points, and (**c**) the narrow scan energy spectrum of the Ni peak with fitting spectra of Ni and NiO.

4. Conclusions

This study investigated whether activated carbon (AC) could be replaced with other additives such as TiO_2 and SiO_2 for the enhanced cementation of Co^{2+} and Ni^{2+} using aluminum (Al) in sulfate and chloride solutions. In summary, the Co^{2+} and Ni^{2+} recovery efficiencies using Al in sulfate and chloride solutions were almost 0% because of the presence of an Al oxide layer on an Al surface. The adsorption of Co^{2+} and Ni^{2+} occurred when using only AC, while it did not occur when using only TiO_2 and SiO_2. When using an AC/Al-mixture or TiO_2/Al-mixture, the Co^{2+} and Ni^{2+} recovery efficiencies from sulfate and chloride solutions were enhanced compared to using Al, AC, TiO_2, and SiO_2/Al-mixture. From the results of AES analysis, Co and Ni were mostly deposited as zero-valent forms on TiO_2 attached to Al surface. This work establishes that using a conductor (AC) or a semiconductor (TiO_2) could enhance the recovery of Co^{2+} and Ni^{2+} by Al-based cementation even under mild conditions (e.g., low temperature, 25 °C; mild pH conditions, pH 4–5; no Cl^- or a low concentration). Moreover, it is expected that other conductive materials could also be used for the removal and/or recovery of metal ions using Al.

Supplementary Materials: The following are available online at https://www.mdpi.com/2075-4701/11/2/248/s1, Table S1: The concentration of Al, Ti, and Si ions after cementation experiment of Co^{2+} in sulfate solution at initial pH 4.0 at 25 °C for 24 h, Table S2: The concentration of Al, Ti, and Si ions after cementation experiment of Ni^{2+} in sulfate solution at initial pH 4.0 at 25 °C for 24 h, Table S3: The concentration of Al, Ti, and Si ions after cementation experiment of Co^{2+} in chloride solution at initial pH 4.0 at 25 °C for 24 h, Table S4: The concentration of Al, Ti, and Si ions after cementation experiment of Ni^{2+} in chloride solution at initial pH 4.0 at 25 °C for 24 h, Figure S1: The activity–pH diagram for 1 mM Co^{2+} species with 0.1 M SO_4^{2-} at 25 °C (created using the GWB Professional Ver. 12.0.3 software), Figure S2: The activity–pH diagram for 1 mM Ni^{2+} species with 0.1 M SO_4^{2-} at 25 °C (created using the GWB Professional Ver. 12.0.3 software), Figure S3: The activity–pH diagram for 1 mM Co^{2+} species with 0.1 M Cl^- at 25 °C (created using the GWB Professional Ver. 12.0.3 software), Figure S4: The activity–pH diagram for 1 mM Ni^{2+} species with 0.1 M Cl^- at 25 °C (created using the GWB Professional Ver. 12.0.3 software).

Author Contributions: Conceptualization, S.C. and S.J.; methodology, S.C., S.J. and N.H.; formal analysis, S.C., S.J., I.P., M.I. and N.H.; investigation, S.C.; writing—original draft preparation, S.C.; writing—review and editing, S.C., S.J., I.P., M.I. and N.H.; supervision, N.H.; project administration, N.H.; funding acquisition, S.J. All authors have read and agreed to the published version of the manuscript.

Funding: This study was financially supported by the Japan Society for the Promotion of Science (JSPS) grant-in-aid for Research Activity start-up (grant numbers: 19K24378).

Data Availability Statement: Data available on request due to restrictions, as the research is ongoing.

Conflicts of Interest: The authors declare no conflict of interest.

References

1. Nelson, A.; Wang, W.; Demopoulos, G.P.; Houlachi, G. Removal of cobalt from zinc electrolyte by cementation: A critical review. *Miner. Process. Extr. Metall. Rev.* **2000**, *20*, 325–356. [CrossRef]
2. Demirkran, N.; Künkül, A. Recovering of copper with metallic aluminum. *Trans. Nonferrous Met. Soc. China* **2011**, *21*, 2778–2782. [CrossRef]
3. Silwamba, M.; Ito, M.; Hiroyoshi, N.; Tabelin, C.B. Recovery of Lead and Zinc from Zinc Plant Leach Residues by Concurrent Dissolution-Cementation. *Metals* **2020**, *10*, 531. [CrossRef]
4. Choi, S.; Yoo, K.; Alorro, R.D.; Tabelin, C.B. Cementation of Co ion in leach solution using Zn powder followed by magnetic separation of cementation-precipitate for recovery of unreacted Zn powder. *Miner. Eng.* **2020**, *145*. [CrossRef]
5. Farahmand, F.; Moradkhani, D.; Sadegh Safarzadeh, M.; Rashchi, F. Optimization and kinetics of the cementation of lead with aluminum powder. *Hydrometallurgy* **2009**, *98*, 81–85. [CrossRef]
6. Abdel-Aziz, M.H.; El-Ashtoukhy, E.S.Z.; Bassyouni, M. Recovery of Copper from Effluents by Cementation on Aluminum in a Multirotating Cylinder-Agitated Vessel. *Metall. Mater. Trans. B Process Metall. Mater. Process. Sci.* **2016**, *47*, 657–665. [CrossRef]
7. Abdollahi, P.; Yoozbashizadeh, H.; Moradkhani, D.; Behnian, D. A Study on Cementation Process of Lead from Brine Leaching Solution by Aluminum Powder. *OALib* **2015**, *2*, 1–6. [CrossRef]
8. Boisvert, L.; Turgeon, K.; Boulanger, J.; Bazin, C.; Houlachi, G. Recovery of Cobalt from the Residues of an Industrial Zinc Refinery. *Metals* **2020**, *10*, 1553. [CrossRef]
9. Li, W.; Cochell, T.; Manthiram, A. Activation of aluminum as an effective reducing agent by pitting corrosion for wet-chemical synthesis. *Sci. Rep.* **2013**, *3*, 1–7. [CrossRef]
10. Park, I.; Tabelin, C.B.; Seno, K.; Jeon, S.; Ito, M.; Hiroyoshi, N. Simultaneous suppression of acid mine drainage formation and arsenic release by Carrier-microencapsulation using aluminum-catecholate complexes. *Chemosphere* **2018**, *205*, 414–425. [CrossRef]
11. Silwamba, M.; Ito, M.; Hiroyoshi, N.; Tabelin, C.B.; Fukushima, T.; Park, I.; Jeon, S.; Igarashi, T.; Sato, T.; Nyambe, I.; et al. Detoxification of lead-bearing zinc plant leach residues from Kabwe, Zambia by coupled extraction-cementation method. *J. Environ. Chem. Eng.* **2020**, *8*, 104197. [CrossRef]
12. Annamalai, V.; Murr, L.E. Effects of the source of chloride ion and surface corrosion patterns on the kinetics of the copper-aluminum cementation system. *Hydrometallurgy* **1978**, *3*, 249–263. [CrossRef]
13. Artamonov, V.V.; Moroz, D.R.; Bykov, A.O.; Artamonov, V.P. Experimental studies of cementation of tin in a dispersed form. *Russ. J. Non-Ferrous Met.* **2013**, *54*, 128–131. [CrossRef]
14. Ekmekyapar, A.; Tanaydin, M.; Demirkiran, N. Investigation of copper cementation kinetics by rotating aluminum disc from the leach solutions containing copperions. *Physicochem. Probl. Miner. Process.* **2012**, *48*, 355–367. [CrossRef]
15. Djokić, S.S. Cementation of Copper on Aluminum in Alkaline Solutions. *J. Electrochem. Soc.* **1996**, *143*, 1300–1305. [CrossRef]
16. Jeon, S.; Tabelin, C.B.; Park, I.; Nagata, Y.; Ito, M.; Hiroyoshi, N. Ammonium thiosulfate extraction of gold from printed circuit boards (PCBs) of end-of-life mobile phones and its recovery from pregnant leach solution by cementation. *Hydrometallurgy* **2020**, *191*, 105214. [CrossRef]
17. Jeon, S.; Tabelin, C.B.; Takahashi, H.; Park, I.; Ito, M.; Hiroyoshi, N. Enhanced cementation of gold via galvanic interactions using activated carbon and zero-valent aluminum: A novel approach to recover gold ions from ammonium thiosulfate medium. *Hydrometallurgy* **2020**, *191*, 105165. [CrossRef]
18. Sulyman, M.; Namiesnik, J.; Gierak, A. Low-cost adsorbents derived from agricultural by-products/wastes for enhancing contaminant uptakes from wastewater: A review. *Polish J. Environ. Stud.* **2017**, *26*, 479–510. [CrossRef]
19. Negem, M.; Nady, H.; El-Rabiei, M.M. Nanocrystalline nickel–cobalt electrocatalysts to generate hydrogen using alkaline solutions as storage fuel for the renewable energy. *Int. J. Hydrog. Energy* **2019**, *44*, 11411–11420. [CrossRef]

20. Zhang, R.; Xia, B.; Li, B.; Lai, Y.; Zheng, W.; Wang, H.; Wang, W.; Wang, M. Study on the characteristics of a high capacity nickel manganese cobalt oxide (NMC) lithium-ion battery-an experimental investigation. *Energies* **2018**, *11*, 2275. [CrossRef]

21. Lee, B.-R.; Noh, H.-J.; Myung, S.-T.; Amine, K.; Sun, Y.-K. High-Voltage Performance of Li[Ni[sub 0.55]Co[sub 0.15]Mn[sub 0.30]]O[sub 2] Positive Electrode Material for Rechargeable Li-Ion Batteries. *J. Electrochem. Soc.* **2011**, *158*, A180. [CrossRef]

22. Liu, S.; Xiong, L.; He, C. Long cycle life lithium ion battery with lithium nickel cobalt manganese oxide (NCM) cathode. *J. Power Sources* **2014**, *261*, 285–291. [CrossRef]

23. IEA. *Global EV Outlook 2020*; OECD: Paris, France, 2020; ISBN 9789264616226.

24. Zeng, X.; Li, J.; Singh, N. Recycling of spent lithium-ion battery: A critical review. *Crit. Rev. Environ. Sci. Technol.* **2014**, *44*, 1129–1165. [CrossRef]

25. Hext, P.M.; Tomenson, J.A.; Thompson, P. Titanium dioxide: Inhalation toxicology and epidemiology. *Ann. Occup. Hyg.* **2005**, *49*, 461–472. [CrossRef]

26. Nesbitt, H.W.; Bancroft, G.M.; Davidson, R.; McIntyre, N.S.; Pratt, A.R. Minimum XPS core-level line widths of insulators, including silicate minerals. *Am. Mineral.* **2004**, *89*, 878–882. [CrossRef]

27. Chen, H.J.; Lee, C. Effects of the Type of Chelating Agent and Deposit Morphology on the Kinetics of the Copper-Aluminum Cementation System. *Langmuir* **1994**, *10*, 3880–3886. [CrossRef]

28. Gao, X.; Wu, L.; Xu, Q.; Tian, W.; Li, Z.; Kobayashi, N. Adsorption kinetics and mechanisms of copper ions on activated carbons derived from pinewood sawdust by fast H3PO4 activation. *Environ. Sci. Pollut. Res.* **2018**, *25*, 7907–7915. [CrossRef]

29. Karnib, M.; Kabbani, A.; Holail, H.; Olama, Z. Heavy metals removal using activated carbon, silica and silica activated carbon composite. *Energy Procedia* **2014**, *50*, 113–120. [CrossRef]

30. Dil, E.A.; Ghaedi, M.; Ghaedi, A.M.; Asfaram, A.; Goudarzi, A.; Hajati, S.; Soylak, M.; Agarwal, S.; Gupta, V.K. Modeling of quaternary dyes adsorption onto ZnO-NR-AC artificial neural network: Analysis by derivative spectrophotometry. *J. Ind. Eng. Chem.* **2016**, *34*, 186–197. [CrossRef]

31. Burakov, A.E.; Galunin, E.V.; Burakova, I.V.; Kucherova, A.E.; Agarwal, S.; Tkachev, A.G.; Gupta, V.K. Adsorption of heavy metals on conventional and nanostructured materials for wastewater treatment purposes: A review. *Ecotoxicol. Environ. Saf.* **2018**, *148*, 702–712. [CrossRef]

32. Park, I.; Tabelin, C.B.; Seno, K.; Jeon, S.; Inano, H.; Ito, M.; Hiroyoshi, N. Carrier-microencapsulation of arsenopyrite using Al-catecholate complex: Nature of oxidation products, effects on anodic and cathodic reactions, and coating stability under simulated weathering conditions. *Heliyon* **2020**, *6*, e03189. [CrossRef] [PubMed]

33. Murr, L.E.; Annamalai, V. An Electron Microscopic Study of Nucleation and Growth in Electrochemical Displacement Reactions: A Comparison of the Cu/Fe and Cu/AI Cementation Systems. *Metall. Trans. B* **1978**, *9*, 515–525. [CrossRef]

34. Murr, L.E.; Annamalai, V. Characterization of copper nucleation and growth from aqueous solution on aluminum: A transmission electron microscopy study of copper cementation. *Thin Solid Films* **1978**, *54*, 189–195. [CrossRef]

35. Reboul, M.C.; Warner, T.J.; Mayer, H.; Barouk, B. A Ten Step Mechanism for the Pitting Corrosion of Aluminium Alloys. *Corros. Rev.* **1997**, *15*, 471–496. [CrossRef]

36. Andersson, M.; Kiselev, A.; Österlund, L.; Palmqvist, A.E.C. Microemulsion-mediated room-temperature synthesis of high-surface-area rutile and its photocatalytic performance. *J. Phys. Chem. C* **2007**, *111*, 6789–6797. [CrossRef]

37. Inada, M.; Mizue, K.; Enomoto, N.; Hojo, J. Synthesis of rutile TiO2 with high specific surface area by self-hydrolysis of TiOCl2 in the presence of SDS. *J. Ceram. Soc. Jpn.* **2009**, *117*, 819–822. [CrossRef]

38. Scherer, J. Auger Electron Spectroscopy (AES): A Versatile Microanalysis Technique in the Analyst's Toolbox. *Microsc. Microanal.* **2020**, *26*, 1564–1565. [CrossRef]

Article

Hydrometallurgical Recovery of Cu and Zn from a Complex Sulfide Mineral by Fe^{3+}/H_2SO_4 Leaching in the Presence of Carbon-Based Materials

María Luisa Álvarez [1], José Manuel Fidalgo [1], Gabriel Gascó [2] and Ana Méndez [1,*]

1 Department of Geological and Mining Engineering, Universidad Politécnica de Madrid, 28040 Madrid, Spain; marialuisa.alvarez@upm.es (M.L.Á.); jmfa88@gmail.com (J.M.F.)

2 Department of Agricultural Production, Universidad Politécnica de Madrid, Ciudad Universitaria, 28040 Madrid, Spain; gabriel.gasco@upm.es

* Correspondence: anamaria.mendez@upm.es

Abstract: Chalcopyrite, the main ore of copper, is refractory in sulfuric media with slow dissolution. The most commonly employed hydrometallurgical process for the oxidation of chalcopyrite and copper extraction is the sulfuric acid ferric sulfate system The main objective of the present work is to study the use of cheap carbon-based materials in the leaching of copper and zinc from a sulfide complex mineral from Iberian Pyrite Belt (IPB). The addition effect of commercial charcoal (VC) and two magnetic biochars (BM and HM) that were obtained by pyrolysis of biomass wastes was compared to that of commercial activated carbon (AC). The experimental results performed in this work have shown that the presence of carbon-based materials significantly influences the kinetics of chalcopyrite leaching in the sulfuric acid ferric sulfate media at 90 °C. The amount of copper and zinc extracted from IPB without the addition of carbon-based material was 63 and 72%, respectively. The highest amount of extracted zinc (>90%) was obtained with the addition of VC and AC in IPB/carbon-based material ratio of 1/0.25 w/w. Moreover, it is possible to recover more than 80% of copper with the addition of VC in a ratio 1/0.25 w/w. Moreover, an optimization of the properties of the carbon-based material for its potential application as catalyst in the leaching of metals from sulfide is necessary.

Keywords: sulfide; leaching; carbon material; copper; zinc

Citation: Álvarez, M.L.; Fidalgo, J.M.; Gascó, G.; Méndez, A. Hydrometallurgical Recovery of Cu and Zn from a Complex Sulfide Mineral by Fe^{3+}/H_2SO_4 Leaching in the Presence of Carbon-Based Materials. *Metals* **2021**, *11*, 286. https://doi.org/10.3390/met11020286

Academic Editor: Dariush Azizi

Received: 31 December 2020
Accepted: 3 February 2021
Published: 6 February 2021

Publisher's Note: MDPI stays neutral with regard to jurisdictional claims in published maps and institutional affiliations.

1. Introduction

Sulfide minerals are an important chemical form of several metal resources in nature. For example, 90% of global copper resources are sulfide ores, where most of the copper is in the form of covellite (CuS), chalcocite (Cu_2S), chalcopyrite ($CuFeS_2$), and bornite (Cu_5FeS_4). Hydrometallurgical processes were used in the treatment of covellite and chalcocite deposits in a similar way to oxidized forms, due to their solubility in acid medium, but these resources are largely exhausted. On the other hand, chalcopyrite the most abundant copper-bearing resource, accounting for appropriately 70% of the known reserves in the world [1–3], and it is submitted to pyrometallurgical treatment. With the increase in the copper demand, the decline of high-grade ores has extended and, nowadays, deposits with grades around 0.4–0.5% of copper are mined. The exploitation of these reserves by traditional flotation methods followed by pyrometallurgical processes (smelting-converting-electrorefining route) is in the limit of economic viability. In addition, the presence of impurities, like As, the generation of large amount of wastes, like slags, and the control of atmospheric emissions, make the research of more eco-friendly processing technologies necessary [4]. The hydrometallurgical route has the advantages of being able to process low-grade ores, to allow better control of co-products, and to have a lower environmental impact [5]. Nevertheless, chalcopyrite is refractory in sulfuric media with slow dissolution rates [6,7].

139

The most commonly employed hydrometallurgical process for the oxidation of chalcopyrite and extraction of copper is the sulfuric acid ferric sulfate system [8,9]. It is well known that low dissolution rate of chalcopyrite and, thus, low metal recoveries, is mainly due to the formation of a passivation layer on the mineral surface [10–12]. Different authors proposed that the composition of this passive layer is elemental sulfur, copper-rich polysulfide, or iron salts with low electrical conductivities, which may prevent the electrons/ions transfer to chalcopyrite matrix and, thus, hinder its copper dissolution [8,13,14].

Different researches have been performed in acidic sulfuric media to accelerate the kinetics of chalcopyrite leaching. Recent researches have studied the chalcopyrite leaching in sulfuric acid with the presence of chloride and nitrate ions [15,16]. Copper extraction increases in the presence of both ions [15,16]. However, the use of chloride could be limited by environmental problems that are related to high solubility of metallic chlorides and high costs of solvent extraction and electrowinning processes. Castillo-Magallanes et al. [17] studied the addition of some organic compounds, like ethylene glycol ($C_2H_6O_2$) and polysorbates. These authors found that the adsorption of organic agents limit the development of surface phases on chalcopyrite, allowing for a higher extraction of copper. Tehrani et al. [18] proposed that the addition of ethylene glycol increase the chalcopyrite dissolution rate by removing the elemental sulfur layer from the surface. Nevertheless, the addition of these two organic agents does not completely inhibit passivation and, although there are improvements in the leaching process, the effect is limited. Kartal et al. [19] investigated the improvement of chalcopyrite leaching by the addition of 20% vol of tetrachloroethylene (C_2Cl_4). These authors found that tetrachlorethylene-assisted leaching led to 80% of copper extraction before passivation.

Other researchers studied the catalytic effect of inorganic compounds, like pyrite [20], iron powder [21], nanosized silica [22], and, more recently, coal, carbon black, or activated carbon [14,23]. The use of activated carbon enhances the kinetics of sulfide minerals leaching in sulfate media, leading to a substantial increase in copper extraction rate [24]. Nakazawa [14] proposed that the enhanced kinetics of chalcopyrite leaching could be attributed to a decreased redox potential, as well as galvanic interaction between chalcopyrite and carbon matrix. The most extensive use of activated carbon in hydrometallurgy is the recovery of gold from cyanide solutions [25]. Activated carbon could show a great electron donor/acceptor due to its surface functional groups. In addition, they usually have an appropriate electrical conductivity that has been shown to promote a galvanic interaction between sulfide minerals and carbon matrix. Moreover, a well-developed porosity (micro, meso, and macroporous) in these carbon materials may facilitate the accessibility to the redox active sites through dissolved exogenous compounds [26].

Nevertheless, the majority of the carbon materials used in these previous studies were commercial activated carbons or carbon black powders, with low information being available on their physicochemical properties. Furthermore, they are relatively expensive and their use could significantly raise the cost of the industrial process. For this reason, the main objective of the present work is to study the use of cheaper carbon-based materials in the leaching of copper and zinc from a sulfide complex mineral. The addition effect of commercial charcoal and two magnetic biochars that were obtained by pyrolysis of biomass wastes was compared to that of commercial activated carbon. The effect of different mineral/carbon material ratios was studied to reduce the consumption of carbon material. pH, redox potential, copper, and zinc content of leaching solution were monitored during the leaching experiments of sulfide ore samples

2. Materials and Methods

2.1. Materials Selection and Characterization

The sulfide ore concentrate from the massive sulfide deposits of the Iberian Pyrite Belt (IPB) was collected after grinding and flotation processes (80% < 50 μm). The sample was air-dried, crushed, and sieved below 50 μm while using a mortar mill Retsch RM 100. Wavelength X-ray fluorescence (WDXRF) was performed in an ARL ADVANT'XP + sequential model from THERMO (SCAI-Malaga University). Concentration data were obtained using the UNIQUANT Integrated Software. XRD was performed using a Bruker diffractometer model D8 Advance A25.

The four carbon-based materials that were used in this work (Table 1) were: a commercial activated carbon (AC) supplied by Panreac (Spain); a commercial charcoal (VC) supplied by Ibecosol (Spain); and, two magnetic biochar samples (BM and HM) obtained, respectively, from pyrolysis of pruning waste and pruning waste hydrochar. Ingelia (Náquera, Spain) supplied pruning waste and pruning waste hydrochar. The two magnetic biochar were prepared by impregnation with ferric sulfate salts of pruning waste (BM) or corresponding hydrochar (HM), followed by pyrolysis at 500 °C for 5 h [27].

Table 1. Main characteristics of carbon-based materials.

Sample	C (%)	H (%)	N (%)	S (%)	O (%)	H/C	O/C	Ash (%)	pH	Eh (mV)	S_{BET} (m^2/g)
CA	85.72	0.88	0.00	0.00	13.41	0.12	0.12	1.00	8.03	453	1138.96
CV	80.21	3.12	0.92	0.00	15.74	0.47	0.15	14.94	8.31	355	2.18
BM	20.18	0.29	0.60	7.83	0.71	0.17	0.03	70.38	7.33	314	56.86
HM	43.81	0.81	1.14	5.12	8.14	0.22	0.14	40.98	8.22	298	183.63

The four carbon-based materials (AC, VC, BM, and HM) were air-dried, crushed, and then sieved below 100 μm using a mortar mill Retsch RM 100 and characterized according to the following properties: pH, Eh, BET surface (S_{BET}), cation exchange capacity (CEC), pore volume (cm^3/g), ash content (%), and elemental C, H, N, O, and S content (dry basis).

The pH and Eh of samples were determined on aqueous solutions at a concentration of 4 g·L^{-1} of carbon-based material sample in distilled water. pH was measured using a Crison micro pH 2000 and Eh with a pH 60 DHS equipment. C, H, N and S content were determined by dry combustion using a LECO CHNS 932 Analyzer. The ash content was calculated by combustion of samples at 850 °C in a Labsys Setaram TGA analyzer. 20 mg of each sample were heated at a rate of 15 °C min^{-1} up to 850 °C using 30 mL min^{-1} of air. O was obtained by difference as 100% − (%C + %H + %N + %S + %Ash). Atomic ratios H/C and O/C were also calculated.

2.2. Leaching Experiments

A thermostatic bath with stirring GFL 1083 (heating power of 1500 W and the voltage 230 V) was used for the leaching tests. The temperature (90 °C) and stirring speed (250 rpm) were controlled during the leaching process for 96 h. The leaching experiments were carried out in 250 mL ISO borosilicate glass jars.

Leaching agent used was prepared by 0.5 M H_2SO_4 solution with a concentration of 5 g·L^{-1} of Fe (III). The pH and Eh of a 0.5 M H_2SO_4 solution were 0.983 and 423 mV, respectively. The purities of Sigma-Aldrich® sulfuric acid and Fe (III) sulfate hydrated Labkem were 97% and 99.5%, respectively. 2.5 g of IPB was weighed in every jar. 50 mL of leaching agent as added. Except for the control, the ratios IPB/carbon-based material (weight/weight) were 1/1, 1/0.5, and 1/0.25.

1 mL of the supernatant liquor of each leaching experiment was withdrawn at different reaction times (2, 4, 24, 48, 72, and 96 h). The sampling was carried out, as follows: first, the stirring was stopped to let the sample stand and favors its decantation. After this short period of time, 1 mL of the supernatant solution was removed and, then, filtered and transferred to a 25 mL graduated flask and make up to volume with distilled water. In order to compensate for this extracted mL and maintain the same conditions throughout the system, 1 mL of the leaching solution was added. Metal extraction in solution was calculated, as follows:

$$Metal\ extraction\ in\ solution\ (\%) = \frac{metal\ content\ in\ leaching\ solution\ (g)}{initial\ IPB\ metal\ content\ (g)} \times 100 \quad (1)$$

Stirring was stopped after 96 h. The pulp was filtered and the solid waste was washed twice with 50 mL of H_2SO_4 solution pH = 2 in order to recover the adsorbent content of metals. The total Cu and Zn extraction degree was determined while taking their content in the final into account and washed solutions using the following formula:

$$Total\ metal\ extraction\ degree\ (\%) = \frac{total\ metal\ extracted\ (g)}{initial\ IPB\ metal\ content\ (g)} \times 100 \qquad (2)$$

where total metal extracted (g) was metal content in leaching solution at 96 h (g) plus metal content in washed solutions (g).

The pH and Eh of leaching solution were determined at different reaction times (0, 2, 4, 24, 48, 72, and 96 h) using a Crison micro pH 2000 and a pH 60 DHS, respectively. The concentration of Cu and Zn in the leaching and washed solutions was determined using a Perkin Elmer AAnalyst 400 Atomic Absorption Spectrophotometer.

3. Results and Discussion

3.1. Characteristics of Samples

Table 1 summarized the main properties of VC, AC, BM, and HM. In spite of the fact that several researches use carbon materials as catalysts in the leaching of metal sulfides, there are not studies comparing the effect of carbon materials with different properties [14,23,24].

The highest C and O content corresponded to AC and VC. AC followed by BM showed the lowest H/C ratios and, consequently, the highest aromaticity. BM showed the lowest O/C and, consequently, the lowest content on oxygen functional groups. BM and HM showed the lowest values of Eh, whereas CA showed the highest value (455 mV).

The XRD analysis of IPB (Figure 1) showed a 52.6% of chalcopyrite, 8.4% of sphalerite, 32.4% of pyrite, 1.4% of cristobalite, and 5.1% of nacrite (Table 2). Table 3 summarizes the chemical composition of IPB that was obtained by X-ray fluorescence. The main valuable metals were Cu (17.61%) and Zn (6.76%). For this reason, the present work is focused on the recovery of these two metals.

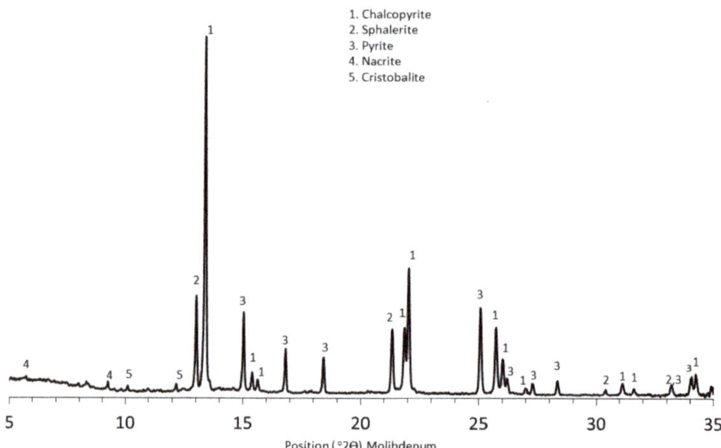

Figure 1. XRD pattern of IPB sample.

Table 2. XRD analysis of Iberian Pyrite Belt (IPB) sample.

Mineral	Content (%)
Chalcopyrite ($CuFeS_2$)	52.6 ± 0.9
Sphalerite (ZnS)	32.2 ± 0.2
Pyrite (FeS_2)	8.4 ± 0.6
Nacrite ($Al_2Si_2O_5(OH)_4$)	5.1 ± 0.1
Cristobalite (SiO_2)	1.4 ± 0.4

Table 3. Chemical composition of IPB sample.

Element	Content (%)
Fe	20.68 ± 0.16
Cu	17.61 ± 0.17
S	13.36 ± 0.17
Zn	6.76 ± 0.11
Si	1.67 ± 0.05
Al	0.664 ± 0.03
Mg	0.413 ± 0.021
Sb	0.229 ± 0.011
As	0.123 ± 0.062
Ca	0.115 ± 0.006
Co	0.0702 ± 0.0035
Ti	0.0329 ± 0.0016
Pb	0.0402 ± 0.0020
K	0.0240 ± 0.0012
P	0.0125 ± 0.0016
Mn	0.0212 ± 0.0011
Se	0.0149 ± 0.008
Sn	0.0116 ± 0.0014
Cd	0.0093 ± 0.0013

3.2. Extraction Degree of Cu and Zn

Figure 2 shows the extraction degree of Cu and Zn after 96 h of IPB leaching in sulfuric/Fe^{3+} solution at 90 °C. The effect of carbon materials (VC, AC, MB, and HM) on the recovery of Cu and Zn was different, depending on metal, carbon-based material, and their ratio with respect to the IPB sample. The amount of Cu and Zn extracted from IPB without the addition of carbon-based material was 63 and 72%, respectively. The highest zinc extraction degree (>90%) was obtained with the addition of VC and AC in the IPB/carbon-based material ratio of 1/0.25. The highest Cu extraction degree (85%) was obtained after the addition of VC in a ratio of 1/0.25. In general, the addition of BM and HM significantly decreased the extraction degree of Cu from IPB sample. Only an increment in extraction of Zn was observed with a IPB/BM ratio of 1/1.

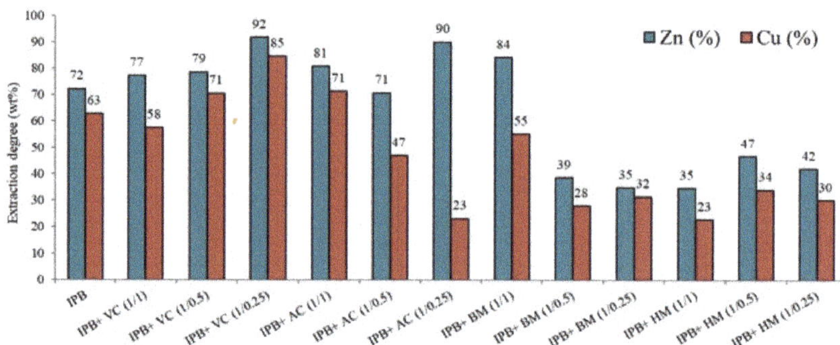

Figure 2. Extraction degree (wt%) of Zn and Cu.

The experimental results showed that the characteristics of a carbon-based material seem to play an important role in the recovery of Cu and Zn from complex sulfide mineral. This means that it is necessary to possess a better knowledge of the mechanism of reaction in the presence of carbon-based materials and the optimization of the properties of the carbon-based material for its potential application as catalyst in the leaching of metals from sulfide ores. Figure 3 shows the Cu and Zn extracted in the leaching solution at different reaction times (2, 4, 24, 48, 72, and 96 h). It is observed that the extraction of both metals fluctuates along time, indicating that an amount of the dissolved Cu or Zn could be adsorbed on the surface of mineral and, especially, on the carbon-based material. Adsorption is more important in the case of Zn. The comparison of Figures 2 and 3 shows that, at 2 h, the Zn concentration in the leaching solution was similar to the extraction degree of Zn. At longer reaction times, the concentration of Zn in the solution decreases. This effect could be due to its adsorption on the surface of the mineral or the carbon-based adsorbent.

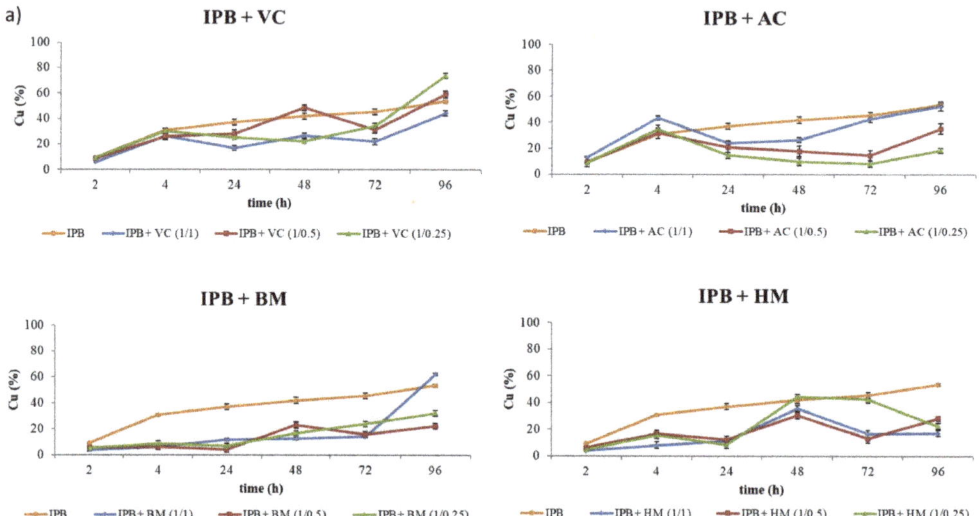

Figure 3. *Cont.*

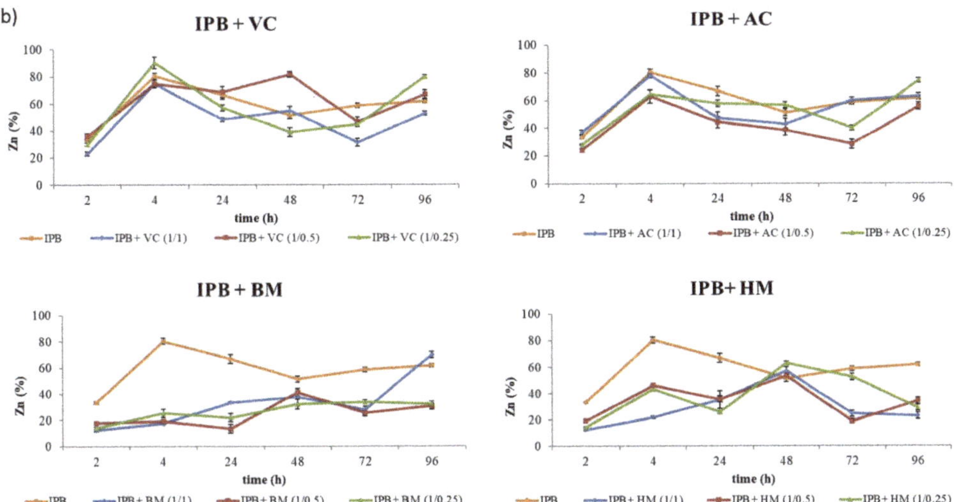

Figure 3. Evolution of Cu (**a**) and Zn (**b**) dissolved at different reaction times (h).

3.3. Eh Evolution during Leaching Experiments

Nakazawa [14] used carbon black in sulfuric acid media at 50 °C and proposed that the enhanced kinetics of chalcopyrite leaching could be attributed to dissolution reactions at the low redox potential in addition to the galvanic interaction between chalcopyrite and carbon black. Cordoba et al. [28] concluded that the redox potential is a key factor in the leaching of chalcopyrite and proposed a critical potential of approximately 450 mV. A high potential of the leaching favors the rapid precipitation of ferric as jarosite and the corresponding passivation of chalcopyrite. Hiroyoshi et al. [29] showed that the leaching rate of chalcopyrite by Fe^{3+}/H_2SO_4 solutions greatly depends on the redox potential and that there is a maximum leaching rate at an optimum redox value. Previous works [30] pointed out that the redox potential has to be low enough for chalcocite formation and high enough for the subsequent chalcocite oxidation. Kamentani and Aaoki [31] investigated the effect of the redox potential in the range of 300 and 650 mV on chalcopyrite leaching at 90 °C. They found that the leaching rate increased with an increase in the suspension potential, until it reached a maximum at 400–430 mV. Thereafter, the leaching rate decreased. Córdoba et al. [28] found that chalcopyrite leaching was remarkably enhanced at low redox potential, which suggested that chalcopyrite dissolves through the intermediate formation of covellite. Figure 4 shows the Eh evolution in the different leaching systems along reaction time. The initial Eh (V) value varies between 641 for IPB and 612 mV for IPB+AC (ratio 1/1). An important decrease on the Eh (V) was observed after 2 h of leaching. The reduction was higher in samples that were treated with AC in spite of this carbon-based materials showing the highest Eh (V) among the materials used in this work (Table 1). The redox potential fell with decreasing Fe^{3+} concentrations, probably due to AC participating in the reaction of sulfide mineral leaching, decreasing the Fe^{3+}/Fe^{2+} ratio according to Nakazawa [14]. Leaching systems with the three AC ratios (1/1, 1/0.5, and 1/0.25) showed similar Eh (V) values along reaction time; however, different leaching percentage of Cu were obtained. Therefore, the Eh (V) was not the only factor that controlled the effect of carbon-bases material on the leaching of metals from sulfide minerals.

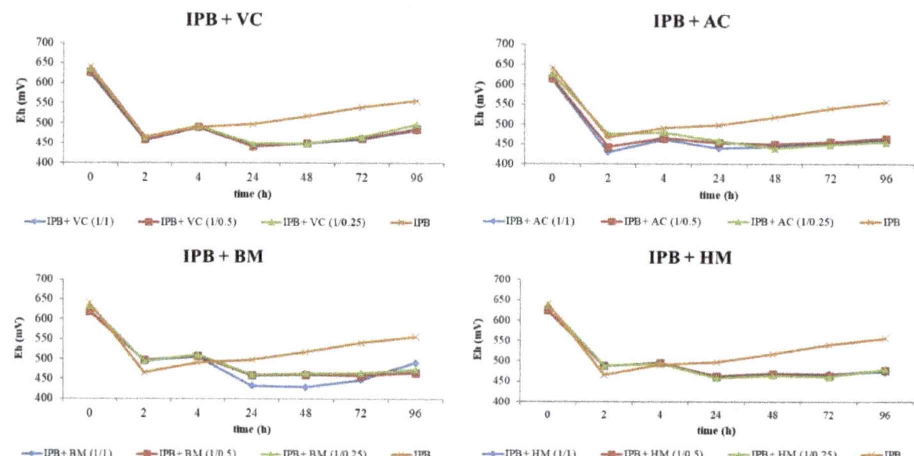

Figure 4. Evolution of Eh (mV) at different reaction times (h).

3.4. pH Evolution during Leaching Experiments

Figure 5 shows the pH evolution in the different leaching systems along reaction time. In spite of carbon-based materials showing alkaline pH (Table 1), their addition to leaching systems significantly decreased the pH of the leaching solution. The highest pH reduction was observed at 2 h of leaching reaction and especially for BM and HM. After 2 h, the slightly increased pH remains, in general, lower than the pH without the addition of carbon-based material. This effect could be due to the reaction of surface functional groups of carbon-based adsorbents with the oxidizing agent (Fe^{3+}/H_2SO_4). These can react with different oxygenated groups generating organic acids and releasing protons to the medium. In addition, the addition of carbon-based adsorbents decreased the Eh of the leaching system (Figure 4) that favors the generation of H_2S, which then was oxidized to S by Fe^{3+} [9]. Córdoba et al. [32] studied the passivation of chalcopyrite in the presence of ferric sulfate solutions at different pH and Eh. They concluded that low pH values (especially <1) of the leaching solution have a negative effect on the chalcopyrite dissolution. The high reduction in the pH of leaching solution after the addition of BM and HM that was observed at 2 and 4 h (Figure 5) could be the reason for an important reduction of dissolved Cu (Figure 4).

In summary, it has been observed that carbon-based materials greatly influence the dissolution of Cu and Zn from chalcopyrite and sphalerite minerals in Fe^{3+}/H_2SO_4 media at 90 °C. The amount of Cu and Zn extracted from IPB without the addition of carbon-based material was, 63 and 72%, respectively. The highest amount of extracted Zn (>90%) was obtained with the addition of VC and AC in a IPB/carbon-based material ratio of 1/0.25. It is possible to recover 85% of copper after the addition of VC in 1/0.25. In general, the addition of carbon-based adsorbents decreases the Eh and pH of the leaching solution. An optimization of the properties of the carbon-based material and experimental conditions used in the presence of carbon-based adsorbent is necessary for its potential application as catalyst in the leaching of metals from sulfide minerals.

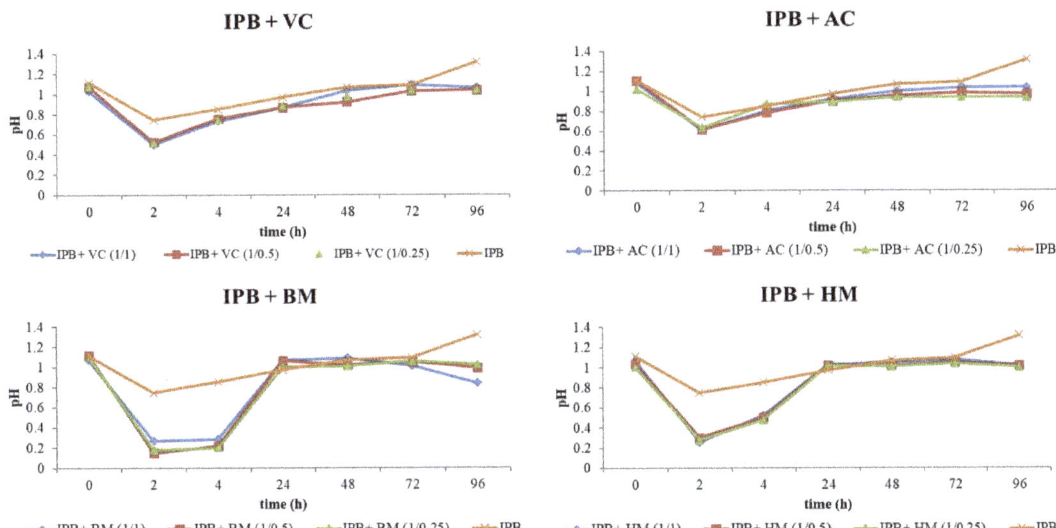

Figure 5. Evolution of pH at different reaction times (h).

4. Conclusions

The main conclusions obtained from the present work were the following:

In general, the addition of carbon-based adsorbents reduces the Eh and pH of leaching systems and probably modifies the reaction mechanisms. The Cu and Zn extracted with Fe^{3+}/H_2SO_4 vary in the presence of carbon-based material. The effect was different, depending on characteristics and the amount of carbon-based material. The addition of commercial charcoal in the ratio 1/0.25 of complex sulfide mineral/charcoal increases the Cu and Zn extracted to 92 and 85%, respectively. Future researches are necessary for knowing the reaction mechanism of chalcopyrite and sphalerite leaching in the presence of carbon-based materials and the properties of carbon-based materials that are involved in the reaction.

Author Contributions: Conceptualization, M.L.Á., J.M.F., G.G. and A.M.; methodology, M.L.Á., J.M.F., G.G. and A.M.; formal analysis, M.L.Á., J.M.F., G.G. and A.M.; M.L.Á., J.M.F., G.G. and A.M writing—original draft preparation; funding acquisition, A.M. All authors have read and agreed to the published version of the manuscript.

Funding: This research was funded by Ministerio de Ciencia, Innovación y Universidades (Spain), grant number RTI2018-096695-B-C31.

Institutional Review Board Statement: Not applicable.

Informed Consent Statement: Not applicable.

Conflicts of Interest: The authors declare no conflict of interest.

References

1. Baba, A.A.; Ayinla, K.I.; Adekola, F.A.; Ghosh, M.K.; Ayanda, O.S.; Bale, R.B.; Sheik, A.R.; Pradhan, S.R. A review on novel techniques for chalcopyrite ore processing. *Int. J. Min. Eng. Miner. Process.* **2012**, *1*, 1–16. [CrossRef]
2. Wang, S. Copper leaching from chalcopyrite concentrates. *JOM* **2005**, *57*, 48–51. [CrossRef]
3. Watling, H.R. The bioleaching of sulphide minerals with emphasis on copper sulphides: A review. *Hydrometallurgy* **2006**, *84*, 81–108. [CrossRef]
4. Karimov, K.A.; Rogozhnikov, D.A.; Kuzas, E.A.; Shoppert, A.A. Leaching kinetics of arsenic sulfide-containing materials by copper sulfate solution. *Metals* **2020**, *10*, 7. [CrossRef]
5. Chagnes, A. Advances in hydrometallurgy. *Metals* **2020**, *9*, 211. [CrossRef]

6. Cortés, S.; Soto, E.E.; Ordóñez, J.I. Recovery of Copper from Leached Tailing Solutions by Biosorption. *Minerals* **2020**, *10*, 158. [CrossRef]

7. Jorjani, E.; Ghahreman, A. Challenges with elemental sulfur removal during the leaching of copper and zinc sulfides and from the residues: A review. *Hydrometallurgy* **2017**, *171*, 333–343. [CrossRef]

8. Watling, H.R. Chalcopyrite hydrometallurgy at atmospheric pressure: 1. Review of acidic sulfate, sulfate–chloride and sulfate–nitrate process options. *Hydrometallurgy* **2013**, *140*, 163–180. [CrossRef]

9. Lorenzo-Tallafigo, J.; Romero-García, A.; Iglesias-Gonzalez, N.; Mazuelos, A.; Romero, R.; Carranza, F. A novel hydrometallurgical treatment for the recovery of copper, zinc, lead and silver from bulk concentrates. *Hydrometallurgy* **2020**, *200*, 105548. [CrossRef]

10. Hackl, R.P.; Peters, E.; King, J.A. Passivation of chalcopyrite during oxidative leaching in sulfate media. *Hydrometallurgy* **1995**, *39*, 25–48. [CrossRef]

11. Li, Y.; Kawashima, N.; Li, J.; Chandra, A.P.; Gerson, A.R. A review of the structure, and fundamental mechanisms and kinetics of the leaching of chalcopyrite. *Adv. Colloid Interface Sci.* **2013**, *197–198*, 1–32. [CrossRef]

12. Nourmohamadi, H.; Esrafili, M.D.; Aghazadeh, V. DFT study of ferric ion interaction with passive layer on chalcopyrite surface: Elemental sulfur, defective sulfur and replacement of M2+(M=Cu and Fe) ions. *Comput. Condens. Matter* **2021**, *26*, e00536. [CrossRef]

13. Munoz, P.B.; Miller, J.D.; Wadsworth, M.E. Reaction mechanism for the acid ferric sulfate leaching of chalcopyrite. *Metall. Mater. Trans. B* **1979**, *10*, 149–158. [CrossRef]

14. Nakazawa, H. Effect of carbon black on chalcopyrite leaching in sulfuric acid media at 50 °C. *Hydrometallurgy* **2018**, *177*, 100–108. [CrossRef]

15. Castellón, C.I.; Hernández, P.C.; Velásquez-Yévenes, L.; Taboada, M.E. An alternative process for leaching chalcopyrite concentrate in nitrate-acid-seawater media with oxidant recovery. *Metals* **2020**, *10*, 518. [CrossRef]

16. Hernández, P.; Gahona, G.; Martínez, M.; Toro, N.; Castillo, J. Caliche and seawater, sources of nitrate and chloride ions to chalcopyrite leaching in acid media. *Metals* **2020**, *10*, 551. [CrossRef]

17. Castillo-Magallanes, N.; Cruz, R.; Lázaro, I. Effect of organic agents on the oxidation process of chalcopyrite in a sulfuric acid solution. *Electrochim. Acta* **2020**, *335*, 136789. [CrossRef]

18. Tehrani, M.E.H.N.; Naderi, H.; Rashchi, F. Electrochemical study and XPS analysis of chalcopyrite dissolution in sulfuric acid in the presence of ethylene glycol. *Electrochim. Acta* **2021**, *369*, 137663. [CrossRef]

19. Kartal, M.; Xia, F.; Ralph, D.; Rickard, W.D.A.; Renard, F.; Li, W. Hydrometallurgy Enhancing chalcopyrite leaching by tetrachloroethylene-assisted removal of sulphur passivation and the mechanism of jarosite formation. *Hydrometallurgy* **2020**, *191*, 105–192. [CrossRef]

20. Dixon, D.G.; Mayne, D.D.; Baxter, K.G.; Galvanox, T.M. A novel galvanically-assisted atmospheric leaching technology for copper concentrates. *Can. Metall. Q.* **2008**, *47*, 327–336. [CrossRef]

21. Sanchez, E.C.; Umetsu, Y.; Saito, F. Effect of iron powder on copper extraction by acid leaching of chalcopyrite concentrate. *J. Chem. Eng. Jpn.* **1996**, *29*, 720–722. [CrossRef]

22. Misra, M.; Fuerstenau, M.C. Chalcopyrite leaching at moderate temperature and ambient pressure in the presence of nanosize silica. *Miner. Eng.* **2005**, *18*, 293–297. [CrossRef]

23. Okamoto, H.; Nakayama, R.; Kuroiwa, S.; Hiroyoshi, N.; Tsunekawa, M. Catalytic effect of activated carbon and coal on chalcopyrite leaching in sulfuric acid solutions. *Shigen Sozai* **2004**, *120*, 600–606. [CrossRef]

24. Nakazawa, H.; Nakamura, S.; Odashima, S.; Hareyama, W. Effect of carbon black to facilitate galvanic leaching of copper from chalcopyrite in the presence of manganese (IV) oxide. *Hydrometallurgy* **2016**, *163*, 69–76. [CrossRef]

25. Medina, D.; Anderson, K.G. A review of the cyanidation treatment of copper-gold ores and concentrates. *Metals* **2020**, *10*, 897. [CrossRef]

26. Yuan, Y.; Bolan, N.; Prévoteau, A.; Vithanage, M.; Kumar, J.; Sik, Y.; Wang, H. Applications of biochar in redox-mediated reactions. *Bioresour. Technol.* **2017**, *246*, 271–281. [CrossRef] [PubMed]

27. Álvarez, M.L.; Gascó, G.; Palacios, T.; Paz-Ferreiro, J.; Méndez, A. Fe oxides-biochar composites produced by hydrothermal carbonization and pyrolysis of biomass waste. *J. Anal. Appl. Pyrolysis* **2020**, *151*, 104893. [CrossRef]

28. Crdoba, E.M.; Munoz, J.A.; Blázquez, M.L.; González, F.; Ballester, A. Leaching of chalcopyrite with ferric ion. Part II: Effect of redox potential. *Hydrometallurgy* **2008**, *93*, 88–96. [CrossRef]

29. Hiroyoshi, N.; Kitayawa, H.; Tsunekawa, M. Effect of solution composition on the optimum redox potential for chalcopyrite leaching in sulfuric acid solutions. *Hydrometallurgy* **2008**, *91*, 144–149. [CrossRef]

30. Hiroyoshi, N.; Miki, H.; Hirajima, T.; Tsunekawa, M. A model for ferrous-promoted chalcopyrite leaching. *Hydrometallurgy* **2000**, *57*, 31–38. [CrossRef]

31. Kametani, H.; Aoki, A. Effect of suspension potential on the oxidation rate of copper concentrate in a sulfuric acid solution. *Metall. Trans. B* **1985**, *16*, 695–705. [CrossRef]

32. Córdoba, E.M.; Muñoz, J.A.; Blázquez, M.L.; González, F.; Ballester, A. Passivation of chalcopyrite during its chemical leaching with ferric ion at 68 °C. *Miner. Eng.* **2009**, *22*, 229–235. [CrossRef]

metals

Article

Hydrometallurgical Leaching of Copper Flash Furnace Electrostatic Precipitator Dust for the Separation of Copper from Bismuth and Arsenic

Michael Caplan [1,*]**, Joseph Trouba** [1]**, Corby Anderson** [1] **and Shijie Wang** [2]

[1] Kroll Institute for Extractive Metallurgy, Mining Engineering Department & George S. Ansell Department of Metallurgical and Materials Engineering, Colorado School of Mines, Golden, CO 80401, USA; jtrouba@mymail.mines.edu (J.T.); cgandersmines@gmail.com (C.A.)

[2] Rio Tinto Kennecott Utah Copper, Magna, UT 84944, USA; Shijie.Wang@riotinto.com

* Correspondence: mcaplan@mymail.mines.edu; Tel.: +1-(281)-300-2619

Abstract: Flash furnace electrostatic precipitator dust (FF-ESP dust) is a recycle stream in some primary copper production facilities. This dust contains high amounts of copper. In some cases, the FF-ESP dust contains elevated levels of bismuth and arsenic, both of which cause problems during the electrorefining stages of copper production. Because of this, methods for separation of copper from bismuth and arsenic in FF-ESP dust are necessary. Hydrometallurgical leaching using a number of lixiviants, including sulfuric acid, sulfurous acid, sodium hydroxide, and water, were explored. Pourbaix diagrams of copper, bismuth, and arsenic were used to determine sets of conditions which would thermodynamically separate copper from bismuth and arsenic. The data indicate that water provides the best overall separation between copper and both bismuth and arsenic. Sodium hydroxide provided a separation between copper and arsenic. Sulfurous acid provided a separation between copper and bismuth. Sulfuric acid did not provide any separations between copper and bismuth or copper and arsenic.

Keywords: copper processing; copper leaching; copper bearing dusts

Citation: Caplan, M.; Trouba, J.; Anderson, C.; Wang, S. Hydrometallurgical Leaching of Copper Flash Furnace Electrostatic Precipitator Dust for the Separation of Copper from Bismuth and Arsenic. *Metals* **2021**, *11*, 371. https://doi.org/10.3390/met11020371

Academic Editor: Srecko Stopic

Received: 13 January 2021
Accepted: 19 February 2021
Published: 23 February 2021

1. Introduction

Rio Tinto Kennecott Copper (RTKC) is a primary copper producer based in Utah, U.S. Like many primary copper producers, RTKC utilizes a flash smelting stage in their process [1]. During the flash smelting process, a portion of the input material is captured, as a dust, by the gases generated by the smelting reactions [2–4]. These gases must be processed prior to release in order to meet regulatory emissions requirements. One of the unit operations necessary for gas treatment is to separate the solid dust from the gases. This is accomplished using a waste heat boiler and an electrostatic precipitator (ESP) [4]. The separated material from the ESP is referred to as the flash furnace electrostatic precipitator dust (FF-ESP dust).

The copper content of the FF-ESP dust is high, over 20% by mass. Because of this, it is necessary to reprocess this material for its copper. Commonly, this is done by simply recycling the FF-ESP dust back into the flash furnace. However, in the case of RTKC, the dust contains high concentrations of both bismuth and arsenic, both of which cause problems further downstream in the copper production process and the final copper product [5].

In metallic copper, arsenic introduces a number of detrimental qualities. It reduces conductivity; increases recrystallization temperature, which may result in increased processing costs if annealing is necessary; and causes grain boundary cracking [6]. In the case of bismuth impurity in copper, bismuth causes embrittlement and increases the work hardening rate [6]. This is because bismuth has no practical solubility in copper and therefore segregates to the copper grain boundaries. In some conditions this can result in liquid metal embrittlement and/or grain boundary wetting. Because of these detriments, the chemical specification for most high-quality cathode coppers, including LME Grade A Cathode Copper, follows ASTM B115-10 Grade 1, which limits arsenic content to 5 ppm and bismuth content to 1.0 ppm [7].

Regarding processing, bismuth and arsenic cause anode passivation and reductions in current efficiency [8]. This is especially true of bismuth. Because of this, the FF-ESP dust cannot always be recycled, resulting in otherwise unnecessary copper losses. Additionally, FF-ESP dust is considered a hazardous waste by U.S. regulatory agencies (in large part due to arsenic), and therefore, the inability to process it results in increased waste stream costs [9]. Alternative methods for FF-ESP dust processing, especially methods that separate copper from bismuth and arsenic, are necessary. That said, any process that deals with the RTKC FF-ESP dust must deal with the arsenic present in it. Generally, this is done via fixation and is well understood in the industry [9,10].

Hydrometallurgical leaching is a commonly used method for the selective separation of one or more elements or compounds. A number of papers have reported on the efficacy of leaching copper flue dusts using sulfuric acid in order to recover copper and other metals; however, separations between copper and bismuth or arsenic are not common [11,12]. Copper flue dusts commonly contain water-soluble copper sulfates; because of this, water is also used to recover copper from copper flue dusts. [4,12] Ha et al. have shown that bismuth can be leached from copper flue dusts using sulfuric acid and sodium hydroxide, but copper is also leached during this process [13].

The potential for hydrometallurgical selectivity and separations can be exhibited by Pourbaix diagrams. These diagrams depict electro-potential and pH on the axes and show the regions of thermodynamic stability for known species of a given system.

Figure 1 shows overlaid copper and bismuth Pourbaix diagrams. This provides regions where separation between copper and bismuth is theoretically possible. The same is shown in Figure 2 but for copper and arsenic instead. All Pourbaix diagrams were generated using StabCal software at 25 °C, and molarities are as stated in the captions.

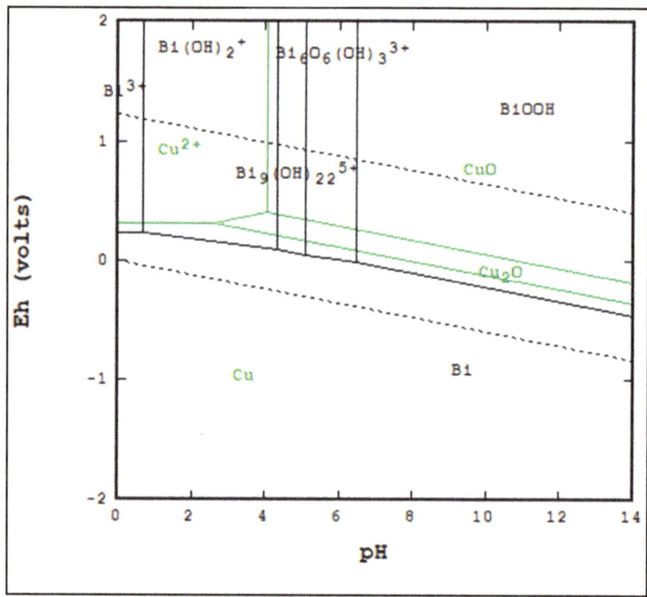

Figure 1. Copper (0.173 M) and bismuth (0.00178 M) Pourbaix diagram overlay.

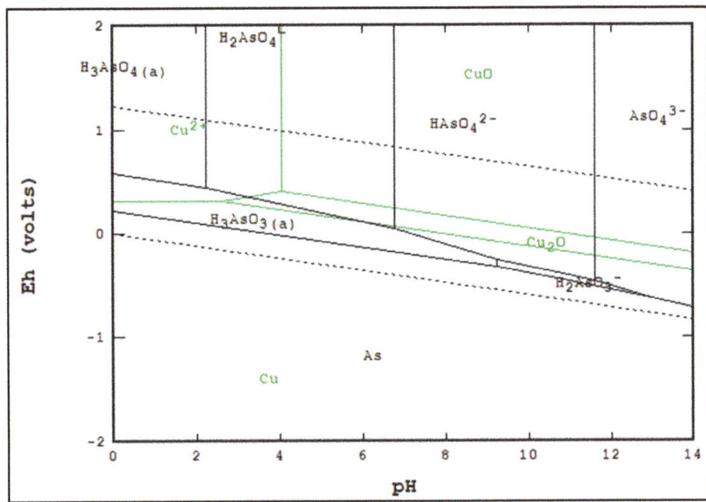

Figure 2. Copper (0.173 M) and arsenic (0.0320 M) Pourbaix diagram overlay.

As shown in this paper, the majority of the RTKC dust is composed of mixed amorphous oxide phases containing the copper, bismuth, and arsenic. Ideally, Pourbaix diagrams are generated using the actual phases present in the system. That said, the phase chemistry and population of the dust is complex and not easily simulated. Because of this, the Pourbaix diagrams used are for pure, ideal phases, and therefore they cannot be used to predict exactly what conditions will provide the thermodynamic possibility for separation, but they can be used as a viable starting point.

From Figure 1, it can be seen that it is thermodynamically possible to separate copper and bismuth in the regions of overlap between $Bi(OH)_2^+$ and CuO or Cu_2O. From Figure 2, it can be seen that it is thermodynamically possible to separate copper and arsenic in the regions of overlap between $HAsO_4^{2-}$ or AsO_4^{3-} and CuO or Cu_2O. By contrasting both figures, it can be seen that it is not thermodynamically likely to separate copper from both arsenic and bismuth in a single leaching stage.

Shown in Figure 3 is a copper Pourbaix diagram with concentration on the vertical axis. Soluble regions are "filled in". It can be seen that as copper concentration increases, so too does the portion of regions that are insoluble. Figure 4 shows the same diagram but for bismuth. The bismuth diagram exhibits the same pattern as copper. Shown in Figure 5 is the 3D arsenic Pourbaix diagram. Unlike copper and bismuth, arsenic solubility increases with arsenic concentration.

Dust processing in general is well suited to hydrometallurgical processing for two primary reasons: the innate ability for hydrometallurgy to be selective, and the environmental benefit of eliminating the emission of toxic metals in gaseous form [14]. Historically, hydrometallurgical developments have enabled high separability and selectivity, such as in the Sherritt Gordon process for the separation and recovery of cobalt, nickel, and copper [15], as well as in the Merrill Crowe cyanidation process for the recovery of gold [14].

In the specific case of copper smelter dusts, a number of studies have been published involving efforts to separate copper from arsenic and bismuth. Ichimura et al., Kovyazin et al., Yang et al., Shanazi et al., and Chen et al. have published studies showing separation of copper from arsenic and/or bismuth via first leaching all three metals and then separating copper from bismuth and arsenic via precipitation methodology [16–19].

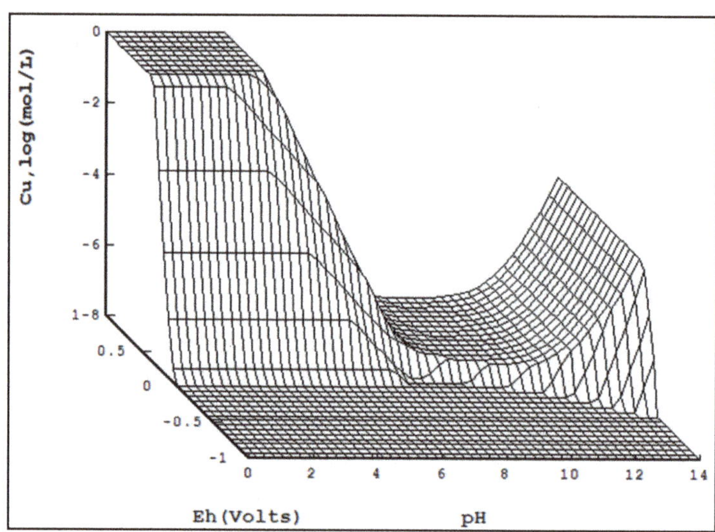

Figure 3. 3D Cu Pourbaix diagram.

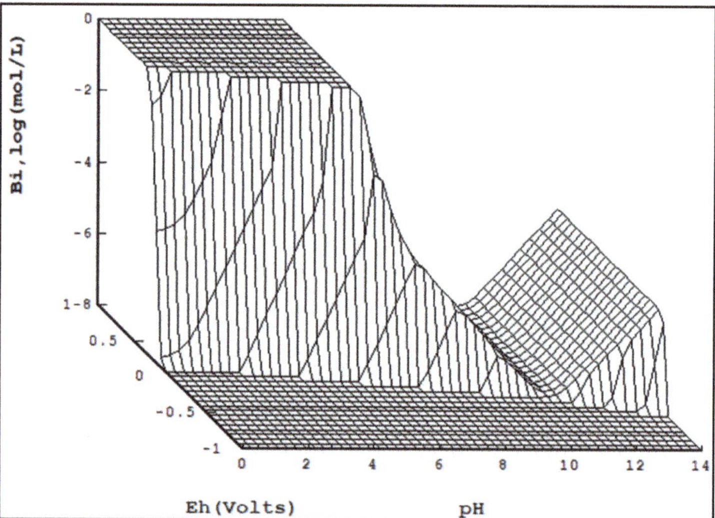

Figure 4. 3D Bi Pourbaix diagram.

This work aims to determine the efficacy of utilizing various hydrometallurgical leaching lixiviants for separating copper from bismuth and arsenic in RTKC FF-ESP dust without a precipitation stage. The lixiviants investigated were sulfuric acid, sulfurous acid, sodium hydroxide, and water. Sulfuric acid is commonly used to recover copper from copper smelter flue dusts and is known to leach arsenic and bismuth as well [19–21]. In this study, it was used to provide a baseline of sorts for recovery. Sulfurous acid was used as it provides similar, but more reductive, conditions compared to sulfuric acid. Li et al. and Guo et al. have shown that alkaline leaching methodologies can be used to recovery arsenic from arsenic-bearing metallurgical dusts [22,23]. Zhang et al. has shown that arsenic can be selectively separated from copper using alkali leaching [21]. Sodium hydroxide was used to confirm the efficacy of alkaline leaching for the separation of arsenic in the unique chemistry of the RTKC dust. Morales et al. has shown that water can provide at least a partial separation between arsenic and copper; because of this, water was used to determine the efficacy of leaching in neutral environments [12].

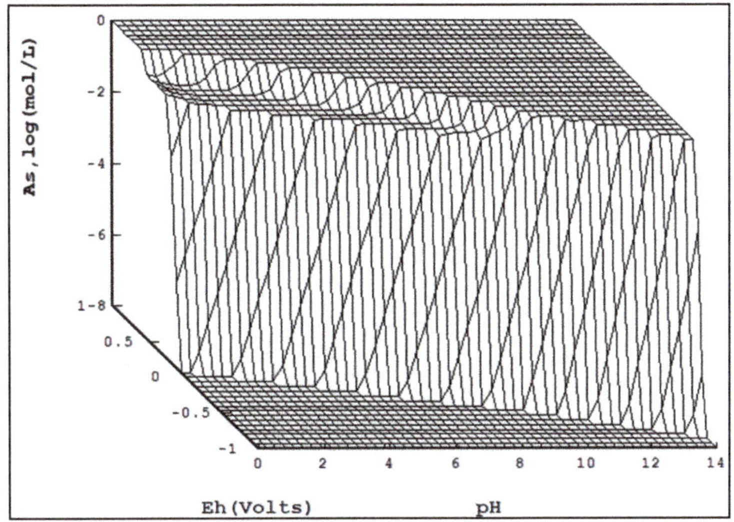

Figure 5. 3D As Pourbaix diagram.

An initial satisfactory separation would occur if copper was transferred to the leachate, with bismuth and arsenic remaining in the residue, or the opposite, with copper in the residue and both bismuth and arsenic in the leachate.

2. Materials and Methods

FF-ESP dust was acquired from RTKC and analyzed for copper, bismuth, and arsenic content using inductively coupled plasma mass spectroscopy (ICP-MS). The ICP-MS results are shown in Table 1. By mass, over 20% of the dust is composed of copper, approximately 5% of the dust is arsenic, and the bismuth concentration is over 7000 ppm. As stated, the concentrations of the arsenic and bismuth are unacceptably high and therefore must be separated from the copper in order for non-problematic electrorefining.

Table 1. Concentration (by mass) of Cu, Bi, and As in FF-ESP dust.

Element	Copper	Bismuth	Arsenic
Concentration	21.99	7443	4.80
Unit	Mass %	ppm	Mass %

In addition to the ICP-MS analysis above, AMICS automated mineralogy was utilized to determine the overall phase composition of the FF-ESP dust. The AMICS work was performed by Eagle Engineering. Table 2 lists the identified phases as well as their fraction of the dust by mass. Figure 6 is an AMICS image, and Table 3 provides a color key for Figure 6. From these, it can be seen that the majority of the dust is composed of mixed amorphous phases containing copper, iron, silicon, and arsenic. Because arsenic and copper are consistently present in the same phases, it is impossible to separate them with physical methods; chemical methods, such as hydrometallurgical leaching, are necessary.

Metals **2021**, *11*, 371

Table 2. AMICS phase composition of FF-ESP dust.

Mineral	Chemistry	Mass %
Barite	$BaSO_4$	0.03
Phase 1	$Cu_{0.23}Fe_{0.24}Si_{0.03}S_{0.07}As_{0.12}O_{0.30}$	22.67
Phase 2	$Cu_{0.19}Fe_{0.12}Si_{0.18}S_{0.01}As_{0.04}O_{0.38}Al_{0.07}$	5.81
$CuFeSiO_4$	$Cu_{0.30}Fe_{0.26}Si_{0.13}O_{0.31}$	3.67
CuFeZnO	$Cu_{0.25}Fe_{0.22}Zn_{0.26}O_{0.27}$	5.65
Phase 4	$Cu_{0.22}Fe_{0.47}S_{0.02}O_{0.29}$	24.42
Phase 5	$Cu_{0.16}Fe_{0.10}Si_{0.14}S_{0.12}As_{0.04}O_{0.44}$	11.40
Phase 6	$Cu_{0.32}Fe_{0.06}Si_{0.20}S_{0.02}As_{0.06}O_{0.32}$	3.93
Phase 7	$Cu_{0.39}Fe_{0.22}Si_{0.02}S_{0.03}As_{0.12}O_{0.24}$	1.51
Phase 8	$Cu_{0.21}Fe_{0.32}Si_{0.04}S_{0.04}As_{0.09}O_{0.30}$	13.04
Phase 9	$Cu_{0.22}Fe_{0.23}Si_{0.07}S_{0.03}As_{0.11}O_{0.31}$	7.85
Quartz	SiO_2	0.01

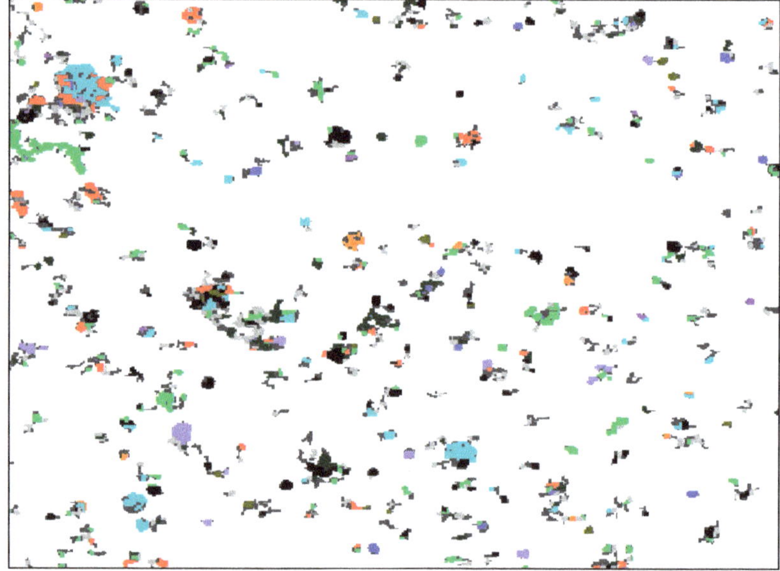

Figure 6. AMICS image of FF-ESP dust.

Table 3. AMICS color key.

Barite		CuFeZnO		Phase 7	
Cu Phase 1		Phase 4		Phase 8	
Cu Phase 2		Phase 5		Phase 9	
$CuFeSiO_4$		Phase 6		Quartz	
-	-	-	-	Lead Oxide	

Table 4 shows the AMICS elemental composition for the FF-ESP dust. When compared to the ICP-MS composition data, there is good agreement between the two for copper and arsenic. The bismuth concentration shown in Table 1 is below the detection limit of AMICS and would not show up in the scan. Given that AMICS composition data are considered "semi-quantitative", only ICP-MS composition data are used for calculations.

Table 4. AMICS elemental composition data for FF-ESP dust.

Element	Mass %
Barium	0.02
Oxygen	35.16
Copper	20.10
Iron	27.66
Silicon	5.40
Sulfur	4.68
Arsenic	5.24
Aluminum	0.41
Zinc	1.32

A 1 L solution containing the requisite concentration of the appropriate lixiviant (ACS grade or higher) was created using deionized water and the lixiviant. This solution was then placed into a baffled flask, heated using a hot plate, and agitated using a magnetic stir bar. The temperature of this solution was measured and controlled via a glass-coated thermocouple. Once the solution reached the appropriate temperature, 50 grams of the FF-ESP dust were added. Upon completion of the leaching time, the leach liquor was separated from the residue using filtration.

The testing conditions for each sulfuric acid, sulfurous acid, caustic, and water design matrix are shown in Table 5, Table 6, Table 7, and Table 8, respectively. The trial conditions for each matrix were generated using Design-Expert 12 software.

Table 5. Test conditions for sulfuric acid trials.

Trial	Time (h)	Temperature (°C)	H_2SO_4 (mL/L)
1	0.5	50	50
2	2	50	25
3	0.5	90	25
4	2	90	50
5	1.25	70	37.5
6	1.25	70	37.5
7	0.5	50	200
8	2	90	100

Table 6. Test conditions for sulfurous acid trials.

Trial	Time (h)	Temperature (°C)	H_2SO_3 (mL/L)
1	0.5	50	50
2	2	50	25
3	0.5	90	25
4	2	90	50
5	1.25	70	37.5
6	1.25	70	37.5
7	0.5	50	150
8	2	90	150

Table 7. Test conditions for caustic trials.

Trial	Time (h)	Temperature (°C)	NaOH (g/L)
1	0.5	50	50
2	2	50	5
3	0.5	90	5
4	2	90	50
5	1.25	70	27.5
6	1.25	70	27.5
7	24	70	30

Table 8. Test conditions for water trials.

Trial	Time (h)	Temperature (°C)
1	0.5	25
2	2	25
3	0.5	90
4	2	90
5	1.25	65
6	1.25	57.5
7	1.5	75

Analysis of the leach liquors was done via atomic absorption spectroscopy (AAS) and ICP-MS. A statistical analysis of the data was completed using Design-Expert 12 software.

3. Results

The recoveries of each element to the leachates, as well as the starting pH, ending pH, and ending Eh values, are shown in Table 9 (the sulfuric acid trials), Table 10 (the sulfurous acid trials), Table 11 (the caustic trials), and Table 12 (the water trials). The starting pH values were calculated using first principles, and random trials were confirmed using titration. The ending pH values were determined via titration. The ending Eh values were determined via Eh probe with a standard hydrogen electrode as reference. The recovery (R) of each element (i) was calculated using the concentration of i in the leachate and FF-ESP dust, the volume (V) of the leachate, and the mass (M) of the dust sample used, as shown in Equation (1).

$$\% \text{ Recovery for element i: } \%R_i = 100 \times ([i_{\text{leachate}}] \times V)/([i_{\text{FF-ESP Dust}}] \times M) \tag{1}$$

Table 9. Elemental recoveries (%) to leachates of sulfuric acid trials, pH data, and Eh data.

Trial	Time (h)	Temperature (°C)	H_2SO_4 (mL/L)	% Recovery			Starting pH	Ending pH	Ending Eh (mV)
				Copper	Bismuth	Arsenic			
1	0.5	50	50	76.4	44.05	61.81	−0.26	0.59	461.3
2	2	50	25	77.21	38.76	51.71	0.04	0.68	454.8
3	0.5	90	25	84.49	42.73	56.82	0.04	0.71	451
4	2	90	50	85.27	48.46	71.28	−0.26	0.51	454.7
5	1.25	70	37.5	83.48	44.36	69.03	−0.14	0.58	459.3
6	1.25	70	37.5	82.29	45.06	65.00	−0.14	0.6	459.2
7	0.5	50	200	77.25	32.23	60.45	−0.87	0.32	461.8
8	2	90	100	89.85	44.23	64.12	−0.57	0.49	460.4

Table 10. Elemental recoveries (%) to leachates of sulfurous acid trials, pH data, and Eh data.

Trial	Time (h)	Temperature (°C)	H_2SO_3 (mL/L)	% Recovery			Starting pH	Ending pH	Ending Eh (mV)
				Copper	Bismuth	Arsenic			
1	0.5	50	50	70.42	1.03	25.84	0.17	1.8	263.1
2	2	50	25	69.31	0.21	25.03	0.47	2.03	326.2
3	0.5	90	25	71.41	4.26	26.25	0.47	2.09	321.1
4	2	90	50	76.84	4.65	30.81	0.17	2.02	331.4
5	1.25	70	37.5	71.12	3.28	29.15	0.29	2.14	318.9
6	1.25	70	37.5	70.64	2.59	27.93	0.29	2.13	298.6
7	0.5	50	150	74.78	1.81	29.63	−0.31	1.92	278.4
8	2	90	150	81.75	3.02	56.26	−0.31	1.9	269.8

Table 11. Elemental recoveries (%) to leachates of caustic trials, pH data, and Eh data.

Trial	Time (h)	Temperature (°C)	NaOH (g/L)	% Recovery			Starting pH	Ending pH	Ending Eh (mV)
				Copper	Bismuth	Arsenic			
1	0.5	50	50	0.25	0.07	27.94	14.1	12.95	−32.1
2	2	50	5	43.13	0.01	0.04	13.1	4.14	231.8
3	0.5	90	5	43.4	0.01	1.76	13.1	3.68	280.1
4	2	90	50	0.46	0.04	63.08	14.1	13.11	−33.8
5	1.25	70	27.5	0.09	0.02	30.13	13.84	12.79	25.2
6	1.25	70	27.5	0.09	0.01	25.63	13.84	12.81	−22.4
7	24	70	30	0.35	0.01	45.21	13.88	12.7	31.2

Table 12. Elemental recoveries (%) to leachates of water trials, pH data, and Eh data.

Trial	Time (h)	Temperature (°C)	% Recovery			Starting pH	Ending pH	Ending Eh (mV)
			Copper	Bismuth	Arsenic			
1	0.5	25	64.14	0.17	7.86	7	1.8	435.6
2	2	25	66.64	0.14	7.2	7	1.89	436.1
3	0.5	90	66.57	0.05	4.86	7	1.67	435.2
4	2	90	66.35	0.03	5.69	7	1.64	435.7
5	1.25	65	65.61	0.04	5.4	7	1.76	433.9
6	1.25	57.5	64.89	0.03	5.67	7	1.79	433.8
7	1.5	75	68.69	0.13	5.33	7	2.06	400.8

4. Discussion

Given that copper is the primary product of RTKC, either a high or a low copper recovery to the leachate is desired. This allows for the majority of the copper to be processed from either the leachate or residue. By comparing the recovery values shown above, it can be seen that sulfuric acid provides the highest copper recovery to the leachate. Unfortunately, sulfuric acid did not provide a separation between copper and either bismuth or arsenic.

Sodium hydroxide provided the lowest copper recovery to the leachate, leaving over 99% of the copper in the residue in all but the low alkalinity trials. Like sulfuric acid, caustic leaching did not provide a separation between copper and bismuth. That said, trial 4 of the caustic leach matrix did provide a moderate separation between copper and arsenic. It provided essentially no copper recovery and a 63% arsenic recovery. That said, higher arsenic recoveries were expected based on existing studies on alkaline leaching of arsenic in metallurgical dusts, which have shown arsenic recoveries of over 90% at the highest as well as many trials with recoveries over 80% [22,23]. The difference in arsenic recovery may be attributed to the exact type of alkaline leaching used, as the referenced studies utilized either pressure NaOH leaching or NaOH–Na_2S leaching, or may simply be due to the difference in the materials leached.

By comparing trial 4 to the other alkaline trials, it becomes apparent that higher alkalinities, higher temperature, and longer leach times provide higher arsenic recovery without increases in recovery of copper or bismuth. This is partially supported by the response surfaces for caustic leaching (below) generated using Design-Expert 12.

It should be noted that caustic trials 2 and 3, which were only slightly alkaline relative to the other caustic trials, were the only caustic trials to leach a non-near-zero amount of copper. It is hypothesized that the sulfur in the FF-ESP dust reacts with water when leached and generates sulfuric acid. The acid then dissolves the copper present in the dust, resulting in the higher copper recoveries. This is supported by the pH data associated with caustic trials 2 and 3, which show that the leach solution ends substantially more acidic than it starts, and the pH data associated with the water leach trials, which show the same pattern. It is believed that the other caustic trials generated sulfuric acid, but that the generated acid was reacted with the sodium hydroxide before it was able to dissolve copper, and thus the lack of a large shift in pH.

Figure 7 shows copper recovery response surfaces from the caustic leaching design matrix at the minimum and maximum sodium hydroxide concentrations tested. Both surfaces indicate that time and temperature have no meaningful effect on copper recovery, but that sodium hydroxide concentration does. The surfaces indicate that at low sodium hydroxide concentrations (approximately 5 g/L) over 30% of the copper can be recovered, and that at higher sodium hydroxide concentrations, copper recovery can be reduced to almost 0%.

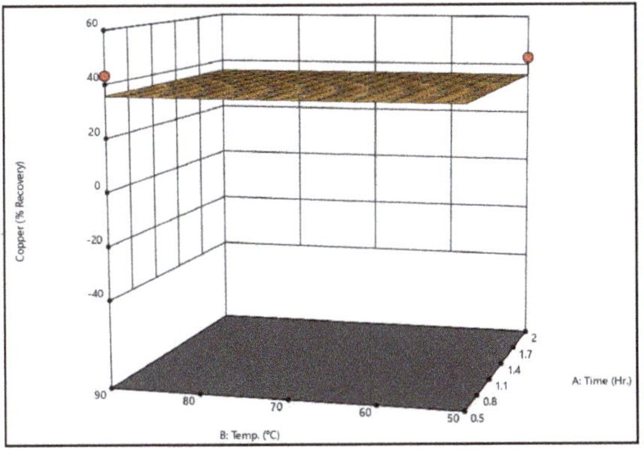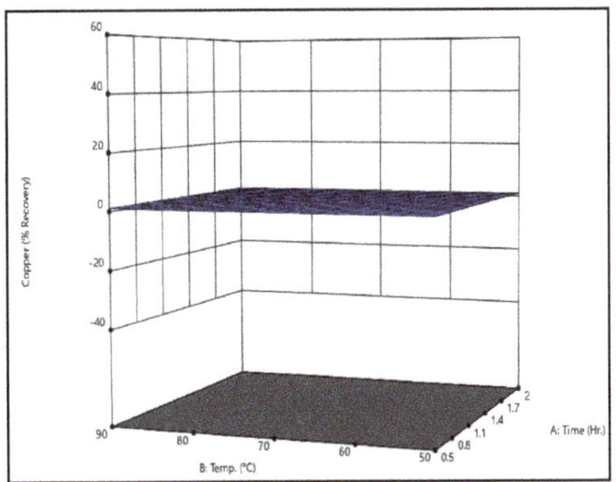

Figure 7. Caustic leaching copper recovery response surfaces: low alkalinity (**left**) and high alkalinity (**right**).

Figure 8 shows bismuth recovery response surfaces from the caustic leaching design matrix at the minimum and maximum temperatures tested. It can be seen that the surfaces are identical, meaning temperature has little to no effect on bismuth recovery. Additionally, it is evident that time has no meaningful effect on bismuth recovery. Only sodium hydroxide concentration has a measurable effect on bismuth recovery; higher sodium hydroxide concentrations are associated with higher bismuth recoveries. That said, it should be noted that increasing sodium hydroxide did not increase bismuth recovery by a meaningful degree within the ranges tested.

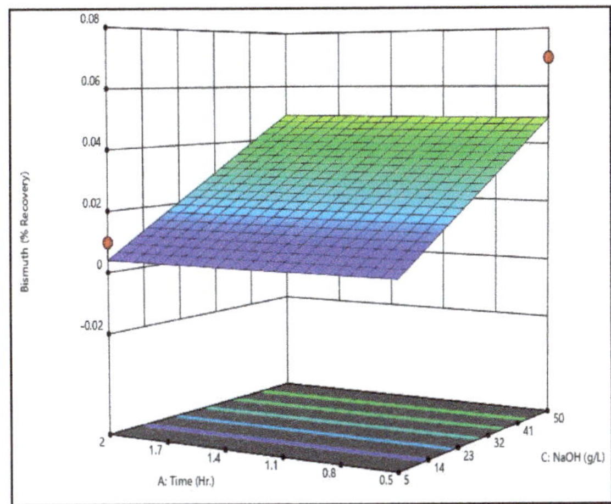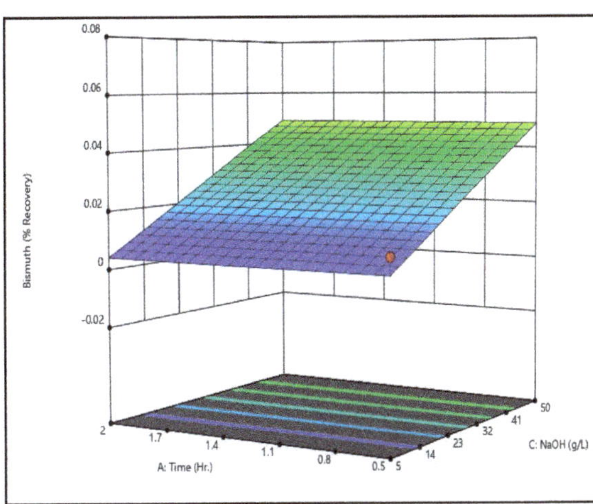

Figure 8. Caustic leaching bismuth recovery response surfaces: low temperature (**left**) and high temperature (**right**).

Figure 9 shows the arsenic recovery response surfaces from the caustic leaching design matrix at the minimum and maximum temperatures tested. Like bismuth recovery, temperature and time had no meaningful effect on arsenic recovery. Only sodium hydroxide concentration had an effect on recovery. Unlike bismuth, however, the effect of sodium hydroxide concentration was substantial; increasing sodium hydroxide concentration from 5 g/L to 50 g/L resulted in an increase of over 40% recovery.

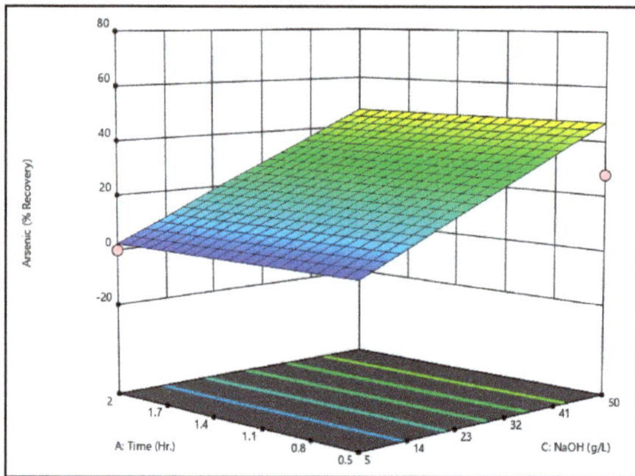 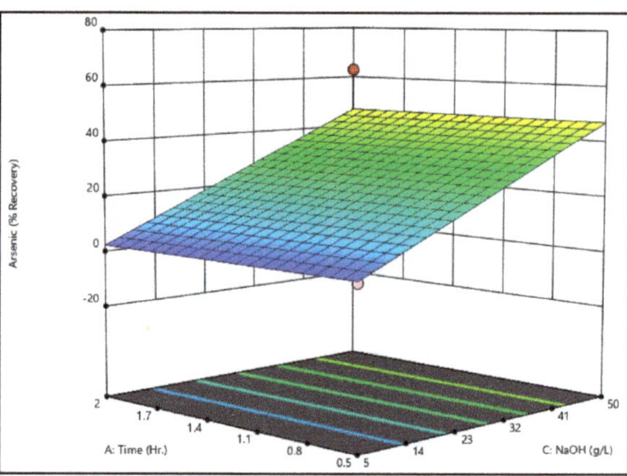

Figure 9. Caustic leaching arsenic recovery response surfaces: low temperature (**left**) and high temperature (**right**).

By comparing the caustic leaching response surfaces to the recovery data shown in Table 11, there is some discrepancy. Some trials exhibited recoveries substantially above or below the generated surfaces (see trial 4's arsenic recovery value, for example). On the surfaces themselves, this can be seen by the dots, which represent actual recovery values at the conditions shown. These discrepancies indicate that there may be some effect of time and temperature on arsenic recovery that is not statistically significant given the dataset. Additional data are necessary to confirm or refute this, however. It should be noted that other studies which have examined alkaline leaching of arsenic in metallurgical dusts have shown that increases in time and/or temperature improve arsenic recovery [22,23].

Both the surfaces and actual recovery values show that increasing sodium hydroxide concentration results in meaningful increases to arsenic recovery without meaningful increases to copper and bismuth recoveries, which remain near zero. Because of this, caustic leaching is a good candidate for separation of arsenic.

As shown in Table 10, sulfurous acid provided the second-highest copper recoveries to the leachate. Additionally, sulfurous acid provided a good separation between copper and bismuth. Sulfurous acid recovered approximately 70–80% of the copper to the leachate while only recovering, at most, 4.65% of the bismuth. A separation between copper and arsenic did occur, but only somewhat, given that between 25% and 30% of the arsenic was recovered for most trials. Lower arsenic recovery is necessary for a satisfactory copper–arsenic separation.

The differences in recovery in bismuth and arsenic between the sulfurous trials and the sulfuric trials may be partially explained by the differences in the ending pH and Eh values for the two lixiviants. The sulfuric trials were more acidic and more oxidizing. In the case of bismuth, the data indicates that the sulfuric trials may have resulted in Bi^{3+} being the dominant phase rather than $Bi(OH)_2^+$, which was the dominant phase in the sulfurous trials based on Figure 1. With arsenic, Figure 2 indicates that H_3AsO_4 or H_3AsO_3 may be the dominant phase for the sulfuric trials, while only H_3AsO_3 may be the dominant phase for the sulfurous trials. That said, neither difference in dominant phase would necessitate a strong separation beyond solubility differences; also, as stated, the Pourbaix diagrams do not utilize the phases present in the RTKC dust.

Figure 10 shows copper recovery response surfaces for the sulfurous acid trials at the maximum and minimum sulfurous acid concentrations tested. The low-acid condition indicates that at low temperatures, time has a negligible effect on copper recovery. The same can be said of temperature for short leaching times. That said, at higher values for either variable, the effect of increasing one of them becomes much more pronounced; increasing either results in increased copper recovery. By comparing the two surfaces, it can be seen that increasing the concentration of sulfurous acid serves to exaggerate this effect.

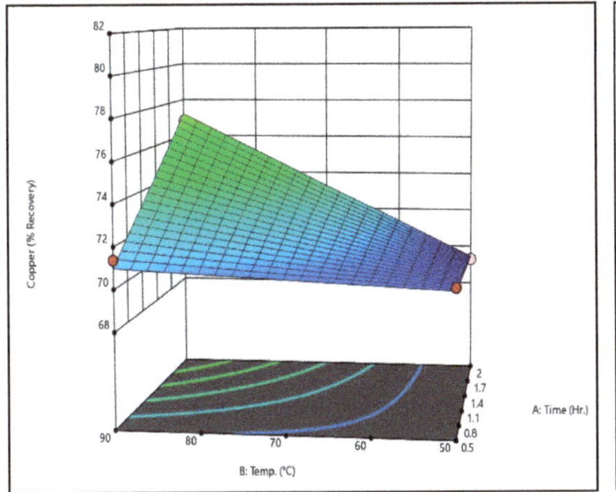 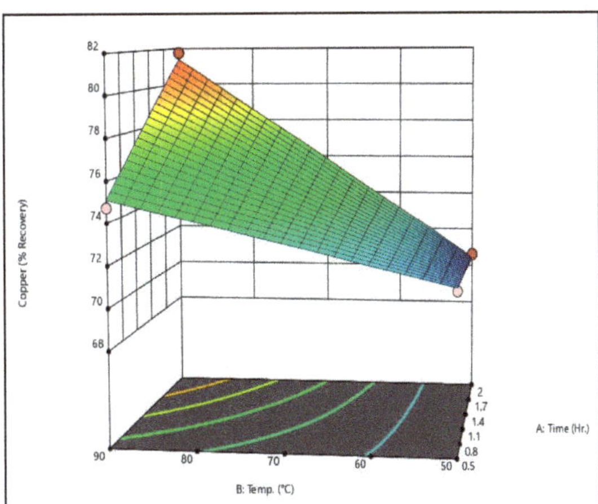

Figure 10. Sulfurous acid copper recovery response surfaces: low acid (**left**) and high acid (**right**).

Figure 11 shows bismuth recovery response surfaces for the sulfurous acid matrix at the minimum and maximum sulfurous acid concentrations tested. The low-acid surface shows that time does not have a substantial impact on recovery, and that higher temperatures are associated with higher bismuth recoveries. When compared with the low-acid surface, the high-acid surface indicates that increased sulfurous acid concentrations serve to reduce the effect of temperature, resulting in a "flatter" surface.

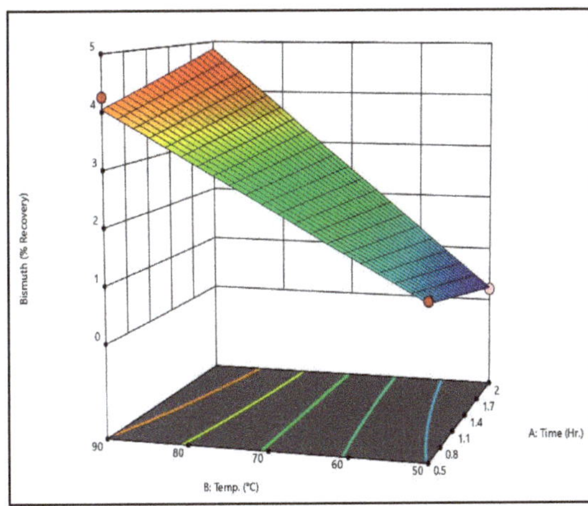 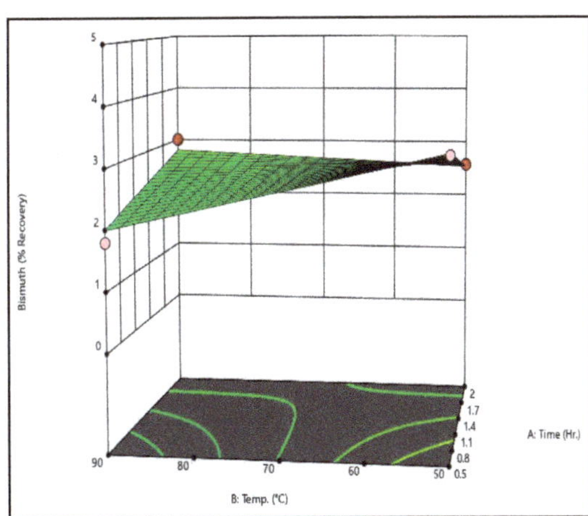

Figure 11. Sulfurous acid bismuth recovery response surfaces: low acid (**left**) and high acid (**right**).

Figure 12 shows arsenic recovery response surfaces for the sulfurous acid matrix at the minimum and maximum sulfurous acid concentrations tested. The low-acid surface shows that time and temperature, individually, have little effect on recovery. That said, increasing both results in increased arsenic recovery. By comparing the two surfaces, it can be seen that increased acid concentration serves to increase this effect so that it is more substantial.

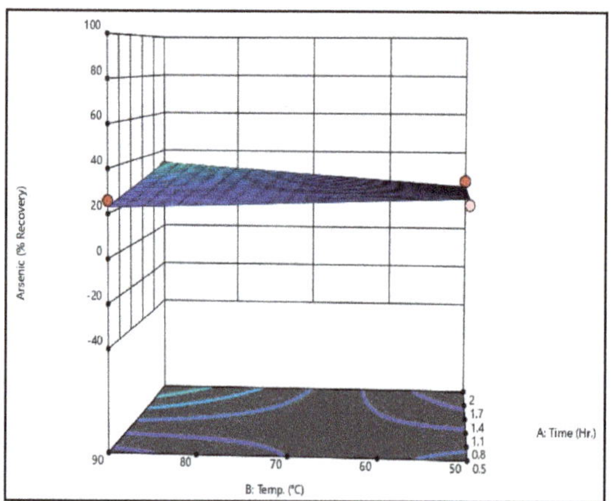
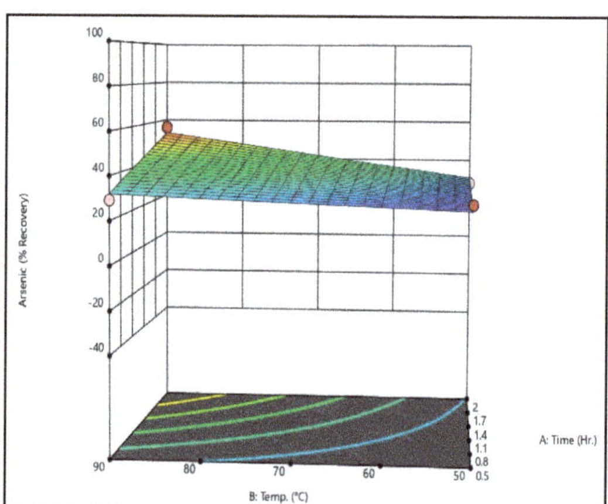

Figure 12. Sulfurous acid arsenic recovery response surfaces: low acid (**left**) and high acid (**right**).

By comparing the sulfurous acid response surfaces and the recovery data in Table 10, it can be said that an improved copper–bismuth separation is possible by increasing the sulfurous acid concentration. Bismuth sees reduced recoveries from increased acid, while copper sees increased recovery. Unfortunately, arsenic also sees improved recoveries from increased sulfuric acid, so sulfurous acid is only a good candidate for separation of bismuth.

In Table 11, it can be seen that water only recovered between 64% and 69% of the copper to the leachate. That said, water did provide an excellent separation between copper and bismuth, recovering essentially none of the bismuth. Additionally, water provided a good separation between copper and arsenic, recovering 7.86% arsenic on the high end. Given that bismuth is known to be more problematic than arsenic, the recovery values for bismuth and arsenic are very promising. Additionally, the recovery data indicate that neither time nor temperature had a substantial impact on the recoveries of copper, bismuth, or arsenic. This is confirmed by the water leach response surfaces below, and (in regard to copper and arsenic) agrees with Morales et al [12].

Figure 13 shows the copper recovery response surface for the water leach matrix. The surface indicates that at high conditions of either time or temperature, increasing the other has little effect on copper recovery. That said, when both are low, copper recovery is reduced, but by less than 5% recovery.

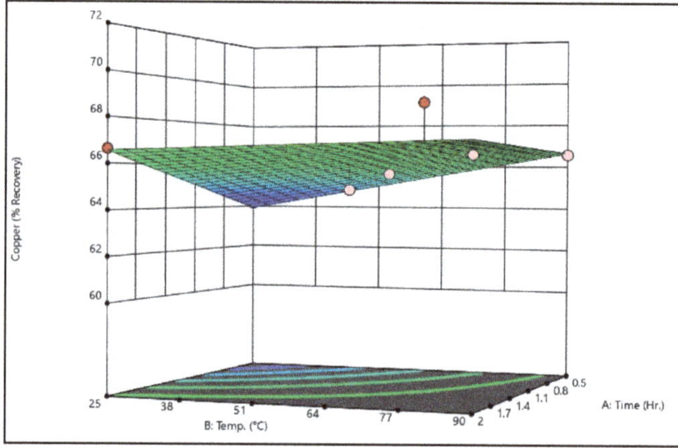

Figure 13. Water leach copper recovery response surface.

Figure 14 shows the bismuth recovery response surface for the water leach matrix. The surface indicates that time has no measurable effect on bismuth recovery and that temperature does. Increases in temperature are associated with reduced bismuth recoveries. It should be noted that the overall range of bismuth recovery is less than 0.2% recovery. Therefore, temperature cannot be said to have a meaningful impact.

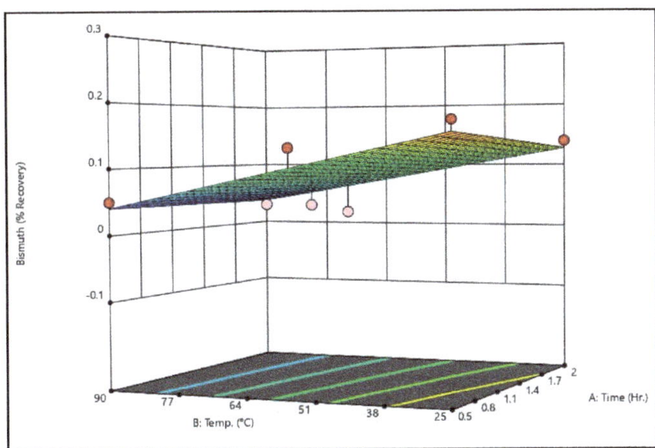

Figure 14. Water leach bismuth recovery response surface.

Figure 15 shows the arsenic recovery response surface for the water leach matrix. Like the bismuth surface, the arsenic surface indicates that time has no measurable effect and that higher temperatures are associated with reduced recoveries. Unlike bismuth, the range of arsenic recoveries is meaningful but still low, at approximately 3%.

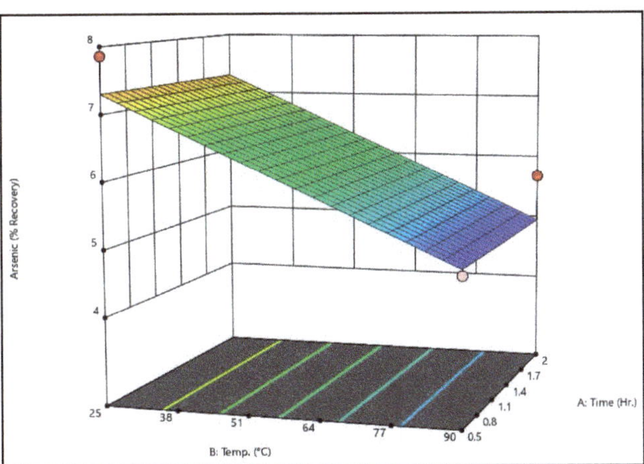

Figure 15. Water leach arsenic recovery response surface.

Generally, the recovery data for the leaching matrices agree with what should occur thermodynamically based on the Pourbaix diagrams. The sulfuric acid tests recovered all three elements, providing no separations between them. This makes sense when the pH and Eh data for sulfuric acid in Table 9 are considered, as the conditions provided by sulfuric acid cause soluble phases of all three elements to form based on the Pourbaix diagrams.

Of the various lixiviants, only water provided a separation between copper and both bismuth and arsenic. When the data in Table 12 and the Pourbaix diagrams are considered, this makes sense, as water provided the thermodynamically correct (that is, fairly neutral and oxidative) conditions to separate copper from both bismuth and arsenic.

As shown in Table 11, the more alkaline trials of the caustic leach matrix were in the region of separation between copper and arsenic based on Figure 2, but not for copper and bismuth based on Figure 1; this agrees with the recovery data for the caustic trials, as only a copper–arsenic separation was achieved.

Interestingly, the conditions shown in Table 10 do not indicate that sulfurous acid should be able to generate a copper–bismuth separation. Based on Figure 1, the stable bismuth phase in the conditions generated by sulfurous acid is $Bi(OH)_2^+$, which is a soluble phase. However, based on the recovery data, little bismuth was actually dissolved by the sulfurous acid. This may be due to the leach time being too short to allow for equilibrium.

Based on the data shown, there are two potential leaching methods for separation of bismuth and arsenic from copper in RTKC FF-ESP dust. One method is to use water, which transfers the majority of the copper to the aqueous phase and leaves behind the vast majority of the bismuth and arsenic in the residue.

Another method is to, first, leach with approximately 50 g/L sodium hydroxide solution. This will leave the almost all of the copper and bismuth in the residue, while transferring potentially over 60% of the arsenic to the leachate. Given that arsenic is not nearly as problematic in electrorefining as bismuth, removing this amount may be satisfactory. Second, the caustic leach residue is leached with sulfurous acid. This should transfer the majority of the copper to the leachate while leaving the vast majority of the bismuth in the residue. A similar two-stage methodology has been proposed by Zhang et al. and shown to be effective [21]. A flowsheet for this process is shown in Figure 16. It should be noted that this was not tested, and therefore does not account for the necessary acid to react with any remaining sodium hydroxide or the possibility of unanticipated insoluble metastable phases during caustic leaching.

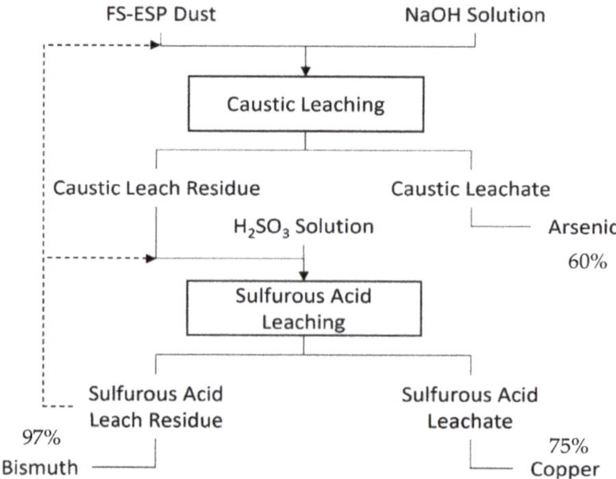

Figure 16. Proposed flowsheet for separation of copper from bismuth and arsenic. Approximate expected recoveries are shown in %.

The benefits of the first method are lower arsenic and bismuth values and a lower operating cost, as virtually no expensive reagents are used. The benefit of the second method is higher copper recovery. Given that copper is the primary product of RTKC, this is an important consideration. It should be noted that whatever copper is not recovered may be leached again some number of times by the same or similar means.

5. Conclusions

Characterization of FF-ESP dust, coupled with thermodynamic analysis and design of experiments for several leaching methods, resulted in the successful separation of copper from bismuth and arsenic in RTKC FF-ESP dust. Water leaching successfully separates both bismuth and arsenic. Caustic leaching successfully separates arsenic via leaching only arsenic under highly alkaline conditions. Sulfurous acid leaching successfully separates bismuth to a higher level than water leaching. It may be possible to get higher separations that what is possible with water leaching by combining caustic leaching and sulfurous acid leaching.

Differences in the expected separations based on the Pourbaix diagrams and the actual separations shown can be likely be attributed to deviations from ideality and interactions between the substances present in the diagrams and the actual materials. Ideally, thermodynamic data for the actual substances present would be generated, as this should allow for a better agreement between the implied separations based on Pourbaix diagrams and experimentation; such thermodynamic data would require either extensive modeling or determination via experimentation.

Continuation of this line of research should include testing of the process shown in Figure 16. Additionally, optimization studies using water and sulfurous acid as lixiviants with wider ranges among the tested variables, and testing for the effect of solid/liquid ratio during leaching, should be conducted with some focus on kinetic modeling. Other methods of alkaline leaching, such as pressurized alkaline leaching, should be pursued to determine if higher arsenic recoveries can be achieved. Moreover, further work regarding the upgrading of copper once in solution should be pursued. Such work should include the separation of copper and iron. This may be done via iron cementation with copper powder, but must be tested.

From an industrial standpoint, once in solution, the copper may be upgraded using a variety of ion-exchange or solvent-extraction methods, and then fed into an electrorefining circuit for production of metallic copper. Bismuth remaining in the leach residue is generally not considered hazardous and should not pose problems regarding disposal, though other elements in the residue, such as arsenic, may. Remaining arsenic could be dealt with via fixation with iron, which is fairly well established in the industry.

Author Contributions: Conceptualization, M.C., J.T., C.A. and S.W.; methodology, M.C and J.T.; investigation, M.C. and J.T.; resources, C.A. and S.W.; writing—original draft preparation, M.C. and J.T.; writing—review and editing, M.C. and C.A.; visualization, M.C.; supervision, C.A. and S.W.; project administration, C.A. and S.W.; funding acquisition, C.A. and S.W. All authors have read and agreed to the published version of the manuscript.

Funding: This research was funded by DOE CMI and Rio Tinto Kennecott.

Acknowledgments: The authors acknowledge Rio Tinto Kennecott Copper, the Colorado School of Mines Kroll Institute for Extractive Metallurgy, and Eagle Engineering. This research was supported by the Critical Materials Institute, an Energy Innovation Hub funded by the U.S. Department of Energy, Office of Energy Efficiency and Renewable Energy, and the Advanced Manufacturing Office.

Conflicts of Interest: The authors declare no conflict of interest.

References

1. Rio Tinto Kennecott. Available online: Riotintokennecott.com (accessed on 16 December 2020).
2. Total Materia. The Copper Flash CC Smelting Process. 2016. Available online: https://www.totalmateria.com/page.aspx?ID=CheckArticle&site=ktn&NM=394 (accessed on 18 November 2020).
3. Yli-Penttila, J.T.; Peuraniemi, E.J.; Jokilaakso, A.; Riihilahti, K.M. Dust formation in flash oxidation of copper matte particles. *Min. Met. Explor.* **1998**, *15*, 41–47.
4. Okanigbe, D.O.; Popoola, A.; Adeleke, A.A. Characterization of copper smelter dust for copper recovery. *Procedia Manuf.* **2017**, *7*, 121–126.
5. Hiskey, J.B.; Chen, T.T. Mechanisms and Thermodynamics of Floating Slimes Formation. In *Proceedings of the Honorary Symposium on Hydrometallurgy, Orlando, FL, USA, 11–15 March 2012*; Elsevier B.V.: Amsterdam, The Netherlands, 2012.

6.   Fountain, C. The Whys and Wherefores of Penalty Elements in Copper Concentrates. In *Proceedings of the MetPlant 2013: Metallurgical Plant Design and Operating Strategies, Perth, Australia15–17 July 2013*; Australasian Institute of Mining and Metallurgy: Carlton, Australia, 2012.

7.   ASTM International. *B115-10 Standard Specification for Electrolytic Copper Cathode*; ASTM International: West Conshohocken, PA, USA, 2016.

8.   Piret, N.L. Optimizing bismuth control during copper production. *JoM* **1994**, *46*, 15–18.

9.   Mehta, A.K. *Investigation of New Techniques for Control of Smelter Arsenic Bearing Wastes*; U.S. EPA: Cincinnati, OH, USA, 1981.

10.   Twidwell, L.G.; Mehta, A.K. Disposal of Arsenic Bearing Copper Smelter Flue dust. *Nucl. Chem. Waste Manag.* **1985**, *5*, 297–303.

11.   Ke, J.-J.; Qiu, R.-Y.; Chen, C.-Y. Recovery of metal values from copper smelter flue dust. *Hydrometallurgy* **1984**, *12*, 217–224.

12.   Morales, A.; Cruells, M.; Roca, A.; Bergo, R. Treatment of copper flash smelter flue dusts for copper and zinc extraction and arsenic stabilization. *Hydrometallurgy* **2010**, *105*, 148–154.

13.   Ha, T.K.; Kwon, B.H.; Park, K.S.; Mohapatra, D. Selective leaching and recovery of bismuth as Bi2O3 from copper smelter converter dust. *Sep. Purify. Technol.* **2015**, *142*, 116–122.

14.   Gupta, C.K. *Chemical Metallurgy Principles and Practice*; Wiley: Hoboken, NJ, USA, 2003.

15.   Rosenqvist, T. *Principles of Extractive Metallurgy*; Tapir Academic Press: Trondheim, Norway, 2004.

16.   Ichimura, R.; Tateiwa, H.; Almendares, C.; Sanchez, G. Arsenic Immobilization and Metal Recovery from El Teniente Smelter Dust. In Proceedings of the John E. Dutrizac International Symposium on Copper Hydrometallurgy, Toronto, ON, Canada, 25–30 August 2007.

17.   Kovyazin, A.A.; Kochin, V.A.; Timofeev, K.L.; Krayuhin, S.A. Comprehensive Processing of Fine Metallurgical Dust. *KnE Mater. Sci.* **2020**, 249–252. [CrossRef]

18.   Yang, T.; Fu, X.; Liu, W.; Chen, L. Hydrometallurgical Treatment of Copper Smelting Dust by Oxidation Leaching and Fractional Precipitation Technology. *JoM* **2017**, *69*, 1982–1986.

19.   Chen, Y.; Liao, T.; Li, G.; Chen, B.; Shi, X. Recovery of bismuth and arsenic from copper smelter flue dusts after copper and zinc extraction. *Miner. Eng.* **2012**, *39*, 23–28.

20.   Liu, W.-F.; Fu, X.-X.; Yang, T.-Z.; Zhang, D.-C.; Chen, L. Oxidation leaching of copper smelting dust by controlling potenital. *Trans. Nonferrous Met. Soc. China* **2018**, *28*, 1584–1861.

21.   Zhang, Y.; Jin, B.; Huang, Y.; Song, Q.; Wang, C. Two-stage leaching of zinc and copper from arsenic-rich copper smelting hazardous dusts after alkali leaching of arsenic. *Sep. Purify. Technol.* **2019**, *220*, 250–258.

22.   Li, Y.; Liu, Z.; Li, Q.; Liu, F.; Liu, Z. "Alkaline oxidative pressure leaching of arsenic and antimony bearing dusts. *Hydrometallurgy* **2016**, *166*, 41–47.

23.   Guo, X.-Y.; Yi, Y.; Shi, J.; Tian, Q.-H. Leaching behavior of metals from high-arsenic dust by NaOH-Na2S alkaline leaching. *Trans. Nonferrous Met. Soc. China* **2016**, *26*, 575–580.

Article

Study on the Extraction and Separation of Zinc, Cobalt, and Nickel Using Ionquest 801, Cyanex 272, and Their Mixtures

Wensen Liu [1,2,3], **Jian Zhang** [3,4,5], **Zhenya Xu** [6], **Jie Liang** [1,*] **and Zhaowu Zhu** [2,3,4,*]

1. School of Chemical and Environmental Engineering, China University of Mining and Technology (Beijing), Beijing 100083, China; liuwensen@ipe.ac.cn
2. Innovation Academy for Green Manufacture, Chinese Academy of Sciences, Beijing 100190, China
3. National Engineering Laboratory for Hydrometallurgical Cleaner Production Technology, Beijing 100190, China; zhangjian01@ipe.ac.cn
4. Key Laboratory of Green Process and Engineering, Institute of Process Engineering, Chinese Academy of Sciences, Beijing 100190, China
5. School of Chemistry and Chemical Engineering, University of Chinese Academy of Sciences, Beijing 100049, China
6. The College of Chemical Engineering, Beijing University of Chemical Technology, Beijing 100028, China; 2018200201@mail.buct.edu.cn
* Correspondence: liangjie@cumtb.edu.cn (J.L.); zhwzhu@ipe.ac.cn (Z.Z.)

Abstract: Both Cyanex 272 (bis (2,4,4-trimethylpentyl) phosphinic acid) and Ionquest 801 (2-ethylhexyl phosphonic acid mono-2-ethylhexyl ester) are commonly used for metal extraction and separation, particularly for zinc, cobalt, and nickel, which are often found together in processing solutions. Detailed metal extractions of zinc, cobalt, and nickel were studied in this paper using Cya-nex 272, Ionquest 801, and their mixtures. It was found that they performed very similarly in zinc selectivity over cobalt. Cyanex 272 performed much better than Ionquest 801 in cobalt separation from nickel. However, very good separation of them was also obtained with Ionquest 801 at its low concentration with separation factors over 4000, indicating high metal loading of cobalt can significantly suppress nickel extraction. Slop analysis proved that two moles of dimeric extractants were needed for one mole extraction of zinc and cobalt, but three moles were needed for the extraction of one mole nickel. A synergistic effect was found between Cyanex 272 and Ionquest 801 for three metal extractions with the synergistic species of *M(AB)* determined by the Job's method.

Keywords: solvent extraction; cyanex 272; ionquest 801; zinc; cobalt; nickel

Citation: Liu, W.; Zhang, J.; Xu, Z.; Liang, J.; Zhu, Z. Study on the Extraction and Separation of Zinc, Cobalt, and Nickel Using Ionquest 801, Cyanex 272, and Their Mixtures. *Metals* **2021**, *11*, 401. https://doi.org/10.3390/met11030401

Academic Editor: Dariush Azizi

Received: 9 February 2021
Accepted: 24 February 2021
Published: 1 March 2021

Publisher's Note: MDPI stays neutral with regard to jurisdictional claims in published maps and institutional affiliations.

1. Introduction

In nickel laterite processing, zinc, cobalt, and nickel often present together in leach solutions due to their similar chemical properties [1–3]. Ionqest 801 and Cyanex 272 are commonly used for their separations [3–6]. Cyanex 272 has been successfully used in a Murrin Murrin Nickel laterite project, where two solvent extraction circuits are used to separate zinc from cobalt and nickel in the first circuit, and, then, cobalt from nickel in the second circuit [7,8]. Although Cyanex 272 performed very well in their separation, its high manufacturing cost and, accordingly, high price drive some practices to turn to other alternatives, such as Ionquest 801 [9], and this is particularly true in China [10–13].

Ionquest 801 has stronger metal extraction capacity than Cyanex 272, but generally has less selectivity for cobalt over nickel [14,15]. The separation factor of cobalt over nickel normally is over 2000 with Cyanex 272 compared to around 150 with Ionquest 801. However, if cobalt loading is high with Ionquest 801, good separation can also be obtained. For example, the cobalt loading increased from 1.55 g/L to 6.92 g/L with Ionquest 801. The separation factor of cobalt over nickel rapidly increased from 106.5 to 858.0 [16], indicating that the metal separation can be significantly affected by the extraction conditions. Ionquest 801 has been used to simultaneously extract cobalt and magnesium from a concentrated nickel sulphated solution [17], even though Cyanex 272 could perform better than Ionquest 801 in cobalt and magnesium separation from nickel [18]. Detailed extraction properties of zinc, cobalt, and nickel are still highly required using Cyanex 272 and Ionquest 801 to serve a real process application.

Cyanex 272 and Ionquest 801 both are organophosphorus acidic extractants with very similar structures. They have a strong synergistic effect with chelating extractants for the extraction of zinc, cobalt, and nickel [19,20]. Another organophosphorus acidic extractant D2EHPA (bis(2-ethylhexyl) phosphoric acid) also shows a strong synergistic effect for nickel extraction with N-bearing chelating reagents [21,22]. The synergistic effect of the mixture of Cyanex 272 and Ionquest 801 for cobalt and manganese has been studied for cobalt and manganese and a maximum synergistic effect of around 3–4 was obtained by Zhao et al. [23]. Using the mixture of Ionquest 801 (P 507) and Cyanex 272 was also used to recover cobalt and nickel from a leach solution by Liu et al. [24], and it was found that their optimised synergistic effect at P 507 to Cyanex 272 ratio of 3:2. However, contrary results were reported for rare earth extraction. For instance, Liu et al. [25] revealed that the mixtures of P 507 and Cyanex 272 have a synergistic effect in heavy rare earth extraction with the extraction species of $RE(HB_2)(HA_2)_2$, while Quinn et al. [26] reported that the mixture of Ionquest 801 and Cyanex 272 has an antagonistic effect in their extractions. This is likely due to the fact that the interaction between them strongly depends on extraction conditions. Therefore, a detailed study is required to verify their interaction mechanism for the metal extractions.

The extraction of zinc, cobalt, and nickel with Cyanex 272 and Ionquest 801 has been widely investigated in these years [27–33]. Most research studies focused on the metal separation properties in an attempt to find potential applications. Few research studies focused on their extraction mechanisms with some discrepancies that might be due to the different testing conditions. For example, the extraction of cobalt and nickel with similar types of extractants Cyanex 272, Ionquest 801, and D2EHPA was reported via a complex species combined directly with two molecules of extractant, but four extractant molecules were involved to explain the relationship of LogD against pH [30]. In contrast, Tait [29] reported that the $LogD$(Co) and $LogD$(Ni) against the Log [Cyanex 272] have a linear relationship with the slopes of 2.0 and 3.1, respectively, suggesting that two molecules of Cyanex 272 participated in each metal extraction for cobalt, but three for nickel. The mechanism of metal extraction by Cyanex 272 and Ionquest 801 needs further study.

Herein, the extraction and separation of zinc, cobalt, and nickel with Cyanex 272, Ionquest 801, and their mixtures were studied in detail. Thermodynamic equilibrium calculations and slop analysis were used to study the metal extraction reactions. The synergistic or antagonistic effect of Cyanex 272 and Ionquest 801 on extraction of these three metal extractions was also discussed.

2. Materials and Methods

2.1. Reagents and Solution Preparation

Cyanex 272 was kindly provided by Cytec Industries (Paterson, NJ, USA) and used as received without further purification. Ionquest 801 was obtained from ChemRex. (Limassol, Cyprus) with >98% purity, and also used as received. ShellSol D70, which is an aliphatic hydrocarbon, supplied by Shell Chemicals (Brisbane, Queensland, Australia), was used as the diluent. The organic solutions were prepared by dissolving extraction reagents into the diluent to desired concentrations. An aqueous feed solution containing 1.0 g/L each of zinc, cobalt, and nickel was prepared by dissolving their corresponding metal sulphates into de-ionized water.

2.2. Metal Extraction pH Isotherms

The determination of metal extraction pH isotherms was carried out in 300 mL of hexagonal glass jars immersed in a water bath to control temperature at 40 °C unless it is indicated. Aqueous and organic solutions each of 100 mL were added into the jar to obtain the A/O ratio of 1:1. Two phase solutions were mixed by an impeller with Φ40 mm six-bottom-bladed disc stirrer equipped to an overhead motor. After the temperature increased to 40 °C, a 200 g/L NaOH solution was used to adjust pH to the desired values. The pH was monitored using a ROSS Sure Flow pH probe (model 8127BN, Thermo Fisher Scientific, Waltham, MA, USA) connected to a Hanna portable pH meter (model HI9125, Hanna Instruments, Woonsocket, RI, USA). The mixed solutions about 20 mL were taken using a syringe with a plastic extension at each desired pH point after pH maintains constant for 2 min, and then two phases were separated using Whatman 1PS filter paper, which only allowed organic solution to pass through. Aqueous solutions were then filtered again using membrane syringe filters to completely remove entrained organic and analysed by inductively coupled plasma-optical emission spectroscopy (ICP-OES, Optima 5300V, Perkin-Elmer, Waltham, MA, USA). Organic solutions were stripped with 100 g/L H_2SO_4 at an A/O ratio of 1:1 and 40 °C. The loaded strip liquors were then filtered and then analysed by ICP-OES. Mass balance was calculated based on the metal in the feed solution and distributed in the two phases. The analysis results with a mass balance in the range of 95 to 105% were adopted.

3. Results and Discussion

3.1. Metal Extraction pH Isotherms of Cyanex 272 and Ionquest 801

The metal extraction pH isotherms of Cyanex 272 and Ionquest 801 with the aqueous feed solution were determined as shown in Figure 1a,b. By comparison, Ionquest 801 performed stronger for the extraction of all three metals than Cyanex 272. The pH_{50} (pH against half metal extraction) of 0.3 M Ionquest 801 were 1.70, 3.88, and 5.36 for zinc, cobalt, and nickel, respectively (Figure 1b), while with 0.3 M Cyanex 272, they were 2.09, 4.16, and 6.20 for zinc, cobalt, and nickel, respectively (Figure 1a). This has been well concluded and documented elsewhere [14]. The gaps of ΔpH_{50}(Co/Ni) with Cyanex 272 was clearly larger than that with Ionquest 801. However, the gaps ΔpH_{50}(Zn/Co) with Ionquest 801 was similar to or slightly large than with Cyanex 272. These indicate that Cyanex 272 is advantageous in the selectivity of cobalt over nickel compared to Ionquest 801, but is slightly inferior to Ionquest 801 in zinc selectivity over cobalt. However, some processes used it for zinc separation from cobalt [7].

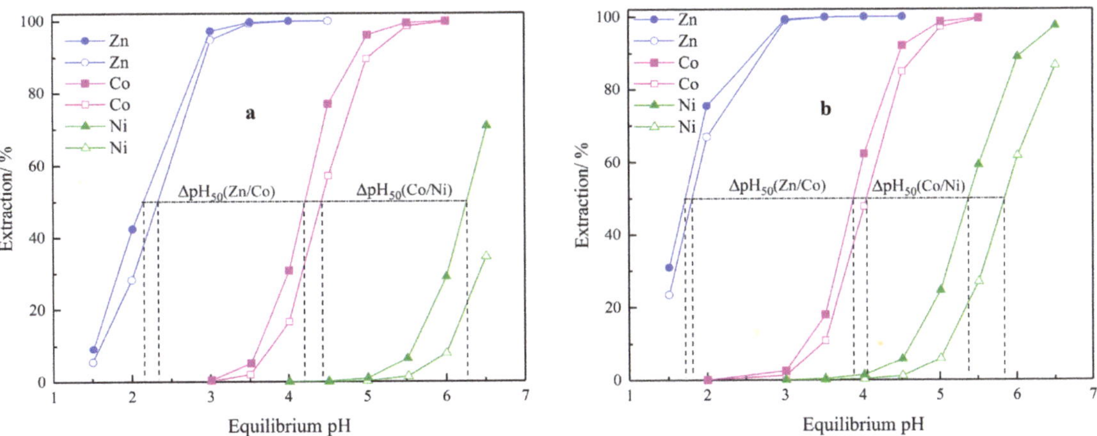

Figure 1. Metal extraction pH isotherm of Cyanex 272 (**a**) and Ionquest 801 (**b**) at an A/O ratio of 1:1 and 40 °C (Organic concentrations: solid-label curve, 0.3 M, and open-label curve, 0.2 M).

Detailed metal extraction pH_{50} and ΔpH_{50}(Co-Zn), ΔpH_{50}(Ni-Co) of Cyanex 272 and Ionquest 801 under various concentrations are obtained in Table 1. All of these results showed that Cyanex 272 has higher pH_{50} and larger ΔpH_{50}(Ni-Co) compared with

Ionquest 801 at the same concentration, suggesting weaker metal extraction and better cobalt selectivity over nickel. The ΔpH_{50}(Co-Zn) of both Cyanex 272 and Ionquest 801 are similar with the latter being slightly larger at the same concentration, indicating slightly better zinc selectivity over cobalt with Ionquest 801 than that with Cyanex 272.

Table 1. Metal extraction pH_{50} and ΔpH_{50} with various concentrations of Cyanex 272 and Ionquest 801.

Cyanex 272 Concentration (*M*)	pH_{50}			ΔpH_{50}	
	Zn	Co	Ni	Co-Zn	Ni-Co
0.1	2.64	5.10	>6.5	2.46	-
0.2	2.32	4.42	>6.5	2.10	-
0.3	2.15	4.21	6.28	2.06	2.07
0.4	2.05	4.15	6.06	2.10	1.91
Ionquest 801 Concentration (*M*)	pH_{50}			ΔpH_{50}	
	Zn	Co	Ni	Co-Zn	Ni-Co
0.1	2.30	4.78	>6.5	2.48	-
0.2	1.80	4.05	5.85	2.25	1.80
0.3	1.71	3.89	5.40	2.18	1.51
0.4	1.55	3.70		2.15	1.35

Cobalt extraction and its separation factors over nickel with various concentrations of Cyanex 272 and Ionquest 801 at different pH values are calculated in Table 2. Organic percentages by metal loading (% of organic concentration occupied by loaded metals) are also calculated in Table 2. For both organic systems, cobalt extraction grew with increasing pH and the extractant concentrations. The $SF_{Co/Ni}$ (separation factors of cobalt and nickel) is high over 1000 under various pH with Cyanex 272 for its all-tested concentrations. The $SF_{Co/Ni}$ also reached high over 4000 at tested pH values with 0.1 M Ionquest 801, which are much comparable with 0.1 M Cyanex 272. This indicates that, when using low Ionquest 801 concentration, very good cobalt separation from nickel can also be obtained. The $SF_{Co/Ni}$ over decreased rapidly to less than 100 with increasing Ionquest 801 concentration to 0.4 M, suggesting that cobalt selectivity of Ionquest 801 is more sensitive to the concentration than that of Cyanex 272. In Table 2, with the increase of the extractant concentration, the metal loaded organic percentage was clearly decreased, leaving more organic free from metal loading. These organic materials tend to extract nickel more with Ionquest 801 than with Cyanex 272.

Table 2. Cobalt extraction, its separation factor over nickel ($SF_{Co/Ni}$), and organic loading percentages with Cyanex 272 and Ionquest 801.

Cyanex 272 Concentration (*M*)	Co Extraction (%)			$SF_{Co/Ni}$			Metal Loaded Organic (%)		
	pH 5.0	pH 5.5	pH 6.0	pH 5.0	pH 5.5	pH 6.0	pH 5.0	pH 5.5	pH 6.0
0.1	37.6	69.4	88.6	2973	6640	7163	42.9	53.3	59.3
0.2	89.5	98.6	99.7	4341	4725	4149	29.9	31.9	33.0
0.3	96.0	99.4	99.9	2514	2479	1888	20.4	21.3	23.8
0.4	97.5	99.6	99.9	2314	1630	1126	17.1	18.0	21.0
Ionquest 801 Concentration (*M*)	Co-Extraction (%)			$SF_{Co/Ni}$			Metal Loaded Organic (%)		
	pH 5.0	pH 5.5	pH 6.0	pH 5.0	pH 5.5	pH 6.0	pH 5.0	pH 5.5	pH 6.0
0.1	64.1	85.3	96.4	4938	7313	6103	51.9	59.0	62.8
0.2	97.2	99.4	99.9	549	459	279	32.5	36.2	41.5
0.3	98.5	99.6	>99.9	207	179	98	23.6	27.3	30.5
0.4	99.0	99.8	>99.9	104	87	48	21.0	24.5	25.1

3.2. Metal Extraction Analysis

Thermodynamic equilibriums for these metal extractions by Cyanex 272 and Ionquest 801 were analysed in this study to further clearly understand the metal extraction reactions. The extraction of all three metals with Cyanex 272 and Ionquest 801 is expressed in Equation (1) considering that both extractants present as a dimer.

$$M^{2+} + n\overline{H_2A_2} \rightleftharpoons \overline{M(H_{2n-2}A_{2n})} + 2H^+ \tag{1}$$

$$K_{eq} = \frac{\left[\overline{M(H_{2n-2}A_{2n})}\right]\left[H^+\right]^2}{[M^{2+}]\left[\overline{H_2A_2}\right]_f^n} \tag{2}$$

$$\mathrm{Log}D(M) = \mathrm{Log}K_{eq} + n\mathrm{Log}[\overline{H_2A_2}]_f + 2\mathrm{pH} \tag{3}$$

where M represents the metals of Zn, Co, or Ni. HA represents the extractant of Cyanex 272 or Ionquest 801, hence H_2A_2 is their dimer. The top bar denotes the organic phase. The subscript "f" denotes free extractant concentration (organic free from the metal loading).

Linear relationships of $\mathrm{Log}D(M)$ against $\mathrm{Log}[H_2A_n]_f$ and pH of Cyanex 272 and Ionquest 801 for all three metal extractions were fitted as shown in Figures 2 and 3, respectively, and corresponding n values in Equation (3) are listed in Table 3. For the extraction of zinc and cobalt, n values were close to 2 with both Cyanex 272 and Ionquest 801. However, for nickel extraction, n values were close to 3. These results are similar to those reported by Tait [29] with Cyanex 272. It is suggested that one molecule metal extraction requires two-dimer extractant molecules for zinc and cobalt extraction, but three for nickel extraction with both Cyanex 272 and Ionquest 801. Since nickel extraction requires more extractant molecules for coordination, which is not required for charge equilibrium, interpreting why its extraction occurred at a relatively higher pH compared to the extraction of zinc and cobalt.

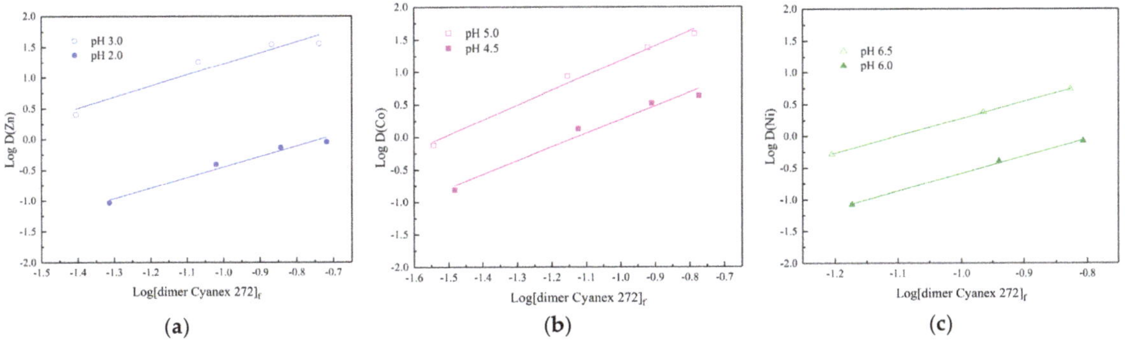

Figure 2. The relationship of $\mathrm{Log}D(M)$ against Log [dimer Cyanex 272]$_f$. (**a**) $\mathrm{Log}D(\mathrm{Zn})$-Log[dimer Cyanex 272], (**b**) $\mathrm{Log}D(\mathrm{Zn})$-Log[dimer Cyanex 272], (**c**) $\mathrm{Log}D(\mathrm{Zn})$-Log[dimer Cyanex 272].

From Equation (3), $\mathrm{Log}D(M)$-$n\mathrm{Log}\,[H_2A_2]_f$ versus pH will give the straight line with a slope of 2. The linear relationships of $\mathrm{Log}D(M)$-$n\mathrm{Log}\,[H_2A_2]_f$ against pH were obtained and shown in Figures 4 and 5. Then, the value used for the calculation is 2 for zinc and cobalt, and 3 for nickel. The line slopes are listed in Table 4. Although some slope values are approaching the integer of 2 for zinc and cobalt extraction with Cyanex 272, many slope values are deviated from 2, ranging from 1.5 to 1.7. This is possibly caused by ionic activity in both aqueous and organic phases.

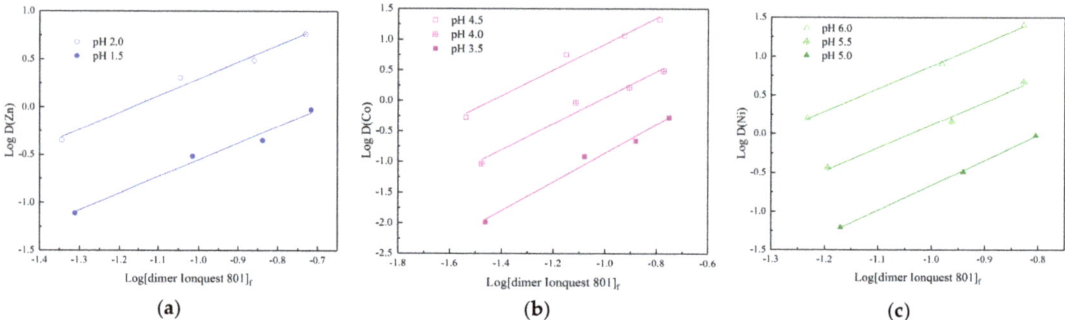

Figure 3. The relationship of Log$D(M)$ against Log [dimer Ionquest 801]$_f$. (**a**) Log$D$(Zn)-Log[dimer Ionquest 801], (**b**) Log$D$(Zn)-Log[dimer Ionquest 801], (**c**) Log$D$(Zn)-Log[dimer Ionquest 801].

Table 3. Slopes of straight lines in Figures 2 and 3, and the corresponding n values in Equation (3).

Metal	pH	n Value	
		Cyanex 272	Ionquest 801
	3.0	1.79	-
Zn	2.0	1.71	1.76
	1.5		1.75
	5.0	2.29	
Co	4.5	2.10	2.12
	4.0	-	2.14
	3.5	-	2.35
	6.5	2.71	-
Ni	6.0	2.74	2.94
	5.5	-	2.96
	5.0	-	3.21

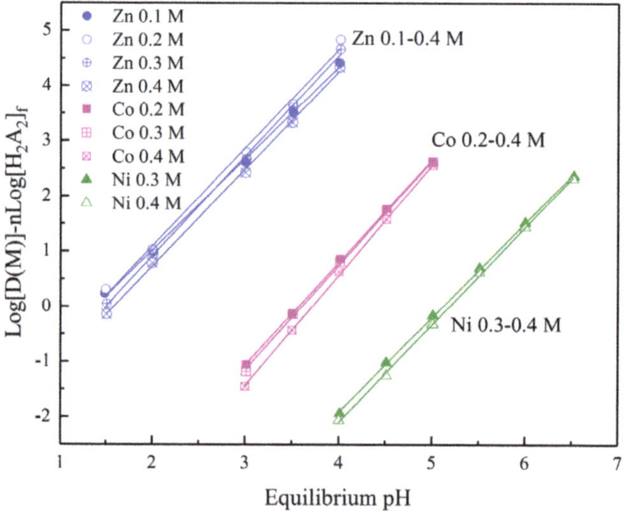

Figure 4. The linear relationship of Log$D(M)$-nLog [H_2A_2]$_f$ and equilibrium pH for the metal extraction with Cyanex 272.

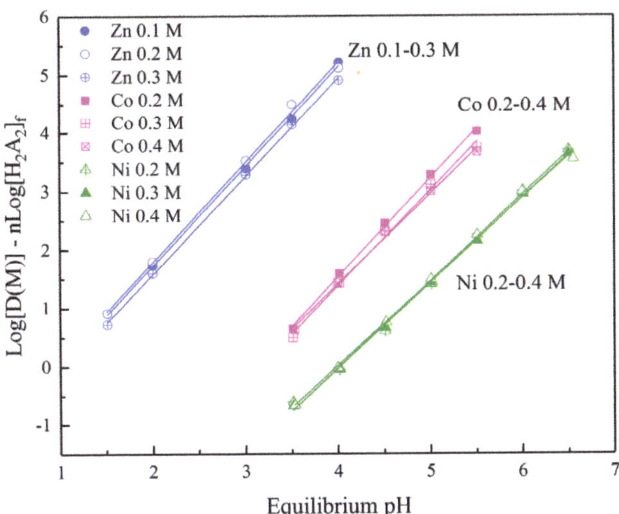

Figure 5. The linear relationship of Log$D(M)$-nLog $[H_2A_2]_f$ and equilibrium pH for the metal extraction with Ionquest 801.

Table 4. Line slope of the linear relationship of $D(M)$-$2n$Log$[HA]_f$ against pH in Figures 4 and 5.

Metal	Cyanex 272 Concentration (M)				Ionquest 801 Concentration (M)			
	0.1	0.2	0.3	0.4	0.1	0.2	0.3	0.4
Zn	1.70	1.79	1.82	1.77	1.70	1.70	1.67	-
Co	-	1.78	1.86	1.99	-	1.69	1.63	1.58
Ni	-	1.68	1.69	1.77	-	1.54	1.49	1.51

3.3. Synergistic Effect of Cyanex 272 and Ionquest 801

Cyanex 272 has a significant advantage in cobalt selectivity over nickel compared to Ionquest 801, but, for zinc separation from cobalt and nickel, Ionquest 801 is more preferred due to its stronger extraction capacity, slightly better selectivity, and, most importantly, its much lower price. If we take both advantages by using each in a separate process for zinc and cobalt extraction, respectively, two extractants might be blended to some extent via phase carryover. A number of investigations have been carried out on the metal extraction using the mixture of Cyanex 272 with Ionquest 801 or D2EHPA, which is another analogue of acidic organophosphorus acid [31,34,35]. No clear synergistic effect on these metal extractions was found. The synergistic effect of Cyanex 272 and Ionquest 801 on the extraction of zinc, cobalt, and nickel was again studied systematically in this study based on the metal extraction pH isotherms. Metal extraction pH isotherms of different compositions of Cyanex 272 and Ionquest 801 were determined (Figure 6). From Figure 6, it is seen that, when the total concentration was maintained constant, metal extractions were basically increased by increasing Ionquest 801 concentration and decreasing Cyanex 272 concentration. This is due to the stronger metal extraction capability of Ionquest 801 than Cyanex 272. However, metal extraction pH isotherms at 0.2 M Cyanex 272 + 0.2 M Ionquest 801 was very close to those at 0.1 M Cyanex 272 + 0.3 M Ionquest 801, which is even stronger by shifting to the right side for the extraction of zinc and cobalt (Figure 6). This should be attributed to their synergistic effect.

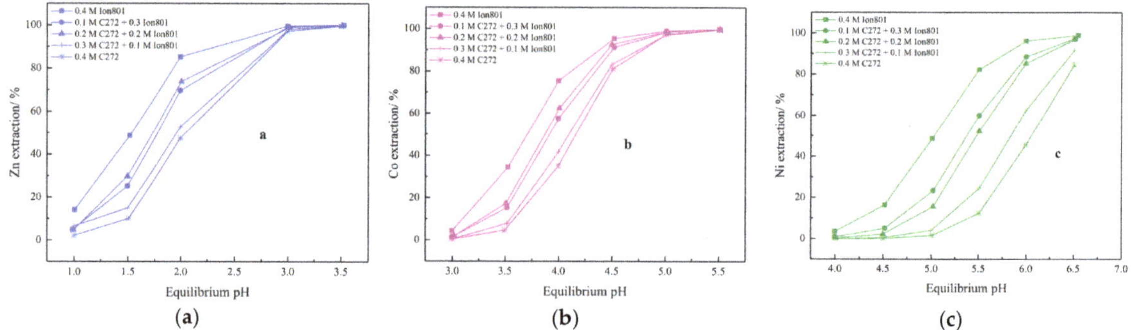

Figure 6. Metal extraction pH isotherms of mixed organic solutions of Cyanex 272 and Ionquest 801 at the total concentration of 0.4 M (in legends: number followed by concentration unit M, C272 is for Cyanex 272, and Ion801 is for Ionquest 801). (**a**) Zn extraction-pH, (**b**) Co extraction-pH, (**c**) Ni extraction-pH.

Half extraction pH_{50} and ΔpH_{50} (Co-Zn and Ni-Co) are shown in Table 5. Clearly, the mixtures all have better zinc and cobalt separation than the Cyanex 272 alone system, but are very similar to the Ionquest 801 alone system. However, the mixed systems are poorer than the Cyanex 272 alone system for cobalt selectivity over nickel. As more Ionquest 801 was used in the system, the performance became poorer. The separation factor of cobalt from nickel ($SF_{Co/Ni}$) dropped rapidly with increasing Ionquest 801 concentration (Table 6), suggesting that lower Ionquest 801 concentration should be used in the mixed system to achieve good cobalt and nickel separation. $SF_{Co/Ni}$ was 300–400 when equal moles of Cyanex 272 and Ionquest 801 were mixed in the extraction system.

Table 5. The half extraction of pH_{50} and ΔpH_{50} (Co-Zn and Ni-Co) with various organic compositions composed of Cyanex 272 and Ionquest 801.

Organic Concentration (*M*)		pH_{50}			ΔpH_{50}	
Cyanex 272	Ionquest 801	Zn	Co	Ni	Co-Zn	Ni-Co
0.4	0	2.05	4.15	6.06	2.10	1.91
0.3	0.1	1.96	4.10	5.84	2.14	1.74
0.2	0.2	1.74	3.89	5.48	2.15	1.59
0.1	0.3	1.79	3.92	5.38	2.13	1.46
0	0.4	1.55	3.70	5.05	2.15	1.35

Table 6. Separation factor of cobalt over nickel ($SF_{Co/Ni}$) under various organic compositions and pH values.

Organic Concentration (*M*)		$SF_{Co/Ni}$			
Cyanex 272	Ionquest 801	pH 4.0	pH 4.5	pH 5.0	pH 5.5
0.4	0	1421	2000	2314	1630
0.3	0.1	696	884	776	759
0.2	0.2	446	555	398	315
0.1	0.3	143	196	217	266
0	0.4	83	107	104	87

The synergistic coefficient of Cyanex 272 and Ionquest 801 under various organic compositions at two appropriate pH values for each metal extraction were calculated based on Equation (4), as shown in Table 7.

$$SC_M = \frac{D_M}{D_{M(C272)} + D_{M(Ion.801)}} \tag{4}$$

where SC_M represents a synergistic coefficient, D_M represents a metal extraction distribution ratio, and C 272 and Ion. 801 are abbreviations of Cyanex 272 and Ionquest 801, respectively. SC_M was clearly larger than 1 at 0.2 M of each Cyanex 272 and Ionquest 801 in the mixture, particularly for nickel extraction, indicating their clear synergistic effect. Under some organic compositions, a slightly antagonistic effect was also observed by $SC_M < 1$, particularly at the organic system consisting of 0.1 M Cyanex 272 and 0.3 M Ionquest 801 for the extraction of zinc and cobalt. Although the reason why the synergistic or antagonistic effect occurred at different organic compositions, it can be generally concluded from Table 7 that, when the concentration of Cyanex 272 is equal to or higher than that of Ionqest 801, a synergistic effect most likely occurs. Otherwise, an antagonistic effect will occur. Based on the Job's method [36], plotting the distribution ratio of $D_{M(S)}$ ($D_{M(S)} = D_M - D_{M(C272)} - D_{M(Ion.801)}$), contributed by the synergistic effect, versus the fraction of Cyanex 272 concentration (Cyanex 272 to the overall concentration) (Figure 7), maximum values were obtained at the fraction of 0.5 for all three metals. It is indicated that the synergistic complex has the structure of one metal molecule combined with each of the Cyanex 272 and Ionquest 801 molecule in the form of *M(AB)* for all three metal extractions.

Table 7. Synergistic coefficient for metal extraction (SC_M) using the mixture of Cyanex 272 and Ionquest 801.

Organic Concentration (M)		SC_{Zn}		SC_{Co}		SC_{Ni}	
Cyanex 272	Ionquest 801	pH = 1.5	pH = 2.0	pH = 4.5	pH = 5.5	pH = 5.5	6.0
0.4	0	1.00	1.00	1.00	1.00	1.00	1.00
0.3	0.1	1.00	0.94	1.34	1.29	4.74	4.03
0.2	0.2	1.10	1.17	1.49	1.80	2.86	3.44
0.1	0.3	0.73	0.73	0.81	0.90	1.03	0.96
0	0.4	1.00	1.00	1.00	1.00	1.00	1.00

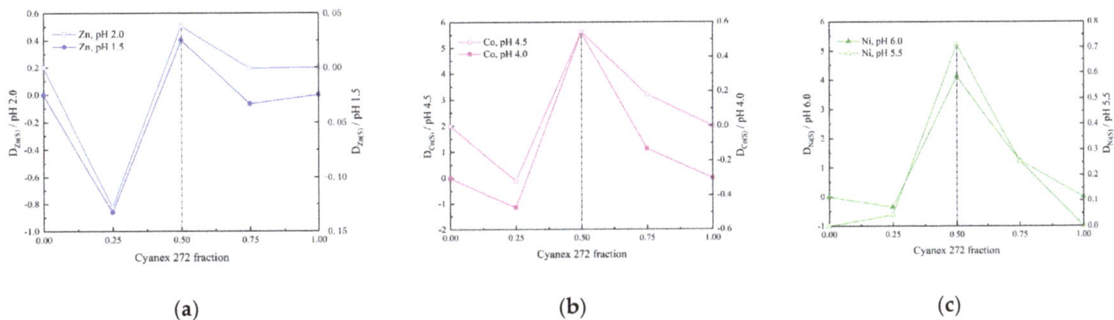

Figure 7. Job's plot of $D_{M(S)}$ versus Cyanex 272 concentration fraction in the mixtures. (**a**) $D_{Zn(s)}$-Cyanex 272 fraction, (**b**) $D_{Co(s)}$-Cyanex 272 fraction, (**c**) $D_{Ni(s)}$-Cyanex 272 fraction.

3.4. Discussion of Cyanex 272 and Ionquest 801 Application

Since Ionquest 801 is stronger for metal extraction than Cyanex 272, lower pH is required for the metal extraction with Ionquest 801 when compared to Cyanex 272. In addition, with its additional advantage of low price, Ionquest 801 should be more preferred to Cyanex 272 in some applications.

- In terms of zinc extraction and separation from cobalt and nickel, Ionquest 801 performed similarly to or even better than Cyanex 272.
- Cyanex 272 is much superior to Ionquest 801 for cobalt and nickel separation, while usually having one magnitude order larger separation factors comparatively. Cyanex 272 would be a preferred selection for cobalt and nickel separation.

- Using low concentration of Ionqest 801 to reduce the free extractant availability during the metal extraction, very good separation of cobalt from nickel is also available with the separation factor over 1000, which is comparable to those using Cyanex 272 as discussed before. Therefore, more rigid concentration control is required if Ionquest 801 is used for cobalt and nickel separation by taking its advantage to lower the cost.
- To separate zinc, cobalt, and nickel in an integral process, if Ionquest 801 is selected in the first solvent circuit to separate zinc and it is followed by Cyanex 272 to separate cobalt, leaving nickel in the raffinate. Ionquest 801 could contaminate Cyanex 272 by the phase carryover. However, as discussed previously, very good cobalt and nickel separation is still available when a small amount of Ionquest 801 mixed in the Cyanex 272 system.

4. Conclusions

Good separation of zinc, cobalt, and nickel from each other can be achieved with Cyanex 272 and Ionquest 801 systems, and both extractants performed similar selectivity for zinc over cobalt with the latter performing slightly better. Although Cyanex 272 has much higher selectivity for cobalt over nickel than that of Ionquest 801, separation factors over thousands were also obtained when low Ionquest 801 concentration or a high A/O ratio was used, which were very similar to that of Cyanex 272, indicating that high cobalt loaded in organic can significantly improve the separation of cobalt from nickel with Ionqest 801.

Slope analysis showed that one molecule metal extraction requires two dimer extractant molecules for zinc and cobalt extraction, but three for nickel extraction with both Cyanex 272 and Ionquest 801. The mixture of Cyanex 272 and Ionquest 801 has a slightly synergistic effect for the extraction of zinc and cobalt, particularly with the equal concentration of each in the mixture. A significant synergistic effect was observed for nickel extraction when Cyanex 272 concentration was higher than the Ionquest 801 concentration. The synergistic species was determined to have the form of *M(AB)*.

Author Contributions: Conceptualization, Z.Z. and J.L. Methodology, Z.Z. Software, W.L. Validation, J.Z. and Z.X. Investigation, Z.Z. Resources, Z.Z. Data curation, J.Z. and Z.X. Writing-original draft preparation, W.L. and Z.Z. Writing-review and editing, W.L., J.L., and Z.Z. Project administration, J.Z. Funding acquisition, Z.Z. All authors have read and agreed to the published version of the manuscript.

Funding: This research was funded by the National Natural Science Foundation of China (51774260, 51804289, 51904286, 21908231), Beijing Natural Science Foundation (2202053), Key Program of Innovation Academy for Green Manufacture, and Chinese Academy of Sciences (Grant No. IAGM-2019-A15), Key Research Program of Frontier Sciences of Chinese Academy of Sciences (Grant No. QYZDJ-SSW-JSC021), and CAS Interdisciplinary Innovation Team.

Institutional Review Board Statement: The study did not involve humans or animals.

Informed Consent Statement: The study did not involve humans or animals.

Data Availability Statement: This study did not report any data and the data presented in this study are available on request from the corresponding author.

Acknowledgments: The authors would like to acknowledge the financial support by all the funders.

Conflicts of Interest: The authors declare no conflict of interest.

References

1. Lewis, A.E. Review of metal sulphide precipitation. *Hydrometallurgy* **2010**, *104*, 222–234. [CrossRef]
2. Zhu, Z.W.; Pranolo, Y.; Zhang, W.S.; Cheng, C.Y. Separation of cobalt and zinc from concentrated nickel sulphate solutions with Cyanex 272. *J. Chem. Technol. Biotechnol.* **2011**, *86*, 75–81. [CrossRef]
3. Mishra, R.K.; Rout, P.C.; Sarangi, K.; Nathsrma, K.C. Solvent extraction of zinc, manganese, cobalt and nickel from nickel laterite bacterial leach liquor using sodium salts of TOPS-99 and Cyanex 272. *Trans. Nonferrous Met. Soc. China* **2016**, *26*, 301–309. [CrossRef]

4. Senapati, D.; Chaudhury, G.R.; Sarma, P.V.R.B. Purification of nickel sulphate solutions containing iron, copper, cobalt, zinc and manganese. *J. Chem. Technol. Biotechnol.* **1994**, *59*, 335–339. [CrossRef]

5. Kursunoglu, S.; Ichlas, Z.T.; Kaya, M. Solvent extraction process for the recovery of nickel and cobalt from Caldag laterite leach solution: The first bench scale study. *Hydrometallurgy* **2017**, *169*, 135–141. [CrossRef]

6. Ichlas, Z.T.; Ibana, D.C. Process development for the direct solvent extraction of nickel and cobalt from nitrate solution: Aluminum, cobalt, and nickel separation using Cyanex 272. *Int. J. Miner. Metall. Mater.* **2017**, *24*, 37–46. [CrossRef]

7. Donegan, S. Direct solvent extraction of nickel at Bulong operations. *Miner. Eng.* **2006**, *19*, 1234–1245. [CrossRef]

8. Crundwell, F.K.; Moats, M.S.; Ramachandran, V.; Robinson, G.; Davenport, W.G. Hydrometallurgical Production of High-Purity Nickel and Cobalt. In *Extractive Metallurgy of Nickel, Cobalt and Platinum Group Metals*; Elsevier: Radarweg, The Netherlands, 2011.

9. Parhi, P.K.; Padhan, E.; Palai, A.K.; Sarangi, K.; Nathsarma, K.C.; Park, K.H. Separation of Co (II) and Ni (II) from the mixed sulphate/chloride solution using NaPC-88A. *Desalination* **2011**, *267*, 201–208. [CrossRef]

10. Li, Y.; Zhang, L.H.; Guo, S.H.; Zhang, L.B. Extraction and separation of cobalt from nickel by P507 microemulsion liquid membrane. *Nonferrous Met. Eng.* **2016**, *6*, 40–44.

11. Liu, M.X.; Shen, Y.Y.; Shao, Q.T.; Wu, Z.M.; Zhong, S.H. Study on mixed phase disengagement of cobalt and nickel by extraction with P 507. *Biol. Chem. Eng.* **2016**, *2*, 1–3.

12. Wang, J.K.; Gao, K.; Xu, W.X.; Wang, Z.J.; Guo, R.; Zhou, Y.; Li, J. Study on extraction and separation of nickel and cobalt with P 507. *Nonferrous Met.* **2018**, *8*, 19–22.

13. Wu, J.H.; Dong, B.; Zhang, X.P.; Ye, F.C.; Wang, H.J.; Ji, H.W.; Guo, F.Y.; Qiu, S.W.; Liu, Z.D. Solvent extraction of Cu, Zn, Co from nickel sulphate solution applying P507. *Nonferrous Met. Sci. Eng.* **2018**, *9*, 19–24.

14. Flett, D.S. Cobalt-Nickel separation in hydrometallurgy: A review. *Chem. Sustain. Dev.* **2004**, *12*, 81–91.

15. Flett, D.S. Solvent extraction in hydrometallurgy: The role of organophosphorus extractants. *J. Organomet. Chem.* **2005**, *690*, 2426–2438. [CrossRef]

16. Thakur, N.V. Extraction studies of base metals (Mn, Cu, Co and Ni) using the extractant 2-ethylhexyl 2-ethylhexyl phosphonic acid, PC 88A. *Hydrometallurgy* **1998**, *48*, 125–131. [CrossRef]

17. Tsakiridis, P.E.; Agatzini-Leonardou, S. Simultaneous solvent extraction of cobalt and magnesium in the presence of nickel from sulfate solutions by Ionquest 801. *J. Chem. Technol. Biotechnol.* **2005**, *80*, 1236–1243. [CrossRef]

18. Zhu, Z.W.; Zhang, J.; Yi, A.F.; Su, H.; Wang, L.N.; Qi, T. Magnesium removal from concentrated nickel solution by solvent extraction using Cyanex 272. *Int. J. Miner. Process. Ext. Metall.* **2019**, *4*, 36–43.

19. Preston, J.S.; du Preez, A.C. Solvent extraction of nickel from acidic solutions using synergistic mixtures containing pyridinecarboxylate esters, Part 1: Systems based on organophosphorus acids. *J. Chem. Technol. Biotechnol.* **1996**, *66*, 86–94. [CrossRef]

20. Cheng, C.Y.; Barnard, K.R.; Zhang, W.S.; Zhu, Z.W.; Pranolo, Y. Recovery of nickel, cobalt, copper and zinc in sulphate and chloride solutions using synergistic solvent extraction. *Chin. J. Chem. Eng.* **2016**, *24*, 237–248. [CrossRef]

21. Sulaiman, R.N.R.; Othman, N. Synergistic green extraction of nickel ions from electroplating waste via mixtures of chelating and organophosphorus carrier. *J. Hazard. Mater.* **2017**, *340*, 77–84. [CrossRef]

22. Sun, Q.; Yang, L.M.; Huang, S.T.; Xu, Z.; Wang, W. Synergistic solvent extraction of nickel by 2-hydroxy-5-nonylacetophenone oxime mixed with neodecanoic acid and bis(2-ethylhexyl) phosphoric acid: Stoichiometry and structure investigation. *Miner. Eng.* **2019**, *132*, 284–292. [CrossRef]

23. Zhao, J.M.; Shen, X.Y.; Deng, F.L.; Wang, F.C.; Wu, Y.; Liu, H.Z. Synergistic extraction and separation of valuable metals from waste cathodic material of lithium ion batteries using Cyanex272 and PC-88A. *Sep. Purif. Technol.* **2011**, *78*, 345–351. [CrossRef]

24. Liu, M.R.; Zhou, G.Y.; Wen, J.K. Separation of divalent cobalt and nickel ions using a synergistic solvent extraction system with P507 and Cyanex272. *Chin. J. Process Eng.* **2012**, *12*, 415–419.

25. Liu, T.C.; Chen, J.; Li, H.L.; Li, K.; Li, D.Q. Further improvement for separation of heavy rare earths by mixtures of acidic organophosphorus extractants. *Hydrometallurgy* **2019**, *188*, 73–80. [CrossRef]

26. Quinn, J.E.; Soldenhoff, K.H.; Stevens, G.W.; Lengkeek, N.A. Solvent extraction of rare earth elements using phosphonic/phosphinic acid mixtures. *Hydrometallurgy* **2015**, *157*, 298–305. [CrossRef]

27. Dreisinger, D.B.; Cooper, W.C. The solvent extraction separation of cobalt and nickel using 2-ethylhexylphosphonic acid mono-2-ethylhexyl ester. *Hydrometallurgy* **1984**, *12*, 1–20. [CrossRef]

28. Devi, N.B.; Nathsarma, K.C.; Chakravortty, V. Separation and recovery of cobalt (II) and nickel (II) from sulphate solutions using sodium salts of D2EHPA, PC 88A and Cyanex 272. *Hydrometallurgy* **1998**, *49*, 47–61. [CrossRef]

29. Tait, B.K. Cobalt-nickel separation: The extraction of cobalt (II) and nickel (II) by Cyanex 301, Cyanex302 and Cyanex 272. *Hydrometallurgy* **1993**, *32*, 365–372. [CrossRef]

30. Sarangi, K.; Reddy, B.R.; Das, R.P. Extraction studies of cobalt (II) and nickel (II) from chloride solutions using Na-Cyanex 272: Separation of Co(II)/Ni(II) by the sodium salts of D2EHPA, PC88A and Cyanex 272 and their mixtures. *Hydrometallurgy* **1999**, *52*, 253–265. [CrossRef]

31. Ahmadipour, M.; Rashchi, F.; Ghafarizadeh, B.; Mostoufi, N. Synergistic effect of D2EHPA and Cyanex 272 on separation of zinc and manganese by solvent extraction. *Sep. Sci. Technol.* **2011**, *46*, 2305–2312. [CrossRef]

32. Begum, N.; Bari, F.; Jamaludin, S.B.; Hussin, K. Solvent extraction of copper, nickel and zinc by Cyanex 272. *Int. J. Phys. Sci.* **2012**, *7*, 2905–2910.

33. Innocenzi, V.; Veglio, F. Separation of manganese, zinc and nickel from leaching solution of nickel-metal hydride spent batteries by solvent extraction. *Hydrometallurgy* **2012**, *50*, 50–58. [CrossRef]
34. Devi, N.B.; Nathsarma, K.C.; Chakravortty, V. Sodium salts of D2EHPA, PC-88A and Cyanex-272 and their mixtures as extractants for cobalt (II). *Hydrometallurgy* **1994**, *34*, 331–342. [CrossRef]
35. Darvishia, D.; Haghshenas, D.F.; Alamdari, E.K.; Sadrnezhaad, S.K.; Halali, M. Synergistic effect of Cyanex 272 and Cyanex 302 on separation of cobalt and nickel by D2EHPA. *Hydrometallurgy* **2005**, *77*, 227–238. [CrossRef]
36. Renny, J.S.; Tomasevich, L.L.; Tallmadge, E.H.; Collom, D.B. Method of Continuous Variations: Applications of Job Plots to the Study of Molecular Associations in Organometallic Chemistry. *Angew. Chem. Int. Ed.* **2013**, *52*, 11998–12013. [CrossRef] [PubMed]

metals

MDPI

Article

Optimization on Temperature Strategy of BOF Vanadium Extraction to Enhance Vanadium Yield with Minimum Carbon Loss

Zhen-Yu Zhou [1,*] and Ping Tang [2,*]

1 Hunan Institution of Science and Technology, No.439, Xueyuan Road, Yueyang 414000, China
2 College of Materials Science and Engineering, Chongqing University, No.174, Shazheng Street, Shapingba District, Chongqing 400044, China
* Correspondence: zhzy@hnist.edu.cn (Z.-Y.Z.); tping@cqu.edu.cn (P.T.)

Abstract: During the vanadium extraction process in basic oxygen furnace (BOF), unduly high temperature is unfavorable to achieve efficient vanadium yield with minimum carbon loss. A new temperature strategy was developed based on industrial experiments. The new strategy applies the selective oxidation temperature between carbon and vanadium (T_{sl}) and the equilibrium temperature of vanadium oxidation and reduction (T_{eq}) for the earlier and middle-late smelting, respectively. Industrial experiments showed 56.9 wt% of V was removed together with carbon loss for 5.6 wt% only in the earlier smelting. Additionally, 30 wt% of vanadium was removed together with carbon loss by 13.4 wt% in middle-late smelting. Applicability analyses confirmed T_{eq} as the high-limit temperature, vanadium removal remains low and carbon loss increased sharply when the molten bath temperature exceeded T_{eq}. With the optimized temperature strategy, vanadium removal increased from 69.2 wt% to 92.3 wt% with a promotion by 23 wt%.

Keywords: vanadium extraction process; vanadium yield; minimum carbon loss; temperature strategy

Citation: Zhou, Z.-Y.; Tang, P. Optimization on Temperature Strategy of BOF Vanadium Extraction to Enhance Vanadium Yield with Minimum Carbon Loss. *Metals* **2021**, *11*, 906. https://doi.org/10.3390/met11060906

Academic Editor: Dariush Azizi

Received: 29 April 2021
Accepted: 28 May 2021
Published: 2 June 2021

Publisher's Note: MDPI stays neutral with regard to jurisdictional claims in published maps and institutional affiliations.

Copyright: © 2021 by the authors. Licensee MDPI, Basel, Switzerland. This article is an open access article distributed under the terms and conditions of the Creative Commons Attribution (CC BY) license (https://creativecommons.org/licenses/by/4.0/).

1. Introduction

Vanadium is a widely used rare metal in many areas such as steel-making, aerospace, and chemical industries [1,2]. It is usually found as a by-product in vanadium-titanium magnetite (VTM), and the most popular method of treating vanadium-bearing hot metal is oxygen blow smelting in converter to form vanadium-enriching slag and semi-steel [3]. Residual vanadium in semi-steel must remain low, mostly under 0.05 wt%. Meanwhile, to ensure subsequent steel-making, carbon content in semi-steel needs to above 3.4 wt% [4]. Therefore, vanadium extraction process demands 'deep devanadium' and 'minimum carbon loss' simultaneously.

The selective oxidation temperature between carbon and vanadium, $T_{sl,}$ a thermodynamic temperature, used to be considered as a key to ensure smelting steps. It related to the transformation from preferential V removal to C [5,6]. Most studies tend to the selective oxidation theory, which required the molten bath temperature never went beyond T_{sl} [4]. However, the demand could not be reached in practice because of the molten bath temperature always exceeded T_{sl} in middle-late smelting. It ranged from 1340 to 1400 °C near the end, and was much higher than corresponding T_{sl}. D.X. Huang's study employed the temperature control strategy based on T_{sl}. It showed that carbon loss must be accepted when the molten bath temperature exceeds T_{sl} for 'deep devanadium', and vanadium removal decreased when the molten bath temperature exceeded T_{sl} in the middle-late smelting. Thus, previous studies mostly focused on vanadium removal, the reduction of (V_2O_3) by C and carbon loss in molten iron was totally ignored [7]. Few studies have concentrated on 'deep devanadium' and 'minimum carbon loss' simultaneously. Further study on a reasonable temperature control strategy is very important to realize these two demands.

In this paper, Industrial experiments were applied to determine the removal characteristics of C and V in various smelting period. Thermodynamic analyses were applied to settle the foundation of new temperature strategy for vanadium extraction. The applicability of new strategy was verified by analyses on final samples and production data.

2. Experimental Procedure

Industrial experiments have taken place, and six heats with similar parameters were performed as one to minimize the experimental errors. The original molten iron composition and temperature are listed in Table 1. The smelting parameters are given in Table 2. The measuring and sampling process is shown in Figure 1.

Table 1. Original compositions and temperatures of molten iron.

Temperature/°C	Composition (wt%)				
	C	V	Si	Mn	Ti
1294 ± 6	4.35 ± 0.05	0.39 ± 0.03	0.11 ± 0.02	0.22 ± 0.02	0.12 ± 0.02

Table 2. Parameters during vanadium extraction smelting.

Furnace Capacity/t	Lance Height/m	Top Flow Rate (Nm³/h)	Bottom Flow Rate (Nm³/h)	Number of Nozzle	Ma	Blowing Time/s	Coolant (Kg/t)
210	1.6~1.9	24,000	1000	4	1.99	~360	41.6 ± 0.6

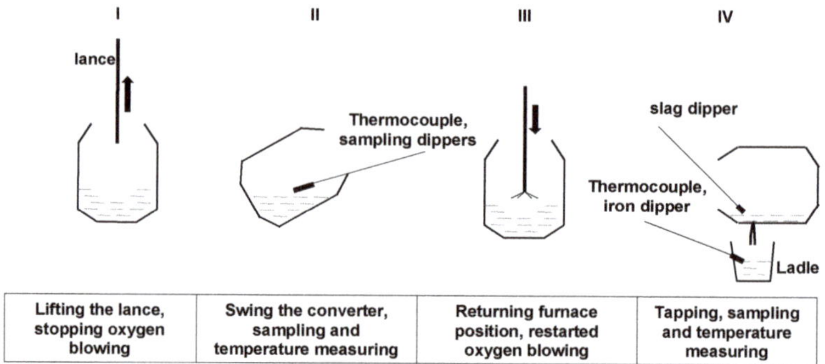

Figure 1. Schematic of sampling and temperature measuring operations.

Coolants were added into molten bath for three times at 60 s, 120 s and 180 s to ensure the molten bath temperature below T_{sl} in the earlier smelting [8]. Sampling and temperature measuring were carried out from 210 s with an interval about 25 ± 5 s. The total smelting lasted around 360 s based on production experience. X-Ray Fluorescence analyses (XRF-1800, Rh-target, 60 KV, 140 mA) were applied to determine the compositions of metal and slag.

3. Results

The thermodynamic temperature T_{sl} for each sample was calculated by equaling $\triangle G$ of reaction (1) to reaction (2), as Equation (3).

$$[C] + (FeO) = [Fe] + CO \quad \Delta G^{\theta} = 98,799 - 90.76T \text{ J/mol} \tag{1}$$

$$\frac{2}{3}[V] + (FeO) = \frac{1}{3}(V_2O_3) + [Fe] \quad \Delta G^{\theta} = -151,376 + 62.33T \text{ J/mol} \tag{2}$$

$$T_{sl} = \cfrac{250,170}{153.09 + 19.147\lg \cfrac{\gamma_{V_2O_3}^{1/3} . x_{V_2O_3}^{1/3} . (\%C) . f_C}{(\%V)^{2/3} . f_V^{2/3} . P_{CO}}} \tag{3}$$

where $x_{V_2O_3}$ was the mole fraction of V_2O_3 in slag and $\gamma_{V_2O_3}$ was the activity coefficient. f_C and f_V were the activity coefficients of carbon and vanadium which referred to Equations (4) and (5). (%C) and (%V) were the mass fractions of carbon and vanadium, respectively. P_{CO} was the partial pressure of CO in the converter.

However, about 55–65% (V_2O_3) in the vanadium slag precipitated by vanadium spinels (FeO.V_2O_3) during smelting [9]. Therefore, the current study employed the value from D.X. Huang based on industrial production data, 10^{-5} [4].

$$\lg f_C = \Sigma e_C^i(\%j) = e_C^C(\%C) + e_C^V(\%V) + e_C^{Si}(\%Si) + e_C^{Mn}(\%Mn) + e_C^{Ti}(\%Ti) + \ldots \tag{4}$$

$$\lg f_V = \Sigma e_V^j(\%j) = e_V^V(\%V) + e_V^C(\%C) + e_V^{Mn}(\%Mn) + e_V^{Si}(\%Si) + e_V^{Ti}(\%Ti) + \ldots \tag{5}$$

where e_C^i was the interaction coefficient to carbon, and e_V^j was the interaction coefficients to vanadium.

The molten bath temperatures, compositions of metal and slag measured in industrial experiments were listed in Tables 3 and 4. Removal ratios of C and V and T_{sl} for each sample were shown in Table 5.

Table 3. The molten bath temperatures and molten iron compositions (wt%) in various time during industrial experiments.

NO.	Time/s	C	V	Si	Mn	Ti	$T_{bt}/(°C)$
0	0	4.31	0.39	0.11	0.22	0.12	1294
1#-1	210	4.07	0.168	0.044	0.096	0.011	1318
1#-2	240	4.0	0.137	0.030	0.060	0.008	1320
1#-3	252	3.82	0.115	0.019	0.054	0.006	1337
1#-4	282	3.60	0.087	0.018	0.046	0.006	1342
1#-5	282	3.66	0.094	0.022	0.049	0.004	1348
1#-6	306	3.59	0.078	0.010	0.034	0.003	1349
2#-1 *	366	3.54	0.051	0.006	0.026	0.001	1376
2#-2 *	366	3.45	0.042	0.009	0.023	0.001	1382
2#-3 *	366	3.51	0.06	0.010	0.030	0.002	1378
2#-4 *	372	3.61	0.053	0.007	0.027	0.001	1394
2#-5 *	378	3.45	0.054	0.005	0.026	0.001	1382
2#-6 *	390	3.49	0.06	0.006	0.024	0.001	1379

* Final samples.

Table 4. Slag compositions (wt%) in various time during industrial experiments.

NO.	Time/s	V_2O_3	FeO	SiO_2	MnO	TiO_2	CaO	MgO
0	0	1.6	90.0	2	2	2	1.5	1.5
1#-1	210	8.2	70.6	4.4	4.4	5.8	1.6	1.4
1#-2	240	10.3	65.5	5.6	5.9	7.8	1.9	1.6
1#-3	252	10.7	62.5	5.9	6.7	8.6	1.9	1.5
1#-4	282	11.1	56.5	7.2	8	9.6	1.7	1.5
1#-5	282	11.5	54.4	8.4	8	9.2	2.0	1.7
1#-6	306	14.0	48.6	9.2	8.5	10.0	1.8	1.9
2#-1 *	366	16.2	32.4	10.2	9.8	10.5	2.0	2.0
2#-2 *	366	17.3	34.5	10.4	10.2	11.2	2.4	1.8
2#-3 *	366	16.5	33.4	9.8	10.4	10.0	1.8	2.2
2#-4 *	372	15.7	35.4	10.2	10.2	11.4	2.0	1.8
2#-5 *	378	16.5	30.9	10.6	11.0	10.8	2.2	2.4
2#-6 *	390	16.5	32.6	10.2	10.6	10.6	2.4	2.0

* Final samples.

Table 5. Removal ratios, T_{sl} and T_{eq} in various time during industrial experiments.

NO.	Time/s	V	V Removal Ratio/wt%	C	C Removal Ratio/wt%	$T_{bt}/(°C)$	$T_{sl}/(°C)$	$T_{eq}/(°C)$
0	0	0.39	-	4.31	0.0	1294	1382	1782
1#-1	210	0.168	56.9	4.07	5.6	1318	1295	1595
1#-2	240	0.137	64.9	4.0	7.2	1320	1285	1566
1#-3	252	0.115	70.5	3.82	11.4	1337	1292	1568
1#-4	282	0.087	77.7	3.60	16.5	1342	1294	1547
1#-5	282	0.094	75.9	3.66	15.1	1348	1294	1542
1#-6	306	0.078	80.0	3.59	16.7	1349	1281	1504
2#-1 *	366	0.051	86.9	3.54	17.9	1376	1260	1407
2#-2 *	366	0.042	89.2	3.45	20.0	1382	1257	1402
2#-3 *	366	0.06	84.6	3.51	18.6	1378	1272	1408
2#-4 *	372	0.053	86.4	3.61	16.2	1394	1257	1415
2#-5 *	378	0.054	86.2	3.45	20.0	1382	1271	1389
2#-6 *	390	0.06	84.6	3.49	19.0	1379	1273	1412

* Final samples.

4. Discussions

4.1. V andC Removal in Various Temperatures

Variations on molten bath temperatures, T_{sl}, V, C and V_2O_3 contents, were shown in Figure 2. The molten bath temperature increased due to the oxidation of dissolved elements. It turned out more rapidly without coolant addition in the middle-late smelting. T_{sl} decreased during vanadium extraction with the decrease in V content and increase in (V_2O_3) in slag, as shown in Figure 2b. The molten bath temperature was controlled below T_{sl} in the earlier smelting and became higher than T_{sl} 23 °C at 210 s.

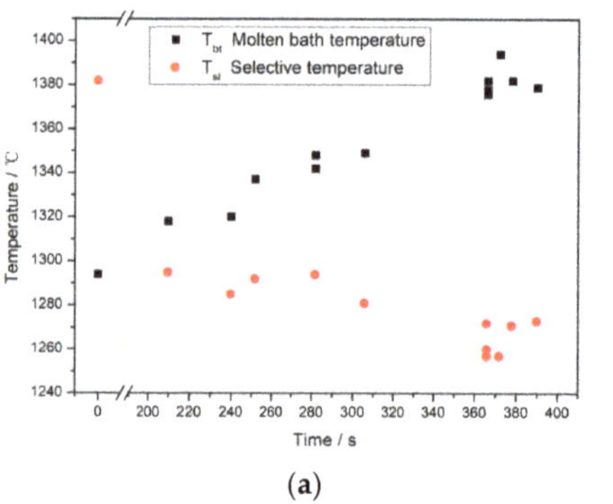

(a)

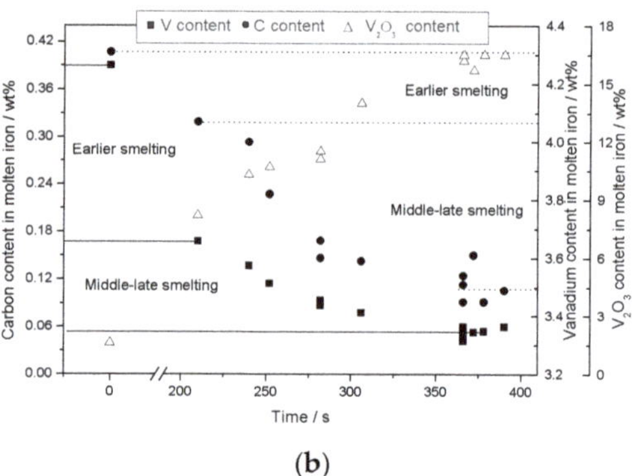

(b)

Figure 2. Variations on V, C, (V_2O_3), the molten bath temperature and T_{sl}. (**a**) variation on T_{bt} and T_{sl}, (**b**) variation on V, C, and (V_2O_3).

The removal ratio of V and C at 210 s were 56.9 wt% and 5.6 wt%. The removal ratios of the final point were 86.3 wt% and 19 wt% (average), as shown in Table 5. As the molten bath temperature exceeded T_{sl} in the middle-late smelting (after 210 s), vanadium oxidation has lost thermodynamic advantage to carbon. Vanadium removal showed a weak tendency compared with the earlier smelting. Therefore, only about 30 wt% V was removed in middle-late smelting. However, carbon removal showed an opposite feature and has no longer been suppressed by vanadium removal. Therefore, about 13.4 wt% C was removed in middle-late smelting, and it was 2.4 times more than the earlier smelting (5.6 wt%), as shown in Table 5. Thus it can be seen, vanadium removal efficiency decreased when the molten bath temperature exceeded T_{sl} in the middle-late smelting. However, carbon removal seemed to climb up rapidly. It was certain to desire a reasonable principle for final temperature control to assure 'enhance vanadium removal' with 'minimum carbon loss'.

4.2. Thermodynamic Analyses of Vanadium Extraction

The present study considered that carbon and vanadium oxidation in the molten bath were mainly indirect [10]. Oxygen was transferred by ferrous oxide (FeO). Carbon monoxide (CO) was considered as the oxidative product of C as it was saturated in molten iron ($\geq$3.4 wt%). In addition, vanadium trioxide (V_2O_3) would be reduced by C, as displayed in Equation (6). The standard Gibbs free energies (ΔG^θ) of reaction (1), (2) and (6) with various temperatures were showed as follows Figure 3.

$$\frac{1}{3}(V_2O_3) + [C] = \frac{2}{3}[V] + CO \quad \Delta G^\theta = 250,170 - 153.09T \text{ J/mol} \tag{6}$$

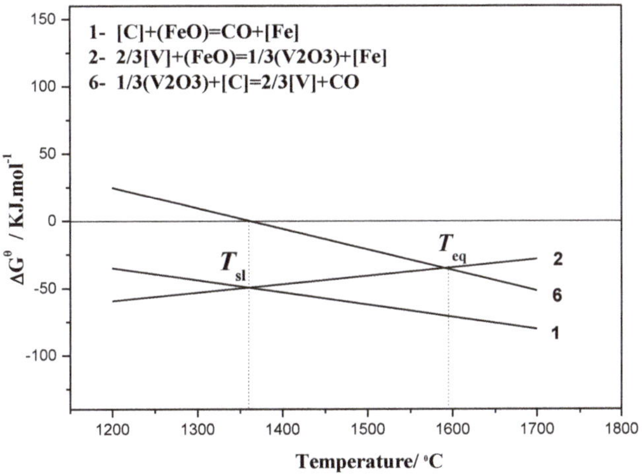

Figure 3. Relationship between ΔG^θ and temperature.

The molten bath temperature exceeded T_{sl} rapidly without coolant addition in the middle-late smelting, as shown in Figure 2a. The removal of C is prior to V in this period, and (V_2O_3) in the slag also was reduced by C, as shown in Figure 3. However, $\triangle G$ of reaction (2) was still less than (6), which meant the thermodynamic driving force of vanadium removal was greater than the reduction. V removal and reduction of (V_2O_3) exist simultaneously.

As the molten bath temperature increasing continuously, there was a temperature-made ΔG of reaction (2) equal to (6). It stood for the thermodynamic equilibrium of vanadium removal and reduction, as shown in Figure 3 [11,12]. In this study, this temperature was described as 'The equilibrium temperature of vanadium oxidation and reduction, T_{eq}'. A coupled reaction represented V removal was deduced by equalizing ΔG of reaction (2) and (6), as displayed in Equation (7). T_{eq} was calculated by the following Equation (8). T_{eq} for each sample was listed in Table 5, and variation on the molten bath temperature and T_{eq} was shown in Figure 4.

$$\frac{4}{3}[V] + (FeO) + CO = \frac{2}{3}(V_2O_3) + [Fe] + [C] \quad \Delta G^\theta = -401,551 + 215.42T \text{ J/mol} \tag{7}$$

$$T_{eq} = \frac{401,551}{215.42 + 19.147 \lg \frac{\gamma_{V_2O_3}^{2/3} \cdot x_{V_2O_3}^{2/3} \cdot [\%C] \cdot f_C}{[\%V]^{4/3} \cdot f_V^{4/3} \cdot a_{FeO} \cdot P_{CO}}} \tag{8}$$

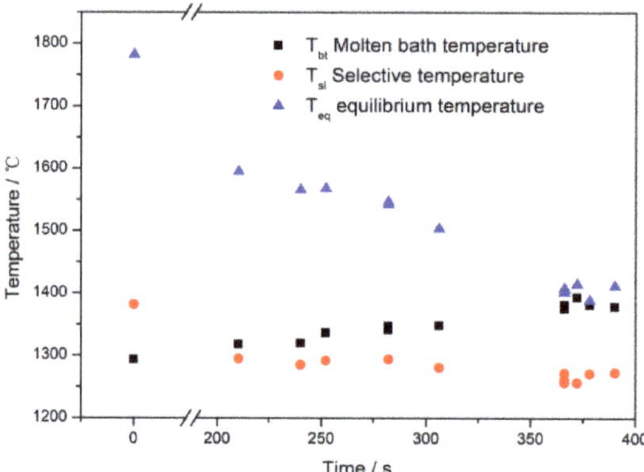

Figure 4. Variations on the molten bath temperature and T_{eq}.

Thus, a new temperature strategy for 'enhance vanadium yield' with 'minimum carbon loss' can be proposed. In the earlier smelting, the molten bath temperature must be controlled below T_{sl} by coolant addition. In middle-late smelting, the molten bath temperature exceeding T_{sl}, T_{eq} was considered as the high-limit temperature. Potential for vanadium removal could be fully exploited and excessive carbon loss was avoided.

4.3. Applicability Analyses of T_{eq} for the Final Temperature

4.3.1. Final Samples

The theoretical residual vanadium in molten iron at the final point was calculated based on Equations (9) and (10) which deduced from Equations (3) and (8), respectively. Six final samples were treated as repetitive measurements in one heat. The Mean Absolute Difference (MAD) between theoretical vanadium contents and measurements were applied to evaluate the applicability of T_{sl} and T_{eq} on temperature strategy, as shown in Figure 5.

$$(\%\text{V})_{\text{sl}} = \left[\frac{\gamma_{V_2O_3}^{1/3} \cdot x_{V_2O_3}^{1/3} \cdot (\%C) \cdot f_C}{10^{(250,170/T - 153.09)/19.147} \cdot f_V^{2/3} \cdot P_{CO}} \right]^{\frac{3}{2}} \tag{9}$$

$$(\%V)_{\text{eq}} = \left[\frac{\gamma_{V_2O_3}^{2/3} \cdot x_{V_2O_3}^{2/3} \cdot (\%C) \cdot f_C}{10^{(401,551/T - 219.422)/19.147} \cdot f_V^{4/3} \cdot a_{FeO} \cdot P_{CO}} \right]^{\frac{3}{4}} \tag{10}$$

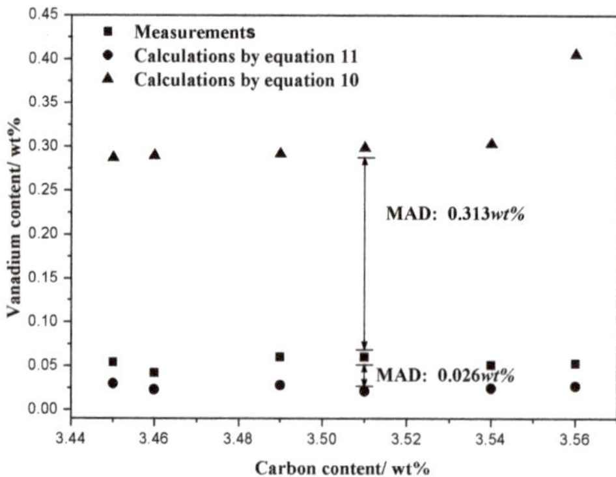

Figure 5. Theoretical vanadium contents and measuring values at the final.

Theoretical vanadium contents was a bit lower than measurements because of the idealized status. The MAD between measurements and theoretical vanadium content calculated based on T_{sl} (Equation (10)) was 0.313 wt% (average). The theoretical values were much higher than measurements, which meant mistakes in calculation foundation. The MAD based on T_{eq} (Equation (10)) was 0.026 wt% (average) and theoretical values were a little lower than measurements. It showed a better fitting with measurements. Thus, it can be seen, T_{eq} should be considered as the thermodynamic basis for describing behaviors in middle-late smelting of vanadium extraction rather than T_{sl}.

4.3.2. Production Data

The temperature strategy was applied in Pansteel. The residual vanadium, molten bath temperature and T_{eq} at the final were also discussed based on production data. The residual vanadium, the temperature differences (ΔT, between T_{eq} and the molten bath temperature) in the final point were shown in Figure 6.

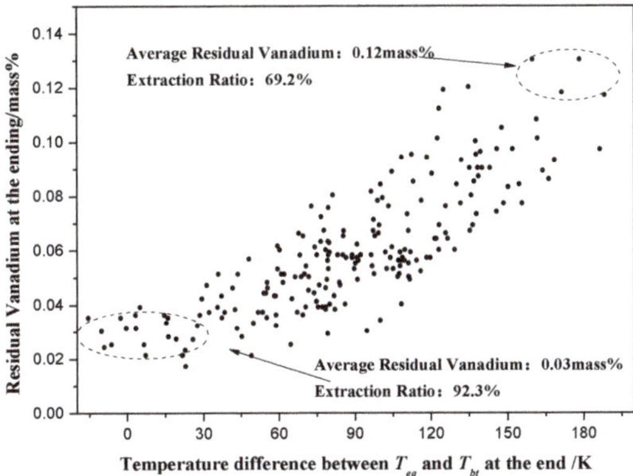

Figure 6. Relationship between the residual vanadium and ΔT at the final.

The temperature differences in most heats were far above zero, which showed the poor understanding on temperature control. The residual vanadium decreased with the decrease in ΔT. It did not decrease further when ΔT approached zero, as shown in Figure 6. The thermodynamic driving force of vanadium removal was bigger than reduction when $\Delta T > 0$, there was still potential on vanadium removal. Therefore, residual vanadium contents decreased with the decrease in ΔT. Vanadium removal and reduction showed fluctuant equilibrium status when the molten pool temperature exceeded T_{eq} (i.e., $\Delta T < 0$).

With the decrease in ΔT, the potential for vanadium removal in middle-late smelting has been fully developed. The vanadium removal ratio has increased from 69.2 wt% to 92.3 wt% with a promotion by 23 wt%. However, the serious carbon loss must be avoided when the molten bath temperature exceeded T_{eq}. T_{eq} must be considered as a maximum temperature limit of the final stage in vanadium extraction.

5. Conclusions

(1) A temperature strategy of vanadium extraction process has been developed to consider the effects of T_{sl} in the earlier smelting and T_{eq} for the high-limit temperature to ensure 'enhance vanadium removal' with 'minimum carbon loss'.

(2) The vanadium removal rate was highly efficient accompanied by slight carbon removal when the molten bath temperature was lower than T_{sl}. The removal rate of vanadium decreased and carbon removal increased when the molten bath temperature exceeded T_{sl}.

(3)   Vanadium removal efficiency remained poor while significant carbon loss was expected when the melt temperature went over T_{eq}, and T_{eq} must be considered as the high-limit temperature in vanadium extraction.

(4)   With the optimized temperature strategy, vanadium removal increased from 69.2 wt% to 92.3 wt% with a promotion by 23 wt% by production data.

Author Contributions: Conceptualization, Z.-Y.Z. and P.T.; methodology, Z.-Y.Z. and P.T.; validation, Z.-Y.Z.; formal analysis, Z.-Y.Z. and P.T.; investigation, Z.-Y.Z. and P.T.; resources, P.T.; data curation, Z.-Y.Z.; writing—original draft preparation, Z.-Y.Z.; writing—review and editing, Z.-Y.Z. and P.T.; visualization, Z.-Y.Z.; supervision, P.T. All authors have read and agreed to the published version of the manuscript.

Funding: This research received no external funding.

Institutional Review Board Statement: Not applicable.

Informed Consent Statement: Not applicable.

Data Availability Statement: Data are contained within the article.

Conflicts of Interest: The authors declare no conflict of interest.

References

1.   Edward, J.; Maxim, I.; Kenneth, M. Reduction of Vanadium(V) by Iron(II)-Bearing Minerals. *Minerals* **2021**, *11*, 316. [CrossRef]
2.   Lidiane, M.A.; Heide, M.; Gaebler, D.J.; Silva, W.E.; Belian, M.F.; Lira, E.C.; Crans, D.C. Acute Toxicity Evaluation of Non-Innocent Oxidovanadium(V) Schiff Base Complex. *Inorganics* **2021**, *9*, 42. [CrossRef]
3.   Zhang, J.H.; Zhang, W.; Xue, Z.L. An Environment-Friendly Process Featuring Calcified Roasting and Precipitation Purification to Prepare Vanadium Pentoxide from the Converter Vanadium Slag. *Metals* **2019**, *9*, 21. [CrossRef]
4.   Huang, D.X. *Vanadium Extracting and Steelmaking*; Metallurgical Industry Press: Beijing, China, 2010; pp. 52–53.
5.   Hong, Y.; Peng, J.; Sun, Z.; Yu, Z.; Wang, A.; Wang, Y.; Liu, Y.-Y.; Xu, F.; Sun, L.-X. Transition Metal Oxodiperoxo Complex Modified Metal-Organic Frameworks as Catalysts for the Selective Oxidation of Cyclohexane. *Materials* **2020**, *13*, 829. [CrossRef] [PubMed]
6.   Condruz, M.R.; Matache, G.; Paraschiv, A.; Badea, T.; Badilita, V. High Temperature Oxidation Behavior of Selective Laser Melting Manufactured IN 625. *Metals* **2020**, *10*, 668. [CrossRef]
7.   Xiang, J.Y.; Xin, W.A.N.G.; Pei, G.S.; Huang, Q.Y.; Lü, X.W. Recovery of vanadium from vanadium slag by composite roasting with CaO/MgO and leaching. *Trans. Nonferrous Metals Soc. China* **2020**, *30*, 3114. [CrossRef]
8.   Penz, F.M.; Schenk, J.; Ammer, R.; Klösch, G.; Pastucha, K.; Reischl, M. Diffusive Steel Scrap Melting in Carbon-Saturated Hot Metal-Phenomenological Investigation at the Solid-Liquid Interface. *Materials* **2019**, *12*, 1358. [CrossRef] [PubMed]
9.   Zhang, S.-Q.; Xie, B.; Wang, Y.; Guan, T.; Cao, H.L.; Zeng, X.L. Reaction of $FeOV_2O_5$ System at High Temperature. *J. Iron Steel Res. Int.* **2012**, *19*, 33. [CrossRef]
10.   Chen, S.K.; Cai, Z.J. An Improved Dynamic Model of the Vanadium Extraction Process in Steelmaking Converters. *Appl. Sci.* **2020**, *10*, 111. [CrossRef]
11.   Zhou, C.G.; Li, J.; Wu, H. Dependence of Temperature and Slag Composition on Dephosphorization at the First Deslagging in BOF Steelmaking Process. *High Temp. Mater. Process.* **2014**, *49*, 24. [CrossRef]
12.   Wang, X.H.; Wang, H.J. Converter practice in China with respect to steelmaking and ferroalloys. *Miner. Process. Extr. Metall.* **2019**, *128*, 46. [CrossRef]

Article

Leaching of Phosphorus from Quenched Steelmaking Slags with Different Composition

Ning-Ning Lv [1,2], Chuan-Ming Du [3,*], Hui Kong [1,2] and Yao-Hui Yu [3]

[1] Key Laboratory of Metallurgical Emission Reduction & Resources Recycling, Anhui University of Technology, Ministry of Education, Maanshan 243002, China; lvning198565@163.com (N.-N.L.); konghui@ahut.edu.cn (H.K.)
[2] School of Metallurgical Engineering, Anhui University of Technology, Maanshan 243032, China
[3] School of Metallurgy, Northeastern University, Shenyang 110819, China; 2010744@stu.neu.edu.cn
[*] Correspondence: duchuanming@smm.neu.edu.cn; Tel.: +86-155-2418-9830

Abstract: Separating P_2O_5 from steelmaking slag is the key to achieving optimum resource utilization of slag. If the P-concentrating $2CaO \cdot SiO_2$–$3CaO \cdot P_2O_5$ solid solution was effectively separated, it can be a potential phosphate resource and the remaining slag rich in Fe_2O_3 and CaO can be reutilized as a flux in steelmaking process. In this study, a low-cost method of selective leaching was adopted, and hydrochloric acid was selected as leaching agent. The dissolution behavior of quenched steelmaking slags with different composition in the acidic solution was investigated and the dissolution mechanism was clarified. It was found that the P dissolution ratio from each slag was higher than those of other elements, achieving an effective separation of P and Fe. The dissolution ratios of P, Ca, and Si decreased as the P_2O_5 content in slag increased. A higher Fe_2O_3 content in slag led to a lower P dissolution ratio. Increasing slag basicity facilitated the dissolution of P from slag. The residue mainly composed of matrix phase and the P_2O_5 content decreased significantly through selective leaching. The P dissolution ratio from slag was primarily determined by the P distribution ratio in the $2CaO \cdot SiO_2$–$3CaO \cdot P_2O_5$ solid solution and the precipitation of ferric phosphate in the leachate. The P-concentrating solid solution was effectively separated from quenched steelmaking slag, even though hydrochloric acid was used as leaching agent.

Keywords: steelmaking slag; phosphorus; leaching; $2CaO \cdot SiO_2$–$3CaO \cdot P_2O_5$

Citation: Lv, N.-N.; Du, C.-M.; Kong, H.; Yu, Y.-H. Leaching of Phosphorus from Quenched Steelmaking Slags with Different Composition. *Metals* **2021**, *11*, 1026. https://doi.org/10.3390/met11071026

Academic Editors: Dariush Azizi and Anna Kaksonen

Received: 26 May 2021
Accepted: 22 June 2021
Published: 25 June 2021

1. Introduction

Phosphorus is a significant nutrient element for animal and plant growth; however, it is one of the most detrimental impurities in the iron and steel industry, and most of the phosphorus is eliminated into slag in the steelmaking process [1]. To improve dephosphorization efficiency and reduce slag generation, hot metal dephosphorization was developed in Japan and widely adopted in steel plants [2]. In this process, dephosphorization and decarburization was conducted in converter, respectively, and dephosphorization slag and converter slag were generated [3]. Due to lower P_2O_5 content, converter slag can be recycled as a flux in dephosphorization process, and then only dephosphorization slag with relatively low basicity is emitted. The hot metal dephosphorization slag normally consists of CaO–SiO_2–FeO–P_2O_5 system, and the industrial operation is mainly carried out in the dicalcium silicate ($2CaO \cdot SiO_2$)-saturated composition range [4,5]. The amount of steelmaking slag is approximately 100~150 kg of per ton of steel [6], while the utilization ratio of steelmaking slag is not high. Large amounts of steelmaking slag are piled up or landfilled directly, causing tremendous waste of valuable components.

As the utilization of iron ores with higher P content, the P_2O_5 content in steelmaking slag is continuously increasing, and then steelmaking slag is regarded an important material to substitute for phosphate rocks [7]. It is well known that $2CaO \cdot SiO_2$ forms a solid solution with tricalcium phosphate ($3CaO \cdot P_2O_5$) at the treatment temperature over a wide composition range [8,9]. This implies that the product of dephosphorization reaction can be concentrated in the $2CaO \cdot SiO_2 - 3CaO \cdot P_2O_5$ (C_2S-C_3P) solid solution, which provides the foundation for P recovery. The P-concentrating C_2S-C_3P solid solution separated from slag can be used as phosphate resource and the remaining slag rich in Fe_tO and CaO can be reutilized as a flux in the ironmaking and steelmaking process, achieving the comprehensive utilization of steelmaking slag.

Various studies have been conducted on the removal of phosphorus from steelmaking slag. Li et al. [10] used centrifugal separation to remove C_2S-C_3P solid solution from the molten slag at high temperature according to the density differences for different mineral phases. Kubo et al. [11] and Lin et al. [12] studied the removal of nC_2S-C_3P solid solution from steelmaking slag according to the differences in the magnetic properties of mineral phases. Recently, some researchers focused on the P recovery from steelmaking slag using acid leaching. Numata et al. [13] reported that in the case of Fe_2O_3-containing slag, the dissolution ratio of each element in the matrix phase was lower than that in the solid solution at various pH conditions. Qiao et al. [14] investigated the dissolution behavior of slag in the buffer solution of $C_6H_8O_7$–NaOH–HCl system and found that most of the P was dissolved while the Fe dissolution ratio was also high. Du et al. [15,16] clarified that Na_2O modification and oxidization of molten slag was beneficial for the dissolution of C_2S-C_3P solid solution from steelmaking slag with high P_2O_5 content in the citric acid solution. Under the optimum conditions, the P dissolution ratio exceeded 85% and the dissolution of Fe was negligible, achieving selective leaching of P. After leaching, most of the P dissolved in leachate can be recovered as calcium phosphates by chemical precipitation, illustrating that acid leaching is an effective and low-cost method to recover P from steelmaking slag [17].

Concerning steelmaking slag with high P_2O_5 content, previous studies primarily studied the selective leaching of P from the furnace-cooled slag in the citric acid solution. Because of slow cooling and the use of organic acid, it resulted in a high treatment cost. In this study, to reduce treatment cost, hydrochloric acid (HCl) was selected as a leaching agent and the quenched slag from dephosphorization process was used. The dissolution behavior of P from quenched steelmaking slags with different composition were investigated. The aim of this study is to achieve an efficient separation of P from steelmaking slag with a simple and low-cost method. It is expected that this will provide theoretical and technical basis for the high value-added utilization of steelmaking slag.

2. Experimental

As reported in previous studies [17,18], the existence of Fe_2O_3 and Na_2O modification was beneficial for the selective leaching of P from slag. In this study, slag composition was simplified, and dephosphorization slags consisting of $CaO-SiO_2-Fe_2O_3-P_2O_5-Na_2O$ system were used. Compared with converter slag, these slags had relatively low slag basicity. Eight kinds of slags with different P_2O_5, Fe_2O_3, and basicity (CaO/SiO_2) were synthesized using reagent-grade $CaCO_3$, SiO_2, Fe_2O_3, $Ca_3(PO_4)_2$, and Na_2SiO_3. The mixed chemical reagents were firstly heated to 1823 K to form a homogeneous liquid slag in a Pt crucible under air. Then, it was cooled to 1673 K at a cooling rate of 3 K/min and held 20 min to precipitate the C_2S-C_3P solid solution. Finally, slag was quickly taken out of the furnace and quenched in water. The synthesized slag was ground and sieved into particles of less than 53 μm. After performing aqua-regia digestion, the element concentration in each slag was determined using inductively coupled plasma atomic emission spectroscopy (ICP-AES) (SPECTRO, Kleve, Germany). Table 1 lists the actual composition of synthesized steelmaking slags. The mineralogical composition and morphology of mineral phases in slag was determined using X-ray diffraction (XRD) (Rigaku Corporation, Tokyo, Japan) analysis and electron probe microanalysis (EPMA) (JEOL, Tokyo, Japan).

Table 1. Actual composition of synthesized slags (mass%).

No.	CaO	SiO$_2$	Fe$_2$O$_3$	P$_2$O$_5$	Na$_2$O	Basicity (C/S)
Slag 1	42.1	24.6	20.4	7.6	5.3	1.71
Slag 2	40.7	23.9	19.6	10.5	5.3	1.70
Slag 3	39.3	23.1	19.1	13.3	5.2	1.70
Slag 4	38.0	22.1	18.2	16.3	5.4	1.72
Slag 5	43.8	25.6	14.9	10.4	5.3	1.71
Slag 6	37.2	21.6	25.4	10.3	5.5	1.72
Slag 7	38.5	26.3	19.8	10.1	5.3	1.46
Slag 8	42.4	22.1	19.9	10.4	5.2	1.92

A Teflon vessel containing 300 mL of distilled water was placed in an isothermal water bath. 1.5 g of slag was added to keep the mass ratio (slag to solution) as 1:200 to cause the slag to fully dissolve, as described in previous study [19]. The slurry was agitated using a rotating stirrer at 200 rpm at room temperature (298 K). The diluted hydrochloric acid (0.4 mol/L) was used as a leaching agent and it was added to the slurry by a pH control and solution addition system. In previous study [18], most of the P could be dissolved from the furnace-cooled slag at pH 4. Hence, the pH of slurry was maintained at 4 to achieve the selective leaching of P. At appropriate intervals, approximately 4 mL of slurry was sampled, and filtered using a syringe filter (<0.45 μm). The concentration of each element in the leachate was analyzed using ICP-AES. After 120 min, the slurry was separated by vacuum filtration and the obtained residue was dried at 373 K. The residue was weighed and analyzed by XRD and EPMA. The chemical composition of residue was determined using the same method for slag analysis.

The dissolution ratio of element X (R_X) from steelmaking slag was calculated using the element concentration in the leachate, as expressed in Equation (1):

$$R_X = \frac{C_X \cdot V \cdot M_{XO}}{m \cdot w_{XO} \cdot M_X} \tag{1}$$

where C_X is the element X concentration in the leachate (mg/L); V is the final leachate volume (L); m is the slag mass (mg); w_{XO} is the oxide XO content in slag; M is molar mass.

3. Results

3.1. Mineralogical Composition

The morphology of mineral phases in synthesized slags with various composition is shown in Figure 1. Each slag principally composed of two mineral phases, and the mass fractions of mineral phases in different slag were obviously different. Table 2 lists the average composition of each mineral phase in slag. The black mineral phase consisting of CaO–SiO$_2$–P$_2$O$_5$ slag system is considered the C$_2$S–C$_3$P solid solution; the grey mineral phase consisting of CaO–SiO$_2$–Fe$_2$O$_3$ slag system is regarded the amorphous matrix phase. The high distribution ratio of P$_2$O$_5$ between the solid solution and the matrix phase indicated that the majority of P$_2$O$_5$ was concentrated in the C$_2$S–C$_3$P solid solution. The enrichment of P$_2$O$_5$ and Fe$_2$O$_3$ in different mineral phases was the basis of P separation by selective leaching. A part of Na$_2$O was also distributed in the C$_2$S–C$_3$P solid solution, which could promote the dissolution of solid solution. As the P$_2$O$_5$ content in slag increased, the P$_2$O$_5$ contents in the solid solution and in the matrix phase both increased. In the case of Slag 4, containing 16.3% P$_2$O$_5$, the solid solution almost consisted of 3CaO·P$_2$O$_5$. As the Fe$_2$O$_3$ content in slag increased, the P$_2$O$_5$ content in the solid solution increased, whereas that in the matrix phase had little change. The P$_2$O$_5$ contents in the solid solution and in the matrix phase both decreased with the increase in slag basicity, but the distribution ratio of P$_2$O$_5$ was still high. If P$_2$O$_5$ was sufficiently concentrated in the C$_2$S–C$_3$P solid solution which could be fully dissolved, separation of P from steelmaking slag could be achieved.

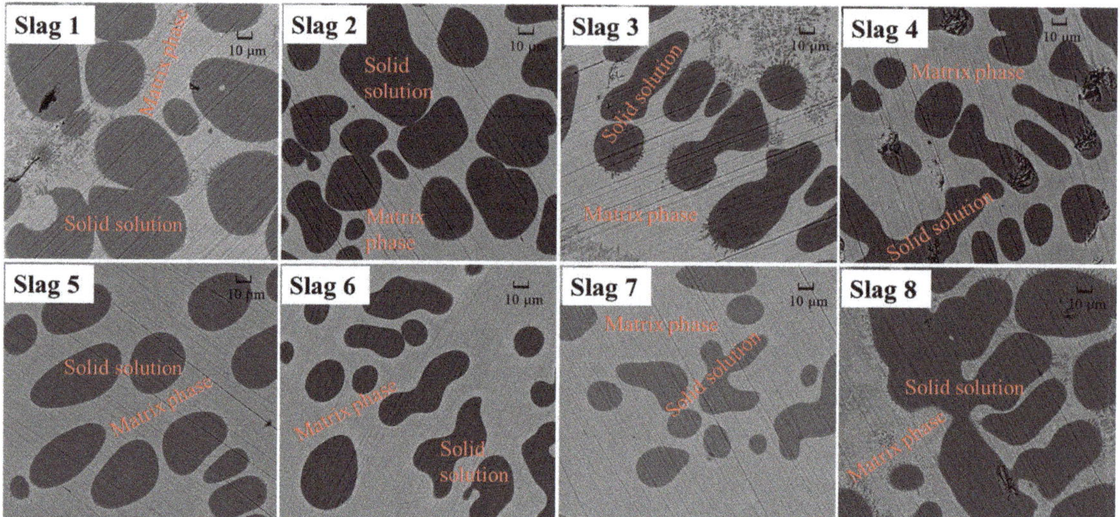

Figure 1. Morphology of mineral phases in the quenched steelmaking slags (EPMA analysis).

Table 2. Average composition of each phase in steelmaking slag (mass%).

		CaO	SiO$_2$	Fe$_2$O$_3$	P$_2$O$_5$	Na$_2$O
Solid Solution	Slag 1	54.8	22.4	1.2	16.2	5.4
	Slag 2	52.3	14.7	1.1	25.1	6.7
	Slag 3	49.9	10.2	1.0	32.1	6.8
	Slag 4	48.0	5.7	0.9	38.5	6.8
	Slag 5	52.6	19.2	1.1	21.2	6.0
	Slag 6	48.1	12.6	1.0	30.2	8.0
	Slag 7	50.7	12.1	1.1	29.4	6.8
	Slag 8	53.7	18.7	1.1	20.8	5.8
Matrix Phase	Slag 1	32.7	28.2	32.5	1.7	4.9
	Slag 2	35.3	28.6	27.8	4.0	4.3
	Slag 3	34.1	29.1	27.0	5.5	4.2
	Slag 4	32.2	30.7	26.2	6.4	4.4
	Slag 5	34.9	31.7	24.9	3.9	4.5
	Slag 6	31.0	26.1	34.5	4.0	4.4
	Slag 7	35.8	30.7	22.5	6.7	4.4
	Slag 8	33.4	26.5	33.1	2.9	4.1

3.2. Leaching Results

The effect of P$_2$O$_5$ content on the change in Ca and P concentrations with leaching time is shown in Figure 2a. The dissolution of each slag primarily occurred in the initial period, resulting in significant increase in Ca and P concentrations. Their concentrations had a little increase after 60 min. With the increase in P$_2$O$_5$ content in slag, the P concentration increased significantly in the leachate, while the Ca concentration decreased. For Slag 4, containing 16.3% P$_2$O$_5$, the Ca and P concentrations reached 507.4 mg/L and 211.8 mg/L, respectively, after 120 min. The calculated dissolution ratios of main elements from slag are shown in Figure 2b. It is worth noting that these elements presented different dissolution behavior. The P dissolution ratio was the highest, and it was approximately 70% in the case of Slag 1. Fe was hard to dissolve, and its dissolution ratio was almost zero. As the P_2O_5 content in slag increased, the dissolution ratios of P, Ca, and Si all decreased, indicating that slag dissolution became difficult. Compared with the leaching results of furnace-cooled slag in the citric acid solution [17], the P dissolution ratio was a little lower, but the dissolution of Fe was negligible, achieving an effective separation of P and Fe as well. The majority of P was dissolved from each slag without a large dissolution of other elements.

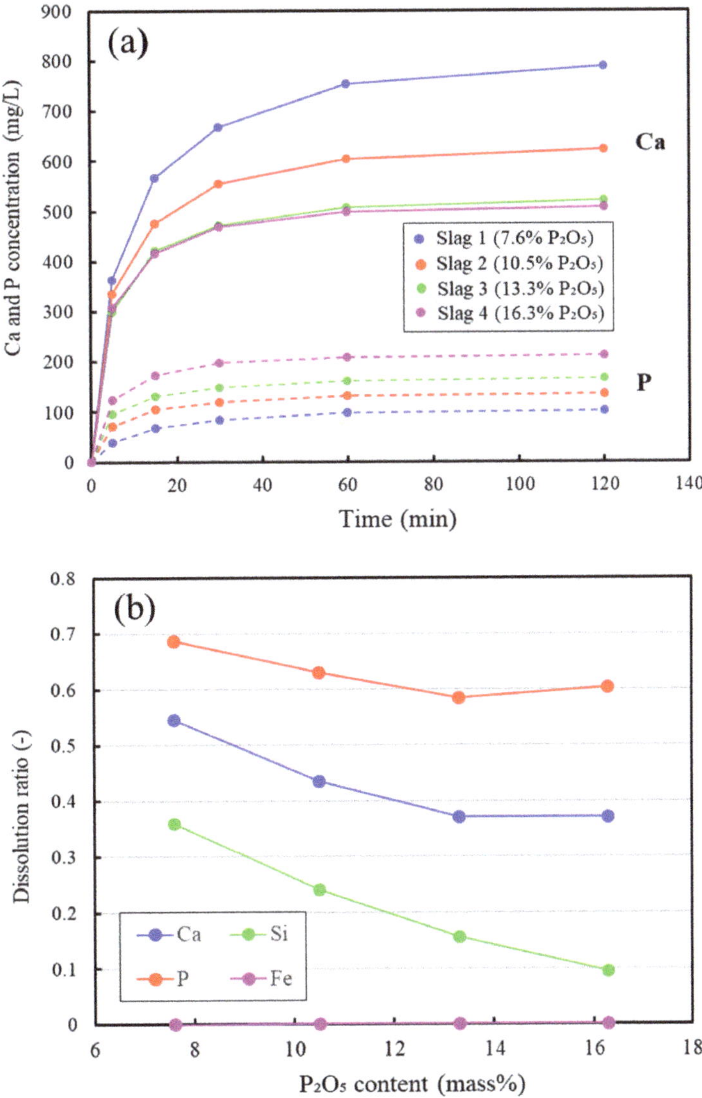

Figure 2. (**a**) Change in the Ca and P concentrations with leaching time; (**b**) Dissolution ratios of main elements from slags with different P_2O_5 contents.

Figure 3a shows the change in Ca and P concentrations with time when slags with different Fe_2O_3 contents were leached. With the increase in Fe_2O_3 content, the Ca concentration in the leachate decreased, whereas the P concentration showed little change, approximately 140 mg/L after 120 min. The dissolution ratios of the main elements from slags with different Fe_2O_3 contents are presented in Figure 3b. The dissolution of Fe was very difficult regardless of Fe_2O_3 content. When the Fe_2O_3 content increased from 14.9% to 19.6%, the Ca and Si dissolution ratios decreased dramatically, while the P dissolution ratio decreased by only 3%. If the Fe_2O_3 content continued to increase, it had a little influence on the dissolution ratio. Although increasing Fe_2O_3 content slightly decreased P dissolution, it significantly suppressed the dissolution of other elements, which was beneficial for selective leaching.

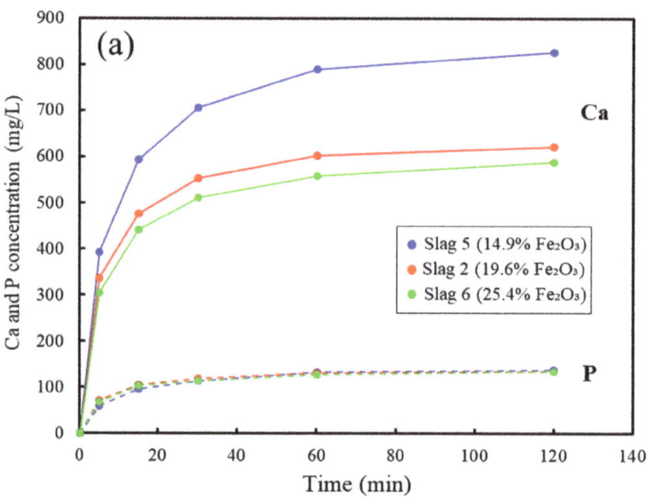

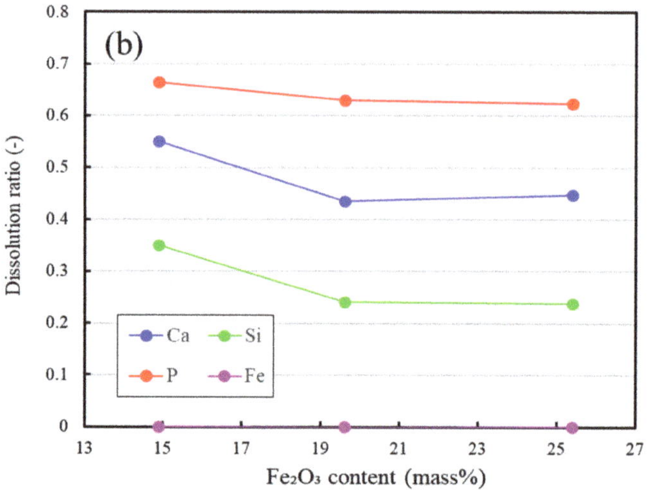

Figure 3. (**a**) Change in the Ca and P concentrations with leaching time; (**b**) Dissolution ratios of main elements from slag with different Fe_2O_3 contents.

The effect of slag basicity on the dissolution behavior of slag is shown in Figure 4. The Ca and P concentrations both increased significantly with slag basicity. The increasing tendency toward dissolution ratios of P, Ca, and Si was almost the same. In the case of low basicity, the dissolution of slag was difficult, resulting in a lower P dissolution ratio. When slag basicity increased to 1.92, the P dissolution ratio reached 77.4% and Fe did not dissolve, illustrating that selective leaching was achieved. The Ca and Si dissolution ratios were 61.7% and 42.2%, respectively. These results suggest that increasing slag basicity facilitates the dissolution of P from slag. Combining with mineralogical composition, it was found that a higher P_2O_5 content in the C_2S-C_3P solid solution caused lower dissolution ratio of each element from slag. To promote P dissolution, the P_2O_5 content in the solid solution should be lowered.

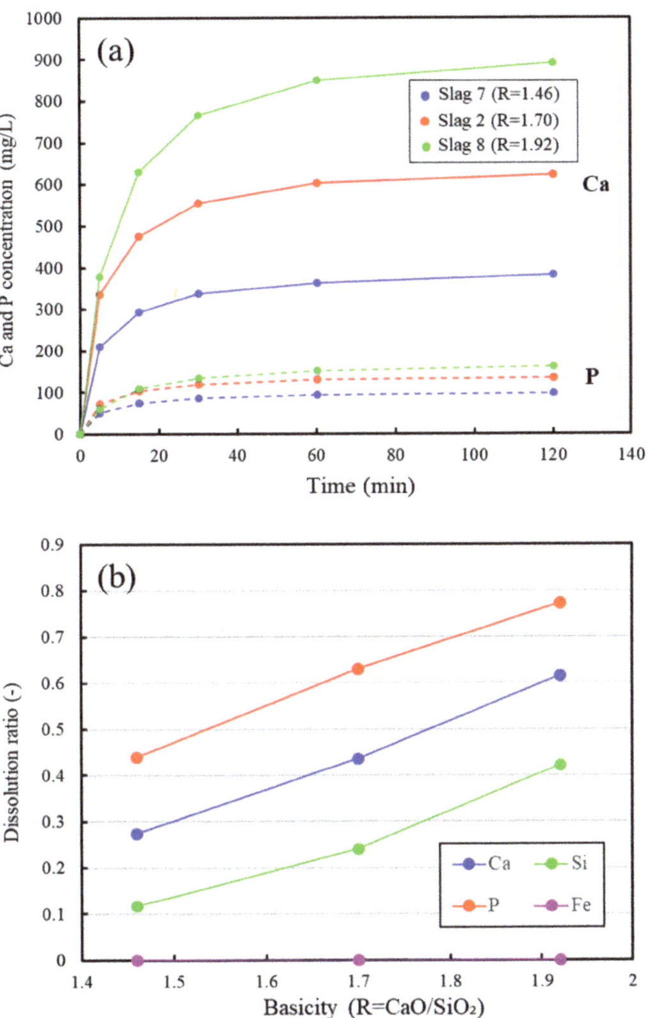

Figure 4. (**a**) Change in the Ca and P concentrations with leaching time; (**b**) Dissolution ratios of main elements from slags with different basicity.

3.3. Characterization of Residue

The XRD patterns of Slag 2 and its residue after leaching is shown in Figure 5. For Slag 2, the characteristic peaks associated with C_2S–C_3P solid solution and the broad peaks of non-crystalline phase were observed, confirming the existence of two mineral phases. Following leaching, the characteristic peaks of C_2S–C_3P solid solution all disappeared, and the broad peaks of non-crystalline phase intensified, illustrating that the P-concentrating phase was fully dissolved. Figure 6 shows the morphology of residue after the leaching of Slag 2. This residue consisted of some irregular particles with single mineral phase. As listed in Table 3, this mineral phase mainly containing CaO, SiO_2, and Fe_2O_3 was similar to the matrix phase. It indicates that only the matrix phase remains in the residue, which is consistent with the XRD results. The average composition of residue is shown in Table 4. Compared with the composition in Table 1, the P_2O_5 content decreased from 10.5% to 4.3%, proving that the C_2S–C_3P solid solution was effectively separated, while the Fe_2O_3 content increased from 19.6% to 31.0%.

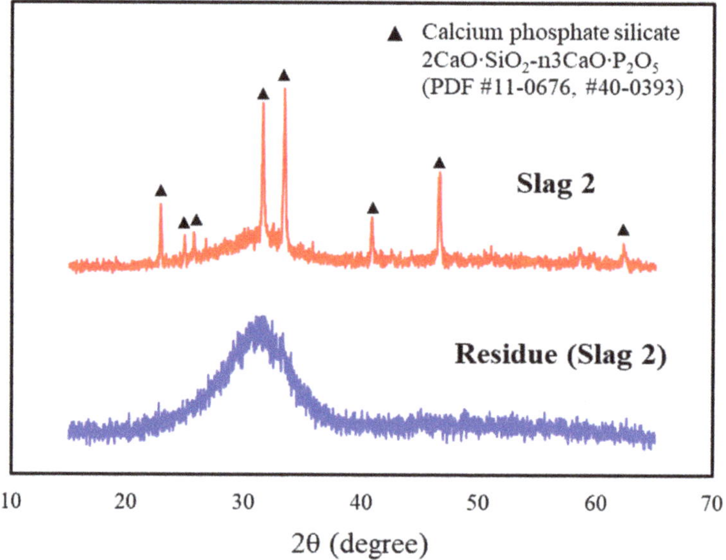

Figure 5. XRD patterns of Slag 2 and its residue after leaching.

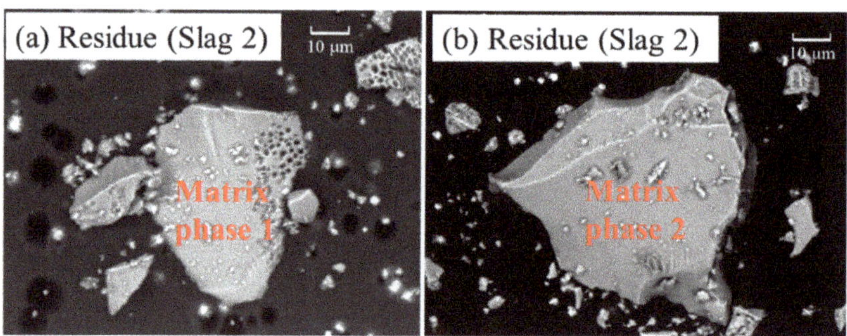

Figure 6. (**a,b**) Morphology of residue surface of Slag 2 (EPMA analysis).

Table 3. Average composition of mineral phases on the residue surface (mass%).

Matrix Phase	Ca	Si	Fe	P	Na	O
1	24.2	12.8	21.1	1.8	3.0	37.1
2	24.2	10.1	17.6	2.1	3.4	42.6

Table 4. Chemical composition of the residue of Slag 2 (mass%).

	CaO	SiO$_2$	Fe$_2$O$_3$	P$_2$O$_5$	Na$_2$O
Residue (Slag 2)	32.9	28.2	31.0	4.3	3.6

Overall, in this process, the P-containing leachate can be used to extract calcium phosphates by chemical precipitation, and the residue consisting of CaO–SiO$_2$–Fe$_2$O$_3$ system has lower P$_2$O$_5$ content, which can be used as a flux in the steelmaking process. These results will promote the comprehensive utilization of steelmaking slag and ease environmental burden.

4. Discussion

To achieve the selective leaching of P, the dissolution behavior and mechanism of P from quenched steelmaking slag should be understood. Several kinds of cations and anions exist in the leachate. Since phosphate ions can precipitate with metallic ions, it is necessary to consider the possibility of phosphate formation. In the aqueous solution, hydroxyapatite (Ca$_{10}$(PO$_4$)$_6$(OH)$_2$) and strengite (FePO$_4$·2H$_2$O) readily precipitate when Ca^{2+} and Fe^{3+} ions coexist with phosphate ions [20]. The precipitation of these phosphates plays a significant role in determining the P concentration in the leachate. Therefore, we investigated the concentration relationship between phosphate ions and metallic ions in the aqueous solution. There are several types of phosphate ions in the aqueous solution depending on pH, including PO$_4^{3-}$, HPO$_4^{2-}$, and H$_2$PO$_4^{-}$ [21]. In thermodynamic calculation, these phosphates ions were considered, and the activity coefficients of ions were assumed to be 1 because their concentrations were relatively low. The solubility lines of FePO$_4$·2H$_2$O and Ca$_{10}$(PO$_4$)$_6$(OH)$_2$ were calculated using the reaction equilibrium constants of Equations (2)–(5) at pH 4, respectively [22,23].

$$FePO_4 \cdot 2H_2O = Fe^{3+} + PO_4^{3-} + 2H_2O \quad logK = -26.07 \tag{2}$$

$$Ca_2(PO_4)_6(OH)_2 + 2H^+ = 10Ca^{2+} + 6PO_4^{3-} + 2H_2O \quad logK = -62.42 \tag{3}$$

$$PO_4^{3-} + H^+ = HPO_4^{2-} \quad logK = 12.36 \tag{4}$$

$$PO_4^{3-} + 2H^+ = H_2PO_4^- \quad logK = 19.56 \tag{5}$$

Figure 7 shows the relationship between P and Fe concentrations in the aqueous solution and experimental results. It was found that the solubility of FePO$_4$·2H$_2$O was very low at pH 4 and large amounts of phosphate and Fe^{3+} ions could not coexist in the aqueous solution. The saturation concentration of P decreased with the increase in Fe concentration. The P concentration in the leachate was high while the Fe concentration was very low, near zero. The points of experimental result all located above the solubility line of FePO$_4$·2H$_2$O, indicating that the P and Fe concentrations were supersaturated and FePO$_4$·2H$_2$O could precipitate. During leaching, the C$_2$S–C$_3$P solid solution dissolved well, and the dissolution of the Fe-containing matrix phase was low, resulting in high concentrations of Ca, Si, and P. As approximately 1.0% Fe$_2$O$_3$ existed in the solid solution, some Fe was also dissolved. However, it was difficult for these Fe^{3+} ions to coexist with phosphate ions, and then Fe^{3+} ions precipitated in the form of FePO$_4$·2H$_2$O. Owing to a small quantity of Fe dissolved from slag, the formation of FePO$_4$·2H$_2$O had little influence on the decrease in P concentration in the leachate. To make phosphate ions exist stably in the leachate, the dissolution of Fe from slag should be suppressed as much as possible.

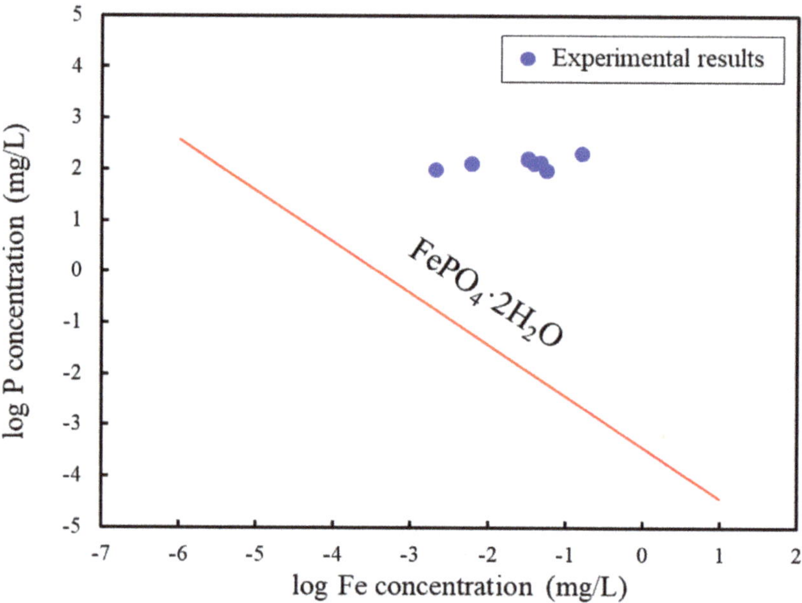

Figure 7. Relationship between P and Fe concentrations in the aqueous solution and experimental results.

Because the Ca concentration was higher than other elements in the leachate, the possibility of $Ca_{10}(PO_4)_6(OH)_2$ formation was evaluated. As shown in Figure 8, it was difficult for the precipitation of calcium phosphates to occur in the aqueous solution at pH 4 unless concentrations on Ca and P were high. In this study, the maximum P concentration was 211.8 mg/L and it was far lower than the saturation concentration. The points of P and Ca concentrations were located far below the solubility line of $Ca_{10}(PO_4)_6(OH)_2$, illustrating that Ca concentration has no effect on phosphate precipitation. The P concentration in the leachate primarily depended on the Fe concentration. Under this condition, most of the P dissolved from slag existed stably in the leachate, proving that selective leaching of P from quenched steelmaking slag with high P_2O_5 content was possible, even in the hydrochloric acid solution.

To evaluate P-selective leaching from each slag, we compared the P distribution ratio in the C_2S–C_3P solid solution with the P dissolution ratio from slag. The mass fractions of solid solution and matrix phase were first calculated using Equations (6) and (7), where α and β are the mass fraction of solid solution and matrix phase, respectively, w^{α}_{XO} and w^{β}_{XO} are the XO content in the solid solution and matrix phase, respectively. The mass fraction was defined as the average results of each oxide. Then, the P distribution ratio in the solid solution (D) was calculated using Equation (8).

$$w_{XO} = \alpha w^{\alpha}_{XO} + \beta w^{\beta}_{XO} \tag{6}$$

$$\alpha + \beta = 1 \tag{7}$$

$$D = \frac{\alpha w^{\alpha}_{P_2O_5}}{w_{P_2O_5}} \tag{8}$$

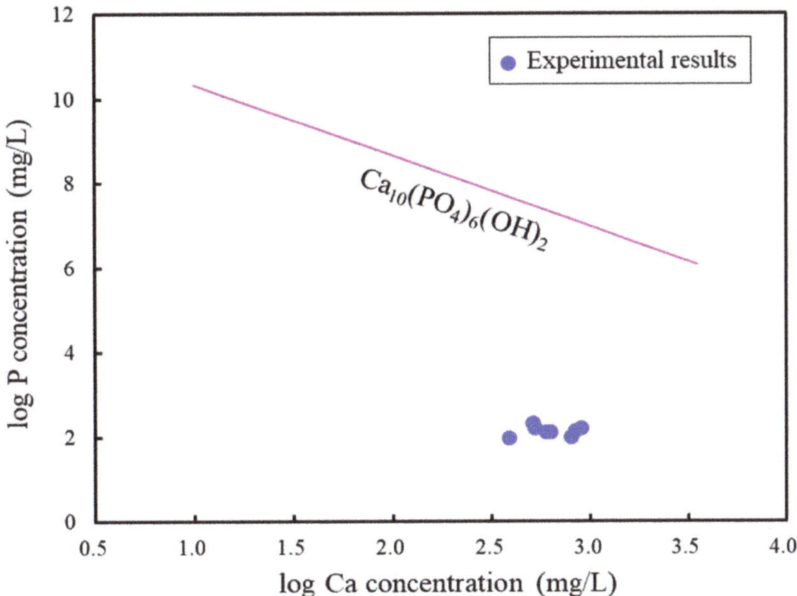

Figure 8. Relationship between P and Ca concentrations in the aqueous solution and experimental results.

As shown in Figure 9, most of the P in slag was distributed in the C_2S–C_3P solid solution. With the P_2O_5 content increased, the P distribution ratio in the solid solution decreased, suggesting that the slag with lower P_2O_5 content facilitated P enrichment. 70.8% of P in Slag 4 containing 16.3% P_2O_5 was concentrated in the solid solution. Increasing Fe_2O_3 content was not beneficial for P enrichment. A higher slag basicity resulted in a higher P distribution ratio in the solid solution. For Slag 7 with low basicity, only half of the P in slag was distributed in the solid solution, which made it difficult to achieve P-selective leaching. In each case, the P dissolution ratio from slag was a little lower than the P distribution ratio in the solid solution. Its variation tendency was the same as that of P distribution ratio. One reason for this was considered to be that a small part of P-concentrating solid solution was not dissolved. The other reason was that a small amount of P dissolved from slag precipitated with Fe^{3+} ions, resulting in a little decrease in P dissolution ratio. Although the P dissolution ratio was not very high, the effective dissolution of P-concentrating solid solution from quenched steelmaking slag was achieved under this condition, similar with the furnace-cooled slags as reported in previous studies [16–18].

To better understand selective leaching, we weighted the mass of remained residue and compared that with the mass fraction of matrix phase. As shown in Figure 10, a large amount of slag was not dissolved under this condition, and the mass fraction of residue was almost equal to that of matrix phase in each slag. This proved that the vast majority of P-concentrating solid solution was dissolved, and the dissolution of matrix phase was difficult, which was consistent with the residue analysis. For Slag 8 with high basicity, the mass fraction of solid solution was high, and then a large amount of slag was dissolved. In summary, the P dissolution ratio from slag was mainly determined by the P enrichment in the C_2S–C_3P solid solution and phosphate precipitation in the leachate. The P-concentrating solid solution was effectively dissolved and separated from quenched steelmaking slag when hydrochloric acid was used as leaching agent.

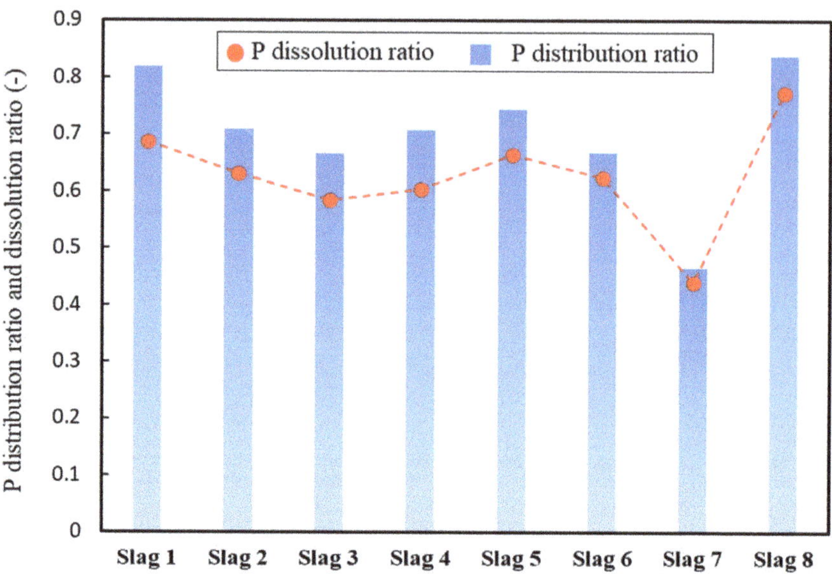

Figure 9. P distribution ratio in the solid solution compared with P dissolution ratio from slag.

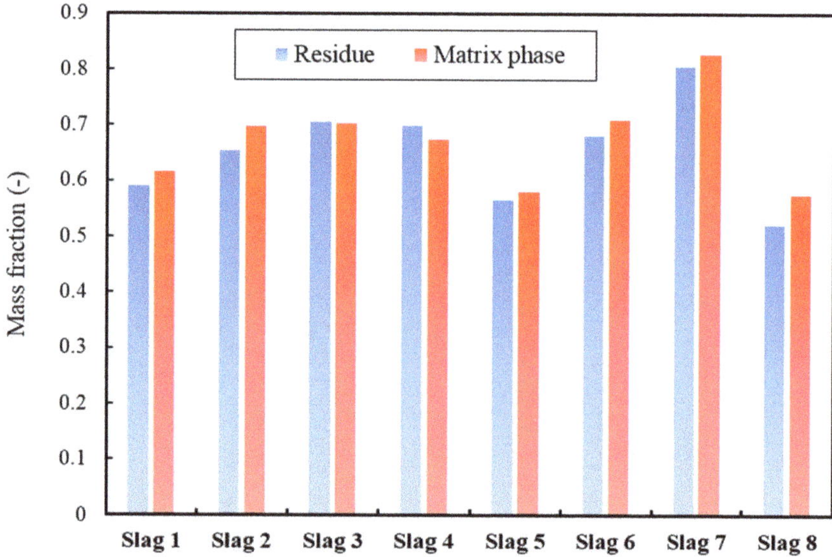

Figure 10. Mass fraction of matrix phase in each slag compared with that of residue.

5. Conclusions

To separate P from steelmaking slag using a simple and low-cost method, selective leaching of C_2S–C_3P solid solution was adopted, and hydrochloric acid was selected as leaching agent. In this study, the dissolution behavior of quenched steelmaking slags with different composition in the acidic solution was investigated. The results obtained are summarized below:

(1) The P dissolution ratio from slag was higher than those of other elements in each case, achieving an effective separation of P and Fe. The C_2S–C_3P solid solution was fully dissolved from slag, and then the residue primarily composed of the matrix phase, causing a significant decrease in P_2O_5 content.

(2) The dissolution ratios of P, Ca, and Si decreased as the P_2O_5 content in slag increased. A higher Fe_2O_3 content in slag resulted in a lower P dissolution ratio. Increasing slag basicity facilitated the dissolution of P from slag.

(3) The P dissolution ratio from slag was mainly determined by the P distribution ratio in the solid solution and the precipitation of ferric phosphate in the leachate. The P-concentrating solid solution was effectively separated from quenched steelmaking slag, even though hydrochloric acid was used as leaching agent, which provided a simple and low-cost method to recover valuable components from steelmaking slag.

Author Contributions: Conceptualization, C.-M.D. and N.-N.L.; methodology, C.-M.D.; investigation, H.K. and Y.-H.Y.; resources, N.-N.L. and C.-M.D.; writing—original draft preparation, N.-N.L.; writing—review and editing, C.-M.D., H.K., and Y.-H.Y.; All authors have read and agreed to the published version of the manuscript.

Funding: This work was Funded by the Open Project Program of Key Laboratory of Metallurgical Emission Reduction & Resources Recycling (Anhui University of Technology), Ministry of Education (No. JKF20-01), Fundamental Research Funds for the Central Universities (No. N2025005), the National Natural Science Foundation of China (No. 52074004), and Director Fund of Anhui Province Key Laboratory of Metallurgical Engineering & Resources Recycling (Anhui University of Technology).

Data Availability Statement: Data are contained within the article.

Conflicts of Interest: The authors declare no conflict of interest.

References

1. Matsubae-Yokoyama, K.; Kubo, H.; Nakajima, K.; Nagasaka, T. A material flow analysis of phosphorus in Japan. *J. Ind. Ecol.* **2009**, *13*, 687–705. [CrossRef]
2. Ogawa, Y.; Maruoka, N. Progress of hot metal treatment technology and future outlook. *Tetsu Hagané* **2014**, *100*, 434–444. [CrossRef]
3. Kitamura, S.; Naito, K.; Okuyama, G. History and latest trends in converter practice for steelmaking in Japan. *Min. Proc. Ext. Met.* **2019**, *128*, 34–45. [CrossRef]
4. Ito, K.; Yanagisawa, M.; Sano, N. Phosphorus distribution between solid $2CaO \cdot SiO_2$ and molten CaO–SiO_2–FeO–Fe_2O_3 slags. *Tetsu Hagane* **1982**, *68*, 342–344. [CrossRef]
5. Kitamura, S.; Yonezawa, K.; Ogawa, Y.; Sasaki, N. Improvement of reaction efficiency in hot metal dephosphorization. *Ironmak. Steelmak.* **2002**, *29*, 121–124. [CrossRef]
6. Gao, X.; Okubo, M.; Maruoka, N.; Shibata, H.; Ito, T.; Kitamura, S.Y. Production and utilization of iron and steelmaking slag in Japan and the application of steelmaking slag for the recovery of paddy fields damaged by Tsunami. *Min. Proc. Ext. Met.* **2015**, *124*, 116–124.
7. Matsubae, K.; Yamasue, E.; Inazumi, T.; Webeck, E.; Miki, T.; Nagasaka, T. Innovations in steelmaking technology and hidden phosphorus flows. *Sci. Total. Environ.* **2016**, *542*, 1162–1168. [CrossRef] [PubMed]
8. Fix, W.; Heymann, H.; Heinke, R. Subsolidus relations in the system $2CaO \cdot SiO_2$-$3CaO \cdot P_2O_5$. *J. Am. Ceram. Soc.* **1969**, *52*, 346–347. [CrossRef]
9. Zhu, B.; Zhu, M.M.; Luo, J.; Dou, X.F.; Wang, Y.; Jiang, H.J.; Xie, B. Distribution behavior of phosphorus in $2CaO \cdot SiO_2$-$3CaO \cdot P_2O_5$ solid solution phase and liquid slag phase. *Metals* **2020**, *10*, 1103. [CrossRef]
10. Li, C.; Gao, J.; Guo, Z. Separation of phosphorus- and iron-enriched phase from CaO-SiO_2-FeO-MgO-P_2O_5 melt with super gravity. *Metall. Mater. Trans. B* **2016**, *47*, 1516–1519. [CrossRef]
11. Kubo, H.; Matsubae-Yokoyama, K.; Nagasaka, T. Magnetic separation of phosphorus enriched phase from multiphase dephosphorization slag. *ISIJ Int.* **2010**, *50*, 59–64. [CrossRef]
12. Lin, L.; Bao, Y.P.; Wang, M.; Jiang, W.; Zhou, H.M. Separation and recovery of phosphorus from P-bearing steelmaking slag. *J. Iron Steel Res. Int.* **2014**, *21*, 496–502. [CrossRef]
13. Numata, M.; Maruoka, N.; Kim, S.; Kitamura, S. Fundamental experiment to extract phosphorous selectively from steelmaking slag by leaching. *ISIJ Int.* **2014**, *54*, 1983–1990. [CrossRef]
14. Qiao, Y.; Diao, J.; Liu, X.; Li, X.S.; Zhang, T.; Xie, B. Dephosphorization of steelmaking slag by leaching with acidic aqueous solution. *JOM* **2016**, *68*, 2511–2519. [CrossRef]

15. Du, C.M.; Gao, X.; Ueda, S.; Kitamura, S. Effect of Na_2O addition on phosphorus dissolution from steelmaking slag with high P_2O_5 content. *J. Sustain. Metall.* **2017**, *3*, 671–682. [CrossRef]
16. Du, C.M.; Gao, X.; Ueda, S.; Kitamura, S. Recovery of phosphorus from modified steelmaking slag with high P_2O_5 content via leaching and precipitation. *ISIJ Int.* **2018**, *58*, 833–841. [CrossRef]
17. Du, C.M.; Gao, X.; Ueda, S.; Kitamura, S. Optimum conditions for phosphorus recovery from steelmaking slag with high P_2O_5 content by selective leaching. *ISIJ Int.* **2018**, *58*, 860–868. [CrossRef]
18. Du, C.M.; Gao, X.; Ueda, S.; Kitamura, S. Effect of Fe^{2+}/T.Fe ratio on the dissolution behavior of P from steelmaking slag with high P_2O_5 content. *J. Sustain. Metall.* **2018**, *4*, 443–454. [CrossRef]
19. Teratoko, T.; Maruoka, N.; Shibata, H.; Kitamura, S. Dissolution behavior of dicalcium silicate and tricalcium phosphate solid solution and other phases of steelmaking slag in an aqueous solution. *High. Temp. Mater. Processes.* **2012**, *31*, 329–338. [CrossRef]
20. Lindsay, W.L.; Moreno, E.C. Phosphate phase equilibria in soils. *Soil. Sci. Soc. Am. J.* **1960**, *24*, 177–182. [CrossRef]
21. Shen, H.Y.; Wang, Z.J.; Zhou, A.M.; Chen, J.L.; Hu, M.Q.; Dong, X.Y.; Xia, Q.H. Adsorption of phosphate onto amine functionalized nano-sized magnetic polymer adsorbents: Mechanism and magnetic effects. *RSC Adv.* **2015**, *5*, 22080–22090. [CrossRef]
22. Markich, S.J.; Brown, P.L. *Thermochemical Data for Environmentally-Relevant Elements*; ANSTO Environment Division: Sydney, NSW, Australia, 1999.
23. Futatsuka, T.; Shitogiden, K.; Miki, T.; Nagasaka, T.; Hino, M. Dissolution behavior of nutrition elements from steelmaking slag into seawater. *ISIJ Int.* **2004**, *44*, 753–761. [CrossRef]

Article

Comparative Studies of Digestion Techniques for the Dissolution of Neodymium-Based Magnets

Mélodie Bonin [1,2], Frédéric-Georges Fontaine [1,3] and Dominic Larivière [1,2,*]

1   Chemistry Department, Université Laval, Québec City, QC G1V 0A6, Canada; melodie.bonin.1@ulaval.ca (M.B.); frederic.fontaine@chm.ulaval.ca (F.-G.F.)
2   Radioecology Laboratory, Université Laval, Québec City, QC G1V 0A6, Canada
3   Canada Research Chair in Green Catalysis and Metal-Free Processes, Université Laval, Québec City, QC G1V 0A6, Canada
*   Correspondence: dominic.lariviere@chm.ulaval.ca; Tel.: +1-418-656-7250

Abstract: The digestion of neodymium (NdFeB) magnets was investigated in the context of recycling rare earth elements (i.e., Nd, Pr, Dy, and Tb). Among more conventional digestion techniques (microwave digestion, open vessel digestion, and alkaline fusion), focused infrared digestion (FID) was tested as a possible approach to rapidly and efficiently solubilize NdFeB magnets. FID parameters were initially optimized with unmagnetized magnet powder and subsequently used on magnet pieces, demonstrating that the demagnetization and grinding steps are optional.

Keywords: NdFeB magnets; critical metals; rare earth elements (REEs); recycling; focus infrared digestion; ICP-OES; electronic waste

Citation: Bonin, M.; Fontaine, F.-G.; Larivière, D. Comparative Studies of Digestion Techniques for the Dissolution of Neodymium-Based Magnets. *Metals* **2021**, *11*, 1149. https://doi.org/10.3390/met11081149

Academic Editor: Bernd Friedrich

Received: 30 June 2021
Accepted: 14 July 2021
Published: 21 July 2021

Publisher's Note: MDPI stays neutral with regard to jurisdictional claims in published maps and institutional affiliations.

1. Introduction

As electronic devices play an increasingly more important role in our lives, society needs to develop strategies to recover valuable metals from end-of-life electronic products. This need is driven by critical metal supply concerns, and by environmental issues with the rapid generation of electronic waste (e-waste) worldwide [1]. The annual global production of e-waste was approximately 53.6 million metric tons (Mt) in 2019 and is expected to increase to 74 Mt by 2030 [1]. Because e-waste contains up to 69 elements from base to precious metals [1], e-waste mining has been proposed as a promising and cost-effective alternative to conventional mining [2]. However, it is estimated that less than 20% of the discarded e-waste is recycled at this time. This low recycling rate is partly attributed to the lack of proper recycling methods for most metals [1]. Therefore, there is a sustainable need to propose new and alternative strategies for e-waste recycling.

Besides precious and base metals, e-wastes also contain rare earth elements (REEs), which are increasingly used in high technology and clean energy applications [3]. REEs are present, for example, in common electronic components such as speakers, hard disk drives, and vibrators [4]. While REEs have relatively low concentrations in most bulk e-waste, the volume of end-of-life electronic devices discarded annually represents a great recycling opportunity. The presence of REEs within these devices comes mostly from neodymium (NdFeB) magnets [5]. NdFeB magnets are typically composed of up to 31 wt% REE [6]. Apart from Nd, which is the main REE, Dy, Pr, and Tb are also added to the magnet in various proportions depending on the application and quality of the magnets required [7]. The possibility of digesting NdFeB magnets into their isolated REE constituents is therefore a critical aspect of an effective recycling strategy.

While the recycling of REEs derived from NdFeB magnets in end-of-life electronic devices is not yet performed commercially, Rademaker et al. [7] emphasized that it would be technically feasible if efficient physical dismantling, separation, hydrometallurgical, and refining methods were available in the future. One challenge associated with REEs is the difficulty of isolating individual REEs from their neighbouring elements in the periodic table [8]. However, most separation techniques for REEs require the dissolution of solid matrices, and NdFeB magnets are no exception. Therefore, digestion of such magnets in a rapid, efficient, and cost-effective manner is an important aspect of a REEs recycling strategy.

Numerous researchers have investigated the dissolution of REEs from NdFeB magnets via various hydrometallurgical approaches (Table 1). Most procedures used variable acid types and concentrations and a two-step sample pre-treatment prior to the acid dissolution. This sample pre-treatment was driven by the need to demagnetize and pulverize the magnet to facilitate its manipulation and dissolution. These steps are time-consuming and potentially hazardous from a chemical and health perspective. For example, as the grinding of magnets leads to small particles, they can be inhaled and deposited within the respiratory system. The exposed magnet surface is increased for smaller particulates which can facilitate the ignition of metallic powders. Hoogerstraete et al. [9] reported a fire after opening a grinding mill containing NdFeB magnets. Sometimes, the heat generated during the grinding process can create a vacuum upon cooling, rendering the opening of the grinding mill challenging. Moreover, even though rare-earth magnets are brittle because they consist of agglomerated particulates (pressed and/or sintered), they are still a hard material and are sufficiently abrasive to damage steel, which can lead to premature wear of the equipment used for e-waste recycling. Grinding NdFeB magnets also requires powerful and resistant grinding equipment [10]. There is therefore a significant interest in assessing the dissolution performances of analytical procedures that do not require grinding and demagnetization processes.

Most dissolution approaches reported in the literature (Table 1) use elevated temperatures to accelerate NdFeB dissolution, usually heated with conventional heat sources such as hotplates and ovens. Recently, Helmeczi et al. [11] reported a rapid digestion of REEs in phosphoric acid by short-wavelength focused infrared radiation (FIR). They reported excellent recoveries and reproducibilities for REEs in various certified reference materials such as OREAS-465 (carbonatite supergene REE-Nb ore), OKA-2 (rare earths and thorium ore), and REE-1 (rare earths, zirconium, and niobium ore). These results suggest that FIR digestion could be a potential alternative to other heat sources for magnet dissolution and, potentially, alleviate previously mentioned sample pretreatments.

Table 1. Hydrometallurgical techniques published on the recycling of REEs from NdFeB magnets.

Magnets Source	Studied Elements	Demagnetization	Particles Size	Pre-Treatment	Leaching Method	Solubilization REEs [a]/Fe (%)	References
Magnetic sludge	Nd, Dy, Fe, B	(Not magnetized)	<250 µm	Drying of the sludge	1 M HNO$_3$ + 0.3 M H$_2$O$_2$, 80 °C, 10 mL/g, 5 min	98, 81/15	Rabatho, et al. (2013) [12]
Manufacturing waste magnets	Nd, Fe, B	350 °C, 15 min	<297 µm	-	3 N H$_2$SO$_4$, 27 °C, 50 mL/g 15 min (ultrasound)	100/100	Lee, et al. (2013) [10]
Manufacturing waste magnets	Nd, Dy, Fe	(Not magnetized)	not mentioned	Roasting at 400 °C, 2 h	1 M acetic acid, 90 °C, 100 mL/g, 400 rpm, 3 h	94, 93/1	Yoon, et al. (2015) [13]
HDD	Nd, Pr	320 °C, time not mentioned	<250 µm	-	2 M H$_2$SO$_4$, 70 °C, 20 mL/g, 15 min	90/not mentioned	München, Bernardes, and Veit (2018) [14]
Wind turbine	Nd, Pr, Dy, Fe, B, Al, Co	310 °C, 60 min	<149 µm	Roasting at 850 °C, 6 h	0.5 M HCl, 10 mL/g, 500 rpm, 5 h	98, 97/<1	Kumari, et al. (2018) [15]
HDD	Nd, Dy, Fe	350 °C, 60 min	<100 µm	900 W microwave, opened vessels, 5 min	0.5 M HCl, 70 °C, 25 mL/g, 900 rpm, 2 h	56/low (not mentioned)	Tanvar, Kumar, and Dhawan (2019) [16]

Table 1. *Cont.*

Magnets Source	Studied Elements	Demagnetization	Particles Size	Pre-Treatment	Leaching Method	Solubilization REEs [a]/Fe (%)	References
HDD	Nd, Dy, Fe	350 °C, 30 min	<500 μm		2 M H_2SO_4, 27 °C, 20 mL/g, 15 min	100/100	Erust, et al. (2019) [17]
HDD	Fe, Nd, Co, Ni	400 °C, 45 min	<420 μm	-	1.3 M $(NH_4)_2S_2O_8$, 75 °C, 50 mL/g, 15 min	99/64	Ciro, et al. (2019) [18]
HDD	Nd, Pr, Dy, Fe	350 °C, 3 h	~2 mm	-	1 M H_2SO_4, 25 °C, 20 mL/g, 90 min	100/100	Kumari, et al. (2020) [19]
HDD	Nd, Pr, Dy, Tb, B, Fe, Al, Cu, Ni, Co	350 °C, 60 min	<250 μm	-	1.6 N HCl or H_2SO_4, 20 mL/g, 5 min	100/100	Present work
HDD	Nd, Pr, Dy, Tb, B, Fe, Al, Cu, Ni, Co	No demagnetization	Coarsely broken	-	1.6 N HCl or H_2SO_4, 20 mL/g, ~30 min for $6 \times 3 \times 2$ mm pieces	100/100	Present work

[a] Presented as "Nd" (or mixed REEs) or "Nd, Dy" percentages.

In this article, we compared the dissolution of intact and pulverized magnets (both magnetized and demagnetized) by focused infrared digestion (FID) to other, more conventional dissolution techniques (microwave digestion, hot plate, and alkaline fusion).

2. Materials and Methods

2.1. Materials and Reagents

Two different samples were used in this study. The first sample, for method comparison and optimization, was prepared from magnets obtained from hard disk drives (HDD) collected in electronic waste bins located on the main campus at Laval University (Quebec City, QC, Canada) and separated from their brackets. The second sample, for the study concerning the dissolution of unaltered magnets, consisted of cylindrical magnets of 0.751 ± 0.006 g (6 mm in diameter and 2 mm in height) manufactured as one single batch, purchased from MagnetsShop (Culver City, CA, USA). These magnets were physically cleaved into similar fractions to enable the acid to penetrate beyond the protective Ni-Cu-Ni coating.

Nanopure water (18.2 MΩ·cm at 25 °C) obtained using a Milli-Q system (Millipore, Bedford, MA, USA) was used to dilute the solutions. Standard solutions (1 g·L^{-1}) of Al, B, Co, Cu, Dy, Fe, Nb, Nd, Ni, Pr, Rh (ISTD), and Tb, purchased from PlasmaCal (SCP Science, Baie d'Urfée, QC, Canada), were used to prepare calibration standards. Concentrated ACS-grade H_2SO_4 (Fisher, Ottawa, ON, Canada), and trace metal grade HCl and HNO_3 (VWR, Mississauga, ON, Canada) were used for sample digestion. Unless stated otherwise, all the confidence intervals represent a confidence level of 95%.

2.2. Demagnetisation and Grinding

Hard disk drive magnets (237 g) were demagnetized by thermal demagnetization based on the procedure proposed by Tanvar et al. [16]; the intact magnets were heated in a muffle furnace at 350 °C in a porcelain crucible for 1 h. Then, the demagnetized magnets were placed in a steel dish with grinding rings and ground with an 8500 Shatterbox mill (SPEX SamplePrep, Metuchen, NJ, USA) during successive cycles lasting 70, 50, and 40 s. Between each cycle, the ground sample was sifted through a 250 μm (mesh #60) brass sieve. Pieces larger than 250 μm were reintroduced to the steel dish for the next cycle. The resulting powder had a final mass of 216 g, representing 91% of the original mass. The difference between the initial and final masses can be explained by the presence of unground particles (13 g) that were larger than 250 μm after the last grinding cycle, and losses during grinding and sieving operations.

The particle size of the ground magnets was analyzed with 53 and 150 μm (mesh #270 and #100) sieves with a W.S. Tyler RX-29 Ro-Tap Sieve shaker (Laval Lab, Laval, QC, Canada) for 15 min, and each collected fraction was weighed. The mass loss from sieving represents less than 0.5% of the total mass used. For all the grinding and sieving steps, the sieves used were W.S. Tyler 12 inch (30.5 cm) brass sieves (Laval Lab, ibid.) with a PM4000 balance (Mettler Toledo, Mississauga, ON, Canada).

2.3. Elemental Analyses

The elemental composition of magnets was determined by inductively coupled plasma–optical emission spectroscopy (ICP-OES, iCAP 7000 Series, Thermo Scientific, Montréal, QC, Canada) equipped with a pneumatic concentric nebulizer. Table 2 presents the ICP-OES operating conditions. X-ray fluorescence (XRF) analyses were also used on some digestion residues to determine their elemental constituents. XRF analyses were performed in 3 replicates of 30 s for each filter (Na-S, Ni-Ag, Cr-Co, and Cl-V) using a MiniPal4 (Malvern Panalytical, Québec, QC, Canada).

Table 2. ICP-OES operating conditions.

Instrumental Parameters	iCAP 7000 Series ICP-OES
RF Power (W)	1150
Plasma gas flow (L/min)	12
Auxiliary gas flow (L/min)	0.5
Nebulizer gas flow (L/min)	0.5
Analysis mode	Radial (Fe, Nd), Axial (others)
Stabilization Time (s)	5
Sample flow rate (mL/min)	1.8
Wavelength (nm)	Al(309.271), B(249.773), Co(228.616), Cu(224.700, 324.754), Dy(353.170), Fe(238.204, 259.940), Nb(309.418), Nd(401.225, 406.109), Ni(216.556, 221.647), Pr(422.535), Tb(350.917)

2.4. Leaching

2.4.1. Total Dissolution of the NdFeB Magnets

To evaluate the elemental composition of ground magnets, four dissolution techniques were used: (1) closed-vessel acid digestion (CVAD, SCP Science, Baie d'Urfée, QC, Canada); (2) microwave (MWD, CEM corporation, Matthews, NC, USA); (3) focused infrared digestion (FID, Coldblock, Niagara Falls, ON, Canada), and (4) alkaline fusion (AF, Malvern Panalytical, Québec, QC, Canada). Table 3 presents the conditions used for each type of dissolution technique. Note that for FID and AF, the temperature ramping was extremely rapid, and boiling and melting temperatures, respectively, were obtained in less than a minute in both cases. The choice of a mixture of concentrated HCl and HNO_3 in an 8-to-2 ratio for acid digestion was based on the operating conditions reported by Berghof [20] for the dissolution of permanent magnets. The choice of 3 M of HNO_3 as a dissolution medium for alkaline fusion was based on the procedure published by Milliard et al. [21] concerning the dissolution of refractory species in environmental matrices.

2.4.2. Leaching Experiments

Experiments on the leaching of elements from the NdFeB powder were conducted with 1 g of ground magnetic material by FID using the design of experiments (DOE) performed using JMP Pro (Version 14.3, SAS Institute, Cary, NC, USA). The parameters used for this optimization process on the FID are presented in Table 4.

Table 3. Operating conditions used for the complete dissolution of ground magnets.

Digestion Technique	CVAD	MWD	FID	AF [a]
Instrument	DigiPrep Jr 12 pos.	Mars 5	6 channels ColdBlock	M4 Fluxer
Amount of ground magnet used (g)	0.5	0.25	1.0	0.1
Volume of acid (mL)	30 mL	10 mL	25 mL	100 mL
Nature of the acid used	$HCl:HNO_3$ (8:2)	$HCl:HNO_3$ (8:2)	$HCl:HNO_3$ (8:2)	3 M HNO_3
Digestion procedure	Ramp *ca.* 30 min, 240 min at 100 °C	Ramp 25 min, hold 15 min, 1600 W at 100%	15 min at 100% power for both lamps	See Milliard et al. [21]

[a] Fusion was performed using a 20:1 (~2 g) ratio of LiT/LiB/LiBr 49.5/49.5/1% flux to sample (Malvern Panalytical, Québec, QC, Canada), which was subsequently dissolved under the acidic conditions presented.

Table 4. Values used during DOE optimization process of the focused infrared digestion of ground magnets.

Nominal Variables	Value 1	Value 2	
Acid type	HCl	H_2SO_4	
Numerical Variables	**Low-Value**	**Mid-Value**	**High-Value**
Acid concentration (N)	1	2	3
Acid-to-sample ratio (mL/g)	10	20	30
Dissolution time (s)	300	450	600
Lamp power (%)	80	90	100

After optimization, the following optimal leaching methodology was used for FID: either 20 mL (1.6 N) or 10 mL (3.2 N) of HCl or H_2SO_4, per gram of ground NdFeB, is necessary to quantitatively (>99.9%) solubilize the rare earth elements in 5 min with a lamp power of 100%. For the trials on unaltered magnets, 7.5 mL of 3.2 N of H_2SO_4 was used for about 0.75 g of magnetic material.

3. Results and Discussion

3.1. Particle Size Distribution of the Ground Demagnetized NdFeB Magnets

The particle size plays an important role in the dissolution process of ground magnets. The particle size distribution of ground NdFeB magnets obtained from HDD are presented in Table 5.

Table 5. Mass percentage composition as a function of the particle size of the ground NdFeB magnet powder obtained from HDD.

Particle Size (µm)	Mass Percentage (%)
<53	42
54–150	34
151–250	24

The particle size distribution shows that most of the ground magnet powder (76%) was reduced to particles smaller than 151 µm. The three fractions were pooled together and used for the remaining trials.

3.2. Elemental Composition of Ground Demagnetized NdFeB Magnets

To properly assess various leaching approaches, the elemental composition of the powdered magnet must be known exactly. Four total dissolution approaches were compared: closed-vessel acid digestion (CVAD), microwave digestion (MW), focused infrared digestion (FID), and alkaline fusion (AF). The elemental composition was determined by ICP-OES. The instrument response was validated by analyzing Al, Cu, Dy, Fe, Nb, Nd, Ni, and Pr in the certified reference material REE-1 (CanmetMINING, Ottawa, ON, Canada). The REE-1 material was prepared for ICP-OES analysis by AF using 0.4 g of the material with 1.2 g of flux. Other parameters were as presented in Table 3. Table 6 presents the elemental composition of powdered NdFeB magnets determined by ICP-OES based on the total dissolution technique used.

Table 6. Elemental composition (%) of the powdered NdFeB sample determined by ICP-OES based on the total digestion used.

Element	Digestion Technique			
	Closed-Vessel Acid (*n* = 6)	Microwave (*n* = 2)	Focused Infrared (*n* = 3)	Alkaline Fusion (*n* = 3)
Fe	64 ± 1	67 ± 2	65 ± 2	67 ± 3
Nd	24.5 ± 0.7	25.6 ± 0.9	26.0 ± 0.6	25 ± 2
Pr	3.4 ± 0.1	3.6 ± 0.1	3.5 ± 0.2	3.3 ± 0.3
Ni	1.8 ± 0.2	N.D.	1.9 ± 0.1	1.7 ± 0.3
Dy	1.57 ± 0.04	1.61 ± 0.08	1.5 ± 0.2	1.5 ± 0.1
B	1.04 ± 0.02	0.93 ± 0.03	1.0 ± 0.1	N.M.
Co	1.02 ± 0.02	1.02 ± 0.01	1.00 ± 0.07	1.0 ± 0.1
Al	0.43 ± 0.02	0.50 ± 0.01	0.5 ± 0.1	0.49 ± 0.07
Nb	0.44 ± 0.08	0.4 ± 0.1	0.223 ± 0.005	0.35 ± 0.02
Cu	0.20 ± 0.02	N.D.	0.17 ± 0.03	0.16 ± 0.02
Tb	0.07 ± 0.02	0.09 ± 0.05	0.09 ± 0.03	0.08 ± 0.01
Total	99 ± 2%	101 ± 4%	100 ± 4%	100 ± 6

N.M.—not measurable, due to the addition of borate flux. N.D.—not determined.

The results obtained from four distinct digestion approaches for ground magnets led to a relatively consistent elemental composition except for Nb. This suggests that the sample used was sufficiently homogeneous to be used for leaching comparisons, which is discussed in the next section. As the digestates were filtered prior to ICP-OES analysis, a black residue was noticeable on the filters used for the samples digested by CVAD and FID. Pre-weighed filters were used to determine the mass fraction of the undissolved magnet powder. After filtration, they were washed with water, dried, and weighed using an analytical balance. The filters were then subjected to elemental analysis by XRF (Figure 1). It was determined that the residue represents 0.40 ± 0.08% (σ = 1 SD) of the ground magnet mass used initially.

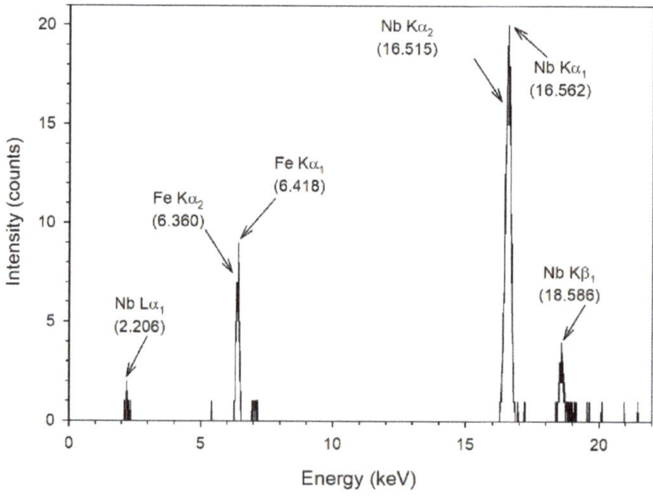

Figure 1. XRF spectrum of the residue present after FID.

The XRF analysis showed that the residue is composed of Nb and Fe. Nb is present in some NdFeB magnets to increase resistance to corrosion and enhance some magnetic properties [22,23]. Its presence in our powdered sample is not unexpected, because of the refractory nature of niobium oxides [24]. We suspect that the presence of Fe is the result of incomplete washings of the filter surface. Except for traces of Nd and Fe, all four digestion techniques were effective to completely dissolve ground demagnetized magnets.

3.3. Leaching of Powdered and Demagnetized Magnets

While complete dissolution of the magnets is mandatory for comparing the digestion techniques and determining the degree of leaching, it is not necessary from the perspective of developing a hydrometallurgical strategy for the recycling of rare earth elements in magnets. Helmeczi et al. [11] recently reported the rapid dissolution of REEs in mineral and environmental matrices using FID. As FID also demonstrated equivalent dissolution performances to other digestion techniques for the complete digestion, this approach was investigated for leaching purposes through a design of experiments (DOE) approach.

3.3.1. Design of Experiments

While the previous digestion procedure, which used a mixture of HCl and HNO_3, was certainly effective to completely dissolve powdered magnets, total digestion is not necessary for the recycling of REEs. The cost of nitric acid and its oxidative characteristics have limited its use in the hydrometallurgical separation of REEs in favor of HCl and H_2SO_4 [25]. Thus, an investigation of the leaching of ground demagnetized magnets was performed in either HCl or H_2SO_4. Five parameters were assessed through a DOE approach (Table 7). A first assessment of the DOE results showed that no parameter had a statistically significant impact on the dissolution. However, by removing either the acid type, dissolution time, or lamp power factors, the results showed that the concentration of acid, the acid-to-sample ratio, and their two-factor interaction were the only significant factors in the leaching of REEs. This suggests that the number of moles of acid was the main parameter for this leaching optimization. Experimentally, it was also determined that the low value used for the dissolution time (i.e., 300 s) set in the DOE was exceedingly sufficient to completely dissolve powdered magnets.

Table 7. Statistical importance of the DOE factors according to a 5- and 4-factor analysis.

Source	Log Worth [a] for 5 Factors	Log Worth [a] for 4 Factors [b]
Concentration	1.734	3.506
Ratio	1.724	3.484
Concentration * Ratio	1.567	3.142
Acid	0.448	0.706
Intensity * Time	0.354	
Intensity * Ratio	0.240	0.360
Time * Acid	0.209	
Time	0.137	
Concentration * Acid	0.137	0.223
Ratio * Acid	0.131	0.215
Intensity * Acid	0.130	0.189
Intensity * Concentration	0.091	0.145
Intensity	0.089	0.127
Time * Ratio	0.064	
Time * Concentration	0.026	

[a] A log worth value of minimum 2 is required for a factor to be considered significant. [b] Dissolution time (Time) factor removed.

To adequately determine the quantity of acid required per gram of powdered magnet to completely dissolve REEs, tests were performed in HCl and H_2SO_4 by varying the number of moles of acid used per gram of magnet. The results are presented for Nd as a representative of the REEs in Figure 2.

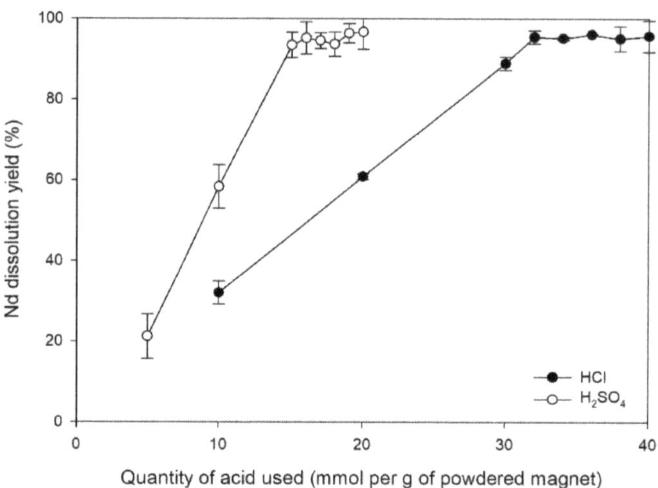

Figure 2. Nd dissolution yield (%) as a function of the amount of acid (HCl or H_2SO_4, in mmol) for 1 g of powdered magnet after a digestion by FID (300 s).

This test highlights the fact that the quantity of acid available to react with the powder is the limiting factor in the dissolution of REEs using FID. The required quantity of H_2SO_4 (16 mmol/g of ground magnet) is exactly half of the needed HCl (32 mmol/g), which is consistent with the balanced redox formulas (Equations (1) and (2)) for both acids. As two atoms of REEs are being oxidized to a trivalent oxidation state, six H^+ cations need to be reduced, which can be found in either three molecules of H_2SO_4 or six of HCl, hence the need for twice as much HCl as H_2SO_4. A similar logic can be applied to iron, one of the main components of the magnet. Based on the composition of the powdered magnet, it was calculated (from composition obtained by FID) that approximately 34 mmol of H^+ was required to oxidize and dissolve the magnet, which is coherent with the quantity found experimentally.

$$2 \text{ REE}_{(metallic)} + 6 \text{ HCl}_{(aq)} \rightarrow 2 \text{ REE}^{3+}_{(aq)} + 6 \text{ Cl}^-_{(aq)} + 3 \text{ H}_{2(g)} + \Delta, \tag{1}$$

$$2 \text{ REE}_{(metallic)} + 3 \text{ H}_2SO_{4(aq)} \rightarrow 2 \text{ REE}^{3+}_{(aq)} + 3 \text{ SO}_4^{2-}_{(aq)} + 3 \text{ H}_{2(g)} + \Delta, \tag{2}$$

Based on these observations, the following final leaching methodology can be proposed for FID: either 20 mL of 1.6 N or 10 mL of 3.2 N of HCl or H_2SO_4 per gram of NdFeB powder will be sufficient to completely solubilize the rare earths.

As stated previously, FID totally leached REEs from magnets, even with the shortest dissolution time tested (5 min). This suggests that the power output associated with the FID is more than sufficient to enable the complete dissolution of REEs. Thus, the Nd dissolution yield was monitored for dissolution times ranging from 60 to 300 s to determine how long, at full lamp power, it would take to achieve complete dissolution. No statistical differences in Nd dissolution yield were noted in the time range selected, except for the trial that ran for only 60 s (94%).

3.3.2. Leaching Performance Comparisons (CVAD, FID)

To determine whether there is a significant advantage in using FID for the leaching of REEs from magnets, the FID approach was compared to closed-vessel acid digestion (CVAD). The digestion of powdered NdFeB magnet was easily achieved in 300 s. As with FID, shorter dissolution times (60 to 300 s) were also investigated with CVAD. However, this technique required longer times—more than 120 s—to achieve complete digestion and quantitative dissolution (84% and 95% for 60 and 120 s, respectively).

Comparisons of the dissolution yields for FID and CVAD are presented in Figure 3. Statistically, both techniques yielded similar results when 10 mL of acid was used. Samples prepared using 20 mL of acid per gram showed lower yields with CVAD than FID. This difference could be explained by the short digestion time used (300 s), which does not allow the larger volume of the solution (20 mL) to be properly heated; this highlights the importance of heat in the rapidity of the digestion process. When performed at room temperature and a contact time of 300 s, the dissolution yields of REEs (i.e., Nd, Pr, Dy, and Tb) reached 80–90% with H_2SO_4 and 30–45% with HCl. As dissolution at room temperature with H_2SO_4 was more effective, further trials used H_2SO_4. However, it should be noted that HCl is also a very suitable acid for such purposes and could be a judicious choice if the subsequent separation scheme is performed in this media.

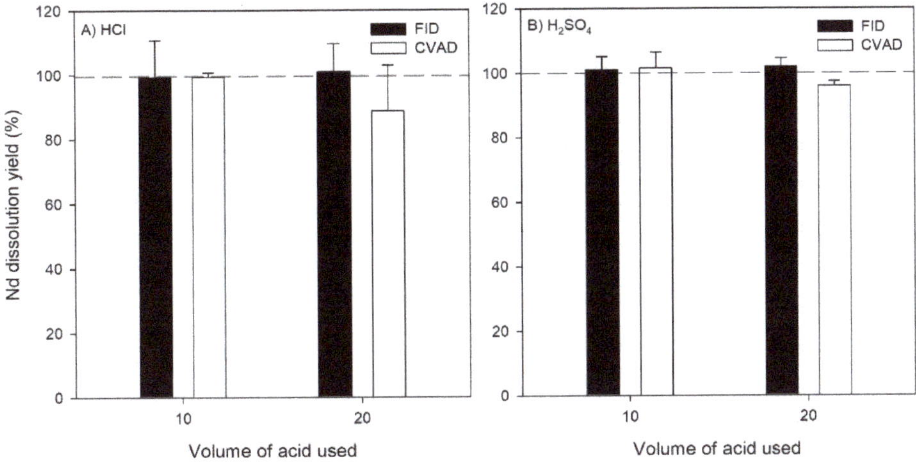

Figure 3. Comparison of FID and CVAD for the digestion of 1 g of magnet powder during 300 s with 10 mL of 3.2 N or 20 mL of 1.6 N of HCl or H_2SO_4.

3.3.3. Magnet Pieces Leaching

To test the effectiveness of FID vs. conventional CVAD on magnet pieces, cylindrical magnets were cut in half and put in with 7.5 mL of 3.2 N of H_2SO_4 for 15 min under various conditions (Table 8). The volume of acid and the molarity were adjusted to correlate with the number of moles required for the complete dissolution of the magnet mass used.

Table 8. Dissolved magnet mass after 15 min of contact time with 3.2 N of H_2SO_4 ($n = 3$).

Conditions	Relative Dissolved Mass [a] (%)
Room temp., without agitation	5 ± 2
Room temp., with agitation [b]	5 ± 2
CVAD	34 ± 5
FID	62 ± 2

[a] The coating of the magnets is not digested with the proposed methods and represents $1.9 \pm 0.2\%$ of the total mass of the magnets. [b] Agitation was performed by the magnets pieces themselves interacting with the magnet field of a stirring plate.

As observed with the ground magnet, once the optimal quantity of acid per gram of magnet is used, the temperature of the acid is a critical parameter for the rapid dissolution of unaltered magnets. These results support the idea that FID, as a powerful heating source, could be an effective method to rapidly dissolve NdFeB magnets for hydrometallurgical recycling.

To determine the required time for the complete dissolution of the magnet pieces by FID and CVAD, the amount of magnet digested was monitored as a function of digestion time (10 to 35 min, Figure 4). The same parameters as previously described were used, but with 15 mL of 3.2 N of H_2SO_4 to avoid the complete evaporation of the acid due to an incomplete condensation of the acidic vapor inside the part of the digestion vessel surrounded by the Peltier cooling block. The temperature of H_2SO_4 was monitored using a thermocouple inserted into the solution while the lamps were on. It showed a working temperature of 103 °C after 1 min, which is close to the expected boiling point of a 2 M solution of H_2SO_4 (102 °C [26]).

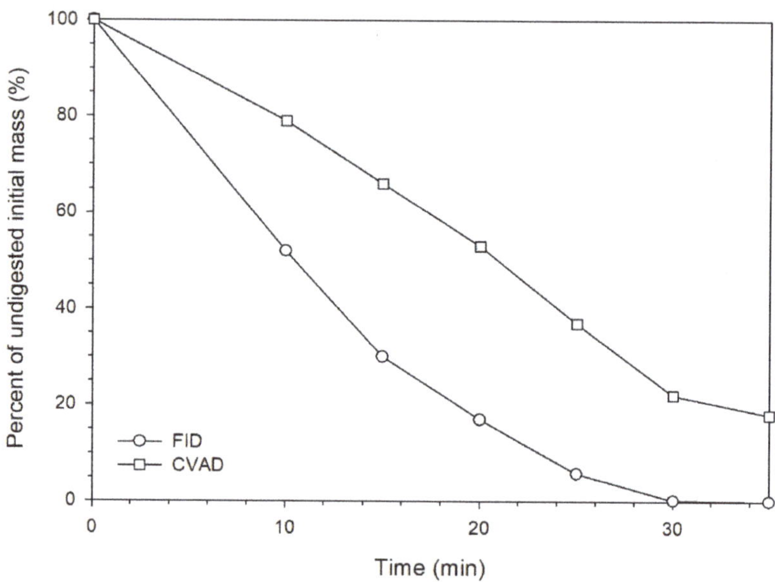

Figure 4. Undissolved magnet mass percentage after digestion in H_2SO_4 by FID and CVAD.

When the remaining magnet mass in the digestion vessel was approximately 10% of its initial value, the efficiency of the digestion tended to decrease for the FID approach. This is likely due to the round shape of the magnets used, which resulted in a smaller contact surface as the digestion progressed. Nonetheless, this experiment demonstrates that dissolution of magnet pieces is faster by FID than by CVAD.

4. Conclusions

The results obtained in this study indicate that focused infrared digestion (FID) could be used as an effective method for the recycling of rare earth magnets, either for the complete dissolution of ground samples (apart from refractory niobium oxides), or for the quantitative dissolution of REEs on magnetized and coarse magnet pieces. The absence of magnetic constituents in the FID unit enables the digestion of the magnets without any prior demagnetization process. The FID method is safer than others because, due to its rapid digestion time, FID does not require crushing of the magnets into fine powders. Skipping the grinding step also facilitates the separation of the undigested Ni-Cu-Ni coating and could potentially help in hydrometallurgical separation later in the recycling process.

Author Contributions: Conceptualization, M.B. and D.L.; methodology, M.B. and D.L.; formal analysis, M.B.; writing—original draft preparation, M.B.; writing—review and editing, F.-G.F. and D.L.; supervision, F.-G.F. and D.L.; project administration, D.L.; funding acquisition, F.-G.F. and D.L. All authors have read and agreed to the published version of the manuscript.

Funding: This research was funded by FRQNT—Team Research Grant (2021 competition)—"Développement d'une filière hydrométallurgique de recyclage des métaux et des terres rares à partir des déchets de téléphones portables et de tablettes électroniques", grant number 284426.

Institutional Review Board Statement: Not applicable.

Informed Consent Statement: Not applicable.

Data Availability Statement: The data presented in this study are available in this article.

Acknowledgments: The authors want to thank Vicky Dodier for her help with the grinding and sieving of the samples and Christa Bedwin for her editorial comments on the manuscript.

Conflicts of Interest: The authors declare no conflict of interest.

References

1. Forti, V.; Cornelis Peter, B.; Kuehr, R.; Bel, G. *The Global E-Waste Monitor 2020: Quantities, Flows, and the Circular Economy Potential*; United Nations University (UNU): Bonn, Germany; United Nations Institute for Training and Research (UNITAR): Geneva, Switzerland; International Telecommunication Union: Geneva, Switzerland; International Solid Waste Association: Rotterdam, The Netherlands, 2020; p. 120.
2. Zeng, X.L.; Mathews, J.A.; Li, J.H. Urban Mining of E-Waste is Becoming More Cost-Effective Than Virgin Mining. *Environ. Sci. Technol.* **2018**, *52*, 4835–4841. [CrossRef] [PubMed]
3. Ferron, C.J.; Henry, P. A review of the recycling of rare earth metals. *Can. Metall. Q.* **2015**, *54*, 388–394. [CrossRef]
4. Lister, T.E.; Wang, P.M.; Anderko, A. Recovery of critical and value metals from mobile electronics enabled by electrochemical processing. *Hydrometallurgy* **2014**, *149*, 228–237. [CrossRef]
5. Lixandru, A.; Venkatesan, P.; Jonsson, C.; Poenaru, I.; Hall, B.; Yang, Y.; Walton, A.; Guth, K.; Gauss, R.; Gutfleisch, O. Identification and recovery of rare-earth permanent magnets from waste electrical and electronic equipment. *Waste Manag.* **2017**, *68*, 482–489. [CrossRef] [PubMed]
6. Yang, Y.; Walton, A.; Sheridan, R.; Güth, K.; Gauß, R.; Gutfleisch, O.; Buchert, M.; Steenari, B.-M.; Van Gerven, T.; Jones, P.T.; et al. REE Recovery from End-of-Life NdFeB Permanent Magnet Scrap: A Critical Review. *J. Sustain. Metall.* **2016**, *3*, 122–149. [CrossRef]
7. Rademaker, J.H.; Kleijn, R.; Yang, Y.X. Recycling as a Strategy against Rare Earth Element Criticality: A Systemic Evaluation of the Potential Yield of NdFeB Magnet Recycling. *Environ. Sci. Technol.* **2013**, *47*, 10129–10136. [CrossRef]
8. Florek, J.; Giret, S.; Juere, E.; Lariviere, D.; Kleitz, F. Functionalization of mesoporous materials for lanthanide and actinide extraction. *Dalton Trans* **2016**, *45*, 14832–14854. [CrossRef]
9. Vander Hoogerstraete, T.; Blanpain, B.; Van Gerven, T.; Binnemans, K. From NdFeB magnets towards the rare-earth oxides: A recycling process consuming only oxalic acid. *RSC Adv.* **2014**, *4*, 64099–64111. [CrossRef]
10. Lee, C.H.; Chen, Y.J.; Liao, C.H.; Popuri, S.; Tsai, S.L.; Hung, C.E. Selective Leaching Process for Neodymium Recovery from Scrap Nd-Fe-B Magnet. *Metall. Mater. Trans. A-Phys. Metall. Mater. Sci.* **2013**, *44A*, 5825–5833. [CrossRef]
11. Helmeczi, E.; Wang, Y.; Brindle, I.D. A novel methodology for rapid digestion of rare earth element ores and determination by microwave plasma-atomic emission spectrometry and dynamic reaction cell-inductively coupled plasma-mass spectrometry. *Talanta* **2016**, *160*, 521–527. [CrossRef]
12. Rabatho, J.P.; Tongamp, W.; Takasaki, Y.; Haga, K.; Shibayama, A. Recovery of Nd and Dy from rare earth magnetic waste sludge by hydrometallurgical process. *J. Mater. Cycles Waste Manag.* **2013**, *15*, 171–178. [CrossRef]
13. Yoon, H.S.; Kim, C.J.; Chung, K.W.; Jeon, S.; Park, I.; Yoo, K.; Jha, M.K. The Effect of Grinding and Roasting Conditions on the Selective Leaching of Nd and Dy from NdFeB Magnet Scraps. *Metals* **2015**, *5*, 1306–1314. [CrossRef]
14. München, D.D.; Bernardes, A.M.; Veit, H.M. Evaluation of Neodymium and Praseodymium Leaching Efficiency from Post-consumer NdFeB Magnets. *J. Sustain. Metall.* **2018**, *4*, 288–294. [CrossRef]
15. Kumari, A.; Sinha, M.K.; Pramanik, S.; Sahu, S.K. Recovery of rare earths from spent NdFeB magnets of wind turbine: Leaching and kinetic aspects. *Waste Manag.* **2018**, *75*, 486–498. [CrossRef]
16. Tanvar, H.; Kumar, S.; Dhawan, N. Microwave Exposure of Discarded Hard Disc Drive Magnets for Recovery of Rare Earth Values. *JOM* **2019**, *71*, 2345–2352. [CrossRef]
17. Erust, C.; Akcil, A.; Tuncuk, A.; Deveci, H.; Yazici, E.Y. A Multi-stage Process for Recovery of Neodymium (Nd) and Dysprosium (Dy) from Spent Hard Disc Drives (HDDs). *Miner. Process. Extr. Metall. Rev.* **2021**, *42*, 90–101. [CrossRef]
18. Ciro, E.; Alzate, A.; López, E.; Serna, C.; Gonzalez, O. Neodymium recovery from scrap magnet using ammonium persulfate. *Hydrometallurgy* **2019**, *186*, 226–234. [CrossRef]
19. Kumari, A.; Jha, M.K.; Pathak, D.D. An innovative environmental process for the treatment of scrap Nd-Fe-B magnets. *J. Environ. Manag.* **2020**, *273*, 7. [CrossRef]
20. Berghof Products + Instruments GMBH. *Microwave Digestion of Permanent Magnets*; Application Note XT4; Eningen, Germany, 2020. Available online: https://www.berghof-instruments.com/en/application/microwave-digestion-of-permanent-magnets/ (accessed on 14 July 2021).
21. Milliard, A.; Durand-Jezequel, M.; Lariviere, D. Sequential automated fusion/extraction chromatography methodology for the dissolution of uranium in environmental samples for mass spectrometric determination. *Anal. Chim. Acta* **2011**, *684*, 40–46. [CrossRef]
22. Yu, L.Q.; Wen, Y.H.; Yan, M. Effects of Dy and Nb on the magnetic properties and corrosion resistance of sintered NdFeB. *J. Magn. Magn. Mater.* **2004**, *283*, 353–356. [CrossRef]

23. Le Breton, J.M.; Teillet, J. Mössbauer and X-ray study of NdFeB type permanent magnets oxidation: Effect of A1 and Nb addition. *J. Magn. Magn. Mater.* **1991**, *101*, 347–348. [CrossRef]
24. Prasad, V.; Baligidad, R.; Gokhale, A. Niobium and Other High Temperature Refractory Metals for Aerospace Applications. *Aerosp. Mater. Mater. Technol.* **2017**, 267–288. [CrossRef]
25. Habashi, F. Extractive metallurgy of rare earths. *Can. Metall. Q.* **2013**, *52*, 224–233. [CrossRef]
26. Washburn, E.W.; West, C.J.; Dorsey, N.E.; National Research Council; International Research Council; National Academy of Science. *International Critical Tables of Numerical Data, Physics, Chemistry and Technology*; The National Academies Press: Washington, DC, USA, 1926.

Article

Viscosity and Structural Investigation of High-Concentration Al₂O₃ and MgO Slag System for FeO Reduction in Electric Arc Furnace Processing

Youngjae Kim [1] and Dong-Joon Min [2,*]

[1]   Mineral Resource Research Division, Korea Institute of Geoscience and Mineral Resources, 124 Gwahak-ro, Yuseong-gu, Daejeon 34132, Korea; youngjae.kim@kigam.re.kr

[2]   Department of Materials Science and Engineering, Yonsei University, Seoul 03722, Korea

*   Correspondence: chemical@yonsei.ac.kr; Tel./Fax: +82-2-2123-2840

Abstract: In the present study, the viscosity of the CaO–SiO_2–FeO–Al_2O_3–MgO slag system was measured for the recovery of FeO in the electric arc furnace (EAF) process using Al dross. Considering the MgO-saturated operational condition of the EAF, the viscosity was measured in the MgO-saturated composition at 1823 K with varying FeO and Al_2O_3 concentrations. An increase in the slag viscosity with decreasing temperature was observed. The activation energy was evaluated, and the change in the thermodynamically equilibrated phase was considered. The changes in the aluminate structure with varying FeO and Al_2O_3 concentrations were investigated by Fourier-transform infrared spectroscopy, which revealed an increase in the [AlO_4] tetrahedral structure with increasing Al_2O_3 concentration. Depolymerization of the aluminate structure was observed at higher FeO concentrations. The Raman spectra showed the polymerization of the silicate network structure at higher Al_2O_3 concentrations. By associations between the silicate and aluminate structures, a more highly polymerized slag structure was achieved in the present system by increasing the Al_2O_3 concentration.

Keywords: CaO–SiO_2–FeO–Al_2O_3–MgO slag system; viscosity; slag structure; silicate structure; aluminate structure; FeO recovery

Citation: Kim, Y.; Min, D.-J. Viscosity and Structural Investigation of High-Concentration Al₂O₃ and MgO Slag System for FeO Reduction in Electric Arc Furnace Processing. *Metals* **2021**, *11*, 1169. https://doi.org/10.3390/met11081169

Academic Editor: Dariush Azizi

Received: 30 June 2021
Accepted: 19 July 2021
Published: 23 July 2021

Publisher's Note: MDPI stays neutral with regard to jurisdictional claims in published maps and institutional affiliations.

1. Introduction

In Korea, steel production in 2020 was 67.1 million tons. Approximately 31% (20.8 million tons) of this was produced by the electrical arc furnace (EAF) process. As approximately 169 kg of EAF slag is produced for each ton of crude steel, the estimated amount of EAF slag generated in 2020 in Korea was 3.5 million tons. With slags from blast furnaces and basic oxygen furnaces, most of the slag can be utilized as raw materials for road construction, backfill, or fertilizers [1]. However, owing to its high concentration of Fe (20–30 wt%), EAF slag is limited in applicability to value-added construction materials [2,3]. EAF slag is typically used as a roadbed or backfill material following an appropriate magnetic separation process [2]. Although several studies have demonstrated the applicability of EAF slag in concrete, road construction materials, and cement without preprocessing to reduce the FeO concentration [4–9], the total amount of EAF slag in these mixtures was limited to obtain the appropriate physical properties.

In order to utilize EAF slag in value-added construction materials and to recover valuable Fe from such slag, the reduction technique called the eco-slag process was proposed for EAF steelmaking [1–3]. Kim et al. [2] suggested a two-stage reduction process of Al reduction by Al dross and direct carbon reduction. In the first stage, Al dross consisting of 30 wt% Al and 70 wt% Al₂O₃ was added approximately 5 min before tapping the steel in the EAF steelmaking process. The addition of 100 kg of Al dross to 90 tons of steel reduced the total Fe content in the EAF slag from 21% to 15%. In the second stage, the tapped EAF slag was transferred to an induction furnace. The EAF slag was agitated using a graphite rod, and further reduction in FeO in the slag was simultaneously performed at 1773 K. After 60 min of operation, the final slag composition was achieved with approximately 5 wt% of FeO. By controlling the cooling rate of the slag, a magnetic and Fe-rich spinel phase forms alongside the amorphous material that was clearly separated [10]. Finally, a suitable slag composition was achieved through crushing and magnetic separation processes for use in ordinary Portland cement [2,3].

During the eco-slag process, the slag composition is significantly changed by adding Al dross and by reducing FeO. The change in slag composition affects the erosion of the EAF refractory. During the EAF process, MgO from the refractory is soluble in the molten slag. As refractory erosion can shorten the service life of the EAF system, MgO saturation in the EAF slag is maintained by the external addition of calcined dolomite or calcined magnesite. Previous studies have investigated the solubility of MgO in $CaO–SiO_2–FeO–Al_2O_3$ systems [11–13]; these studies have shown that MgO solubility in the molten slag system is mainly affected by the equilibrated phase of the slag, such as magnesiowüstite ((Mg, Fe)O) or spinel ($MgAl_2O_4$). In addition, the change in the thermodynamically equilibrated phase affects the ionic state and slag structure of the network-forming oxide [11–13].

The MgO solubility and the viscosity of the EAF slag are mainly affected by changes in the equilibrium phase and its related slag structure. Recently, Lee and Min [14], who studied the activation energy of viscous flow in $CaO–SiO_2–FeO–Al_2O_3–MgO$ systems, reported an abnormal viscosity tendency as the equilibrium phase changed from melilite to di-calcium silicate. Viscosity is a dominant property related to operational conditions, including the slag foaming ability and tapping condition [15,16]. Therefore, understanding the rheological properties with variations in the FeO and Al_2O_3 compositions in the MgO-saturated condition is crucial for practical application of the eco-slag process. Although several studies have investigated the viscosity of molten EAF slag systems [17–21], slags with >10 wt% Al_2O_3 and MgO-saturated compositions have not been studied. In the present study, the viscosity in the high-MgO-concentration region was measured with variations in the FeO and Al_2O_3 contents of the slag, assuming a reduction in FeO by Al dross. In addition, the change in the slag structure was investigated to evaluate the effects of changes in the network structure of oxide melts on the rheological properties of the $CaO–SiO_2–FeO–Al_2O_3–MgO$ system using Raman spectroscopy and Fourier-transform infrared (FT-IR) spectroscopy.

2. Materials and Methods

Prior to the viscosity measurement, MgO solubility in the $CaO–SiO_2–FeO–Al_2O_3$ system at 1823 K was determined by using a thermochemical equilibrium technique [11]. The slag sample was prepared using reagent-grade CaO, SiO_2, Al_2O_3, FeO, and MgO. CaO was obtained by the calcination of $CaCO_3$ at 1273 K for 6 h. The powder was mixed in an agate mortar to obtain a homogeneous mixture. Afterward, approximately 5 g of the powder mixture was placed in a MgO crucible (99% purity) and heated in an electric resistance furnace equipped with $MoSi_2$ heating elements under an Ar atmosphere. The equilibration time was determined as 8 h in a previous study [11]. After 8 h, the samples were removed from the furnace and quenched by blowing Ar gas. The slag was separated from the MgO crucible and ground using a pulverizing ball mill to less than 100 µm for chemical analysis. The slag composition was analyzed using X-ray fluorescence (XRF, S4 Explorer; Bruker AXS, Madison, WI, USA). Table 1 shows the pre- and post-experiment slag compositions. Although the pre-experiment compositions of FeO and Al_2O_3 were 10, 20, 30, and 40 wt%, the post-experiment contents varied because of the different MgO solubilities. For convenience, the pre-experiment concentrations of FeO and Al_2O_3 were used to identify the samples in the present study.

Table 1. Experimental results of MgO solubility for $CaO–SiO_2–FeO–Al_2O_3$ slags at 1823 K.

(wt%)	Pre-Experiment					Post-Experiment				
	CaO	SiO_2	FeO	Al_2O_3	MgO	CaO	SiO_2	FeO	Al_2O_3	MgO
	50	50	0	0	-	41.54	38.97	0.00	0.00	19.49
Initial CaO/SiO_2 = 1.0 Initial FeO = 0 wt%	45	45	0	10	-	36.97	34.86	0.00	7.63	20.55
	40	40	0	20	-	33.28	30.97	0.00	15.33	20.41
	40	40	0	30	-	33.24	29.94	0.00	23.58	13.23
	45	45	10	0	-	38.77	36.25	7.91	0.00	17.07
Initial CaO/SiO_2 = 1.0 Initial FeO = 10 wt%	40	40	10	10	-	34.60	31.77	8.08	7.85	17.70
	35	35	10	20	-	29.96	27.07	9.20	15.72	18.05
	35	35	10	30	-	28.25	25.47	10.13	22.27	13.88
	40	40	20	0	-	34.60	32.76	18.19	0.00	14.46
Initial CaO/SiO_2 = 1.0 Initial FeO = 20 wt%	35	35	20	10	-	29.82	28.18	18.81	7.95	15.24
	30	30	20	20	-	26.03	23.28	19.03	15.22	16.44
	25	25	20	30	-	21.31	19.42	18.34	23.42	17.52
	35	35	30	0	-	31.96	28.77	26.46	0.00	12.82
Initial CaO/SiO_2 = 1.0 Initial FeO = 30 wt%	30	30	30	10	-	27.09	24.43	26.43	8.13	13.92
	25	25	30	20	-	22.11	20.21	26.87	15.56	15.26
	20	20	30	30	-	16.96	15.26	26.08	25.64	16.06

Referring to the MgO-saturated compositions in the $CaO–SiO_2–FeO–Al_2O_3$ slag system at 1823 K, as shown in Table 1, the slag mixture was prepared using reagent-grade CaO, SiO_2, FeO, Al_2O_3, and MgO. Approximately 120 g of the homogeneous powder mixture ground in an agate mortar was placed in a Pt–10% Rh crucible (outer diameter: 41 mm, inner diameter: 40 mm, and height: 65 mm). The crucible was placed in an electric resistance furnace at 1873 K under an Ar atmosphere. After maintaining the conditions for 1 h to achieve thermal equilibrium, the viscosity was measured by using a rotating cylinder method. The viscosity and torque data were recorded each second using a digital viscometer (DV2TLV; Ametek Brookfield, Middleboro, MA, USA) calibrated with silicone oil at room temperature. Figure 1 shows the schematic of the viscosity measurement apparatus. To evaluate the temperature dependency, the viscosity was measured by decreasing the temperature by 25 K at 5 K/min and by maintaining each temperature for 30 min during viscosity measurement.

After the viscosity measurement, the temperature was increased to 1873 K and the crucible was removed from the furnace. The molten slag was quenched on a water-cooled Cu plate. No characteristic X-ray diffraction (XRD) peaks were observed from the quenched sample, indicating that it was in an amorphous state. The obtained sample was crushed and ground to a particle size of less than 100 μm for structural analysis. The intermediate-range order of the slag structure was analyzed using FT-IR spectroscopy (Spectra 100; Perkin-Elmer, Shelton, CT, USA) and Raman spectroscopy (LabRaman HR, Horiba Jobin-Yvon, France). More details of the structural analysis procedure utilizing FT-IR spectroscopy and Raman spectroscopy have been explained elsewhere [22–24].

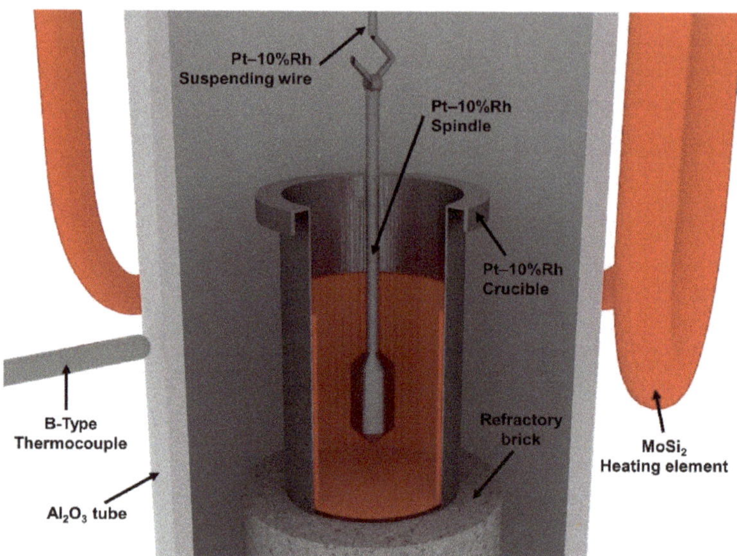

Figure 1. Schematic of the viscosity measurement apparatus.

3. Results and Discussion

3.1. Effect of Temperature on the Viscosity of CaO–SiO₂–FeO–Al₂O₃–MgO Slag

Figure 2 shows the temperature dependence of viscosity in the CaO–SiO_2–FeO–Al_2O_3–MgO system. Typically, slag viscosity decreases with increasing temperature. However, the effect of temperature on the viscosity change varies with slag composition. When both FeO and Al_2O_3 were 0 wt%, the viscosity increased steeply with decreasing temperature. However, when Al_2O_3 was added to the CaO–SiO_2–MgO system, the temperature dependence of viscosity decreased. The addition of Al_2O_3 led to an increase in the temperature dependence of viscosity in the CaO–SiO_2–FeO–MgO system. The relationship between temperature and viscosity can be quantitatively expressed by an Arrhenius-type equation, assuming that viscous shear is a thermally activated process [25]:

$$\eta = \eta_\infty \exp\left(\frac{E}{RT}\right), \tag{1}$$

where η is the viscosity, η_∞ is the pre-exponential constant, R is the ideal gas constant, T is the absolute temperature, and E is the activation energy. From Equation (1), the activation energies of the present slag system were calculated, as shown in Figure 3. The highest activation energy was found in the CaO–SiO_2–MgO ternary slag system. When Al_2O_3 was added to this ternary system, the activation energy initially decreased. However, above 20 wt% Al_2O_3, higher Al_2O_3 concentrations increased the activation energy. In the CaO–SiO_2–FeO–MgO systems, the activation energy increased with increasing Al_2O_3 concentration. As Equation (1) is based on vibrational frequency, the activation energy indicates the energy barrier to be overcome [25]. Turkdogan and Bills described the activation energy for viscous flow as the energy required to move the "flow-unit" from one equilibrium position to another [26]. According to Lee and Min [14], the activation energy was related to the distribution of the network structure and cation–anion interactions. Thus, the activation energy is also affected by the change in the equilibrium phase because the structure of the molten slag is similar to that of the thermodynamic equilibrium phase [14].

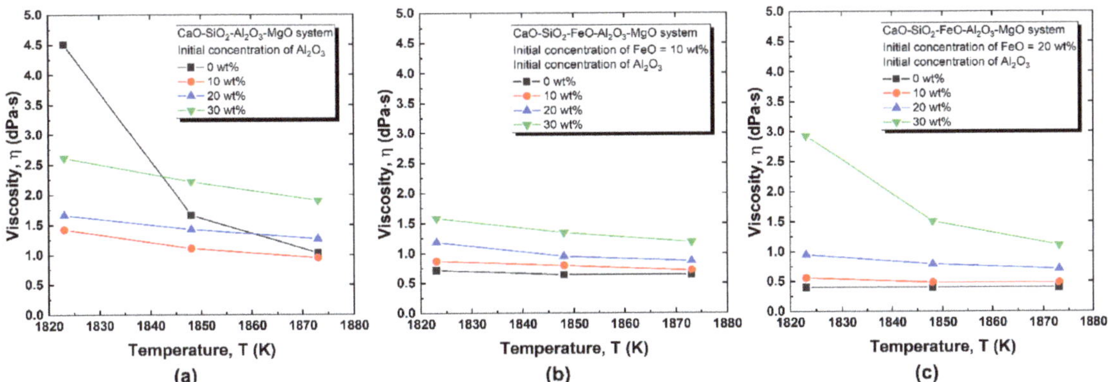

Figure 2. Relationship between viscosity and temperature in the (**a**) CaO–SiO$_2$–Al$_2$O$_3$–MgO system, (**b**) CaO–SiO$_2$–FeO–Al$_2$O$_3$–MgO system with 10 wt% FeO, and (**c**) CaO–SiO$_2$–FeO–Al$_2$O$_3$–MgO system with 20 wt% FeO.

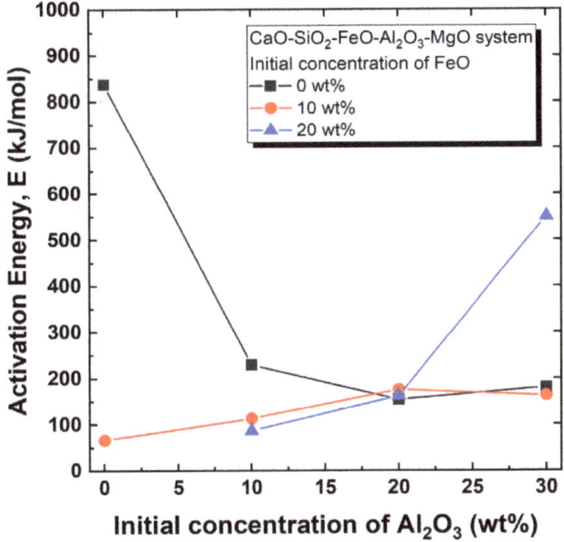

Figure 3. Activation energies of CaO–SiO$_2$–FeO–Al$_2$O$_3$–MgO slag system with varying FeO and Al$_2$O$_3$ concentrations.

Using the thermodynamic calculation software FactSage 8.1 (Thermfact and GTT-Technologies, Montreal, QC, Canada), the thermodynamic equilibrium phases of the molten slags were evaluated. In the CaO–SiO$_2$–MgO ternary system, the determined liquidus temperature was 1823.39 K and the equilibrium phase was merwinite (Ca$_3$MgSi$_2$O$_8$). It can be inferred that this system showed the highest activation energy because merwinite has a rigid structure between cations and silicate anions. The equilibrium phase changed to MgO as Al$_2$O$_3$ was added to the ternary system. As the equilibrium structure was simplified, the activation energy decreased. However, above 20 wt% Al$_2$O$_3$, the equilibrium phase changed to spinel. Due to the high affinity between the Mg cations and aluminate anions, the activation energy was increased.

On the contrary, an increase in the activation energy was observed in the CaO–SiO$_2$–FeO–MgO system as the Al$_2$O$_3$ concentration increased. In order to evaluate the effect of the slag structure on the viscosity, the structural change of the CaO–SiO$_2$–FeO–Al$_2$O$_3$–MgO system was investigated and discussed in the following section.

3.2. *Effect of Slag Structure on the Viscosity of the CaO–SiO$_2$–FeO–Al$_2$O$_3$–MgO Slag*

Figure 4 shows the effect of Al$_2$O$_3$ on the viscosity of the CaO–SiO$_2$–FeO–Al$_2$O$_3$–MgO system at 1873 K. In the CaO–SiO$_2$–MgO system, the viscosity was slightly decreased with the addition of 10 wt% Al$_2$O$_3$. However, the viscosity was simply increased as a the concentration of Al$_2$O$_3$ increased. In the CaO–SiO$_2$–FeO–MgO systems, it is commonly observed that an increase in Al$_2$O$_3$ causes an increase in viscosity. Compared with previous studies that measured viscosity in the CaO–SiO$_2$–Al$_2$O$_3$–MgO system [27,28] or CaO–SiO$_2$–FeO–Al$_2$O$_3$–MgO system [29], the present system showed lower viscosity. The present experiments were carried out in the composition where MgO was saturated at 1823 K. Compared with other studies, the higher MgO concentration resulted in lower viscosity [28]. According to Mysen et al. [30–32], the anionic structure in the aluminosilicate system does not change upon quenching from the molten state. For this reason, the molten slag structure was investigated by analyzing the quenched glass sample. Using FT-IR and Raman spectroscopy, the changes in the network structure with varying Al$_2$O$_3$ and FeO concentrations were evaluated.

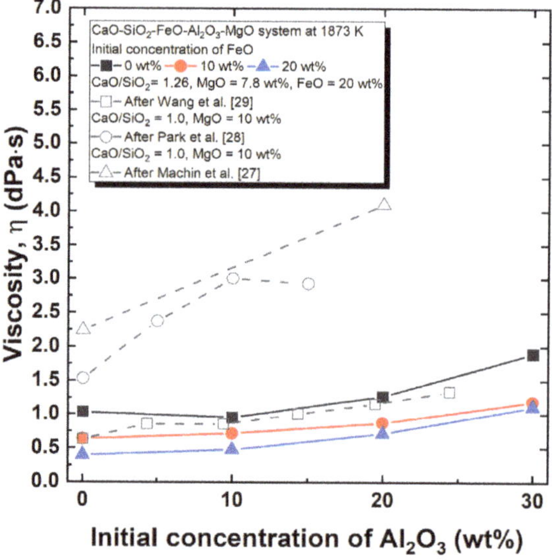

Figure 4. Effect of Al$_2$O$_3$ on the viscosity of CaO–SiO$_2$–FeO–Al$_2$O$_3$–MgO slag system at 1873 K with varying FeO concentrations.

Figure 5 shows the FT-IR transmittance spectra of the slag samples. According to previous FT-IR investigations of slag structures [19,20,33–36], bands indicating the distinct structural units related to the silicate and aluminate structures can be found in three regions: 1200–800 cm^{-1}, 750–630 cm^{-1}, and 630–450 cm^{-1}, corresponding to [SiO$_4$] tetrahedral symmetric stretching vibrations, [AlO$_4$] tetrahedral asymmetric stretching vibrations, and Si–O–Al bending vibrations, respectively. In the silicate network structure, tetrahedral [SiO$_4$] units can be classified depending on the number of bridging oxygens (BOs). As different units have different symmetric stretching vibrations, the absorption band present in the FT-IR spectrum corresponds to the characteristic bonding states of the different units. The number of BOs in the [SiO$_4$] unit is expressed by n in Q_{Si}^n, where 4, 3, 2, and

1 indicate sheets, chains, dimers, and monomers, respectively. Likewise, the number of BOs in the [AlO$_4$] tetrahedral unit is expressed by n in Q^n_{Al}. It is commonly observed that the addition of Al$_2$O$_3$ to the CaO–SiO$_2$–MgO or CaO–SiO$_2$–FeO–MgO system introduces an absorption peak at 750–630 cm^{-1}, indicating the formation of [AlO$_4$] tetrahedral units. When the [AlO$_4$] tetrahedral units form a polymerized network structure or become incorporated into the [SiO$_4$] tetrahedral units, a cation is required for charge balancing [35]. The high affinity between Mg^{2+} and the [AlO$_4$] tetrahedral unit was reported in our previous study [11]. As the viscosity of the CaO–SiO$_2$–FeO–Al$_2$O$_3$–MgO system was measured in the MgO-saturated composition at 1823 K, sufficient Mg^{2+} existed in the molten slag for the charge balance of the aluminate and aluminosilicate network structures. For this reason, the addition of Al$_2$O$_3$ causes the formation of a network structure, and the viscosity monotonically increases with increasing Al$_2$O$_3$ concentration at a fixed initial concentration of FeO, as shown in Figure 4.

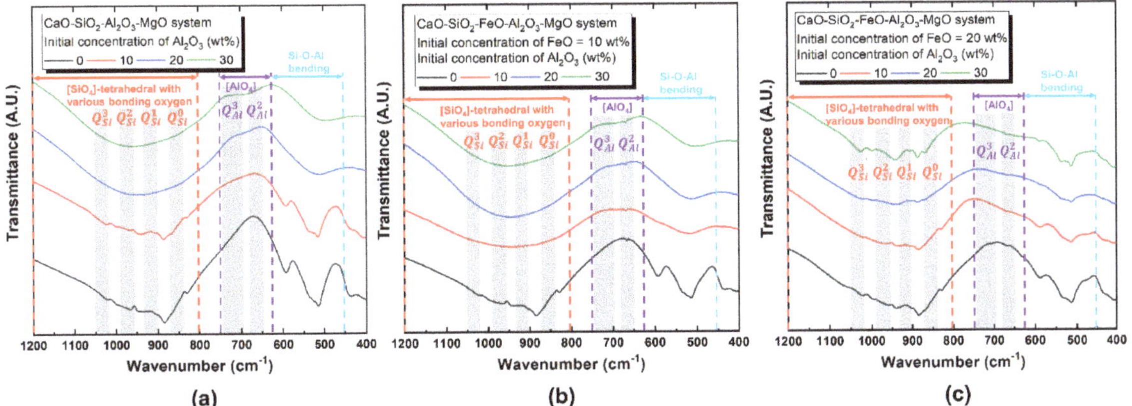

Figure 5. FT-IR transmittance spectra of quenched CaO–SiO$_2$–FeO–Al$_2$O$_3$–MgO slag with varying Al$_2$O$_3$ concentration for FeO contents of (**a**) 0 wt%, (**b**) 10 wt%, and (**c**) 20 wt%.

A decrease in viscosity can be observed in Figure 4 as the FeO concentration increases for a fixed initial Al$_2$O$_3$ concentration. Depending on the number of BOs in the [AlO$_4$] tetrahedral units, two distinct absorption bands can appear in the FT-IR spectra [19]. When the BO number is 3 (Q^3_{Al}), an absorption band is observed in the range 690–750 cm^{-1}. Otherwise, the absorption band observed between 640 and 680 cm^{-1} is attributed to the Q^2_{Al} unit, where the BO number is 2. As shown in Figure 5, a decrease in transmittance at 640–680 cm^{-1} is observed when the initial concentration of FeO is increased at a fixed initial concentration of Al$_2$O$_3$. The increase in Q^2_{Al} units with increasing FeO concentration indicates the depolymerization of the [AlO$_4$] tetrahedral network structure. In the molten oxide system, FeO acts as a network modifier. As Fe^{2+} ions require charge compensation, non-bridging oxygen is formed, which results in depolymerization by reducing the network connectivity.

To quantitatively evaluate the silicate structure changes with varying Al$_2$O$_3$ concentration in the present CaO–SiO$_2$–FeO–Al$_2$O$_3$–MgO system, Raman scattering measurements were performed. Figure 6 shows the original Raman spectra and Raman deconvoluted bands within the 400–1100 cm^{-1} range. Referring to the appropriate references listed in Table 2 [19–21,30,36–42], the Raman spectra were fitted by a Gaussian function and the corresponding structural units of the slag were identified with the aid of Peakfit 4 (Systat Software, San Jose, CA, United States). The relative fractions of the tetrahedral silicate structure units with varying BO numbers Q^n_{Si} were qualitatively evaluated by integrating the areas of the corresponding Gaussian-deconvoluted peaks. As shown in Figure 7, the number of Q^1_{Si} structural units gradually decreased with increasing Al$_2$O$_3$ concentration.

In contrast, the numbers of Q_{Si}^2 and Q_{Si}^3 structural units increased with increasing concentrations of Al_2O_3. The increase in the number of silicate structure units with higher BO numbers indicates the polymerization of the silicate network structure. According to Wang et al. [29] who studied the structure of a $CaO–SiO_2–FeO–Al_2O_3–MgO$ slag system using Raman spectroscopy and magic-angle-spinning nuclear magnetic resonance spectroscopy, a more polymerized silicate network structure was observed with higher Al_2O_3 concentration. The $[AlO_4]$ tetrahedral structural unit can be associated with the $[SiO_4]$ tetrahedral structural unit, thereby increasing the degree of polymerization. Yao et al. [43] also reported silicate network polymerization by the addition of Al_2O_3. When Al_2O_3 functions as a network former for tetrahedral structural units, it can be associated with non-bridging oxygen in the $[SiO_4]$ tetrahedral structural units, thus strengthening the silicate network structure. Therefore, the addition of Al_2O_3 to the $CaO–SiO_2–FeO–Al_2O_3–MgO$ system results in the polymerization of the molten slag system by the formation of an $[AlO_4]$ tetrahedral network structure associated with the $[SiO_4]$ tetrahedral network structure units.

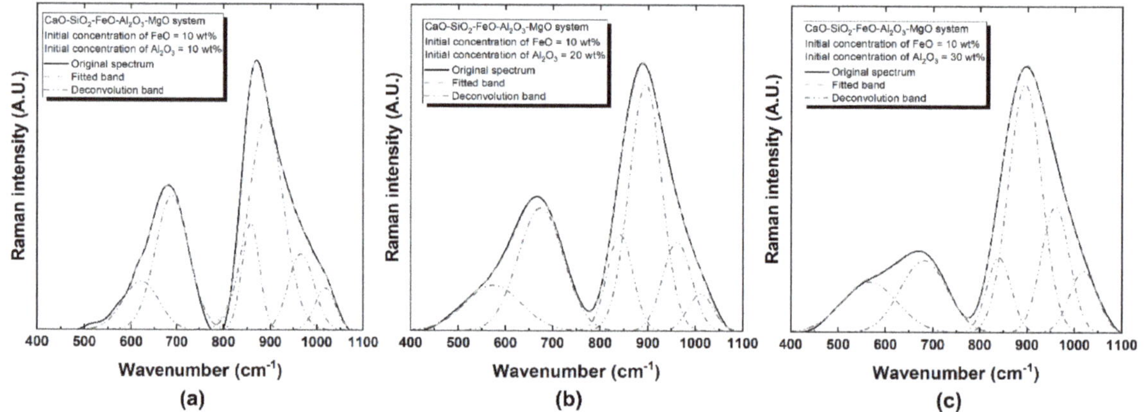

Figure 6. Raman spectra of quenched $CaO–SiO_2–FeO–Al_2O_3–MgO$ slag with 10 wt% initial FeO concentration and varying initial Al_2O_3 concentrations of (**a**) 10 wt%, (**b**) 20 wt%, and (**c**) 30 wt%.

Table 2. Reference Raman peak positions and corresponding assigned aluminate and silicate units.

Reference Position (cm^{-1})	Assignments
500–600 [21,30,37–39]	Symmetric Al–O$^-$ stretching of $[AlO_4]$
630–750 [21,36,37,40,41]	Symmetric Al–O$^-$ stretching of $[AlO_4]$
850–880 [19–21,36,42]	Symmetric Si–O$^-$ stretching of $[Si_2O_4]4^-$ (Q_{Si}^0)
900–930 [19–21,36,42]	Symmetric Si–O$^-$ stretching of $[Si_2O_7]6^-$ (Q_{Si}^1)
950–980 [19–21,36,42]	Symmetric Si–O$^-$ stretching of $[SiO_3]4^-$ (Q_{Si}^2)
1040–1060 [19–21,36,42]	Symmetric Si–O$^-$ stretching of $[Si_2O_5]2^-$ (Q_{Si}^3)

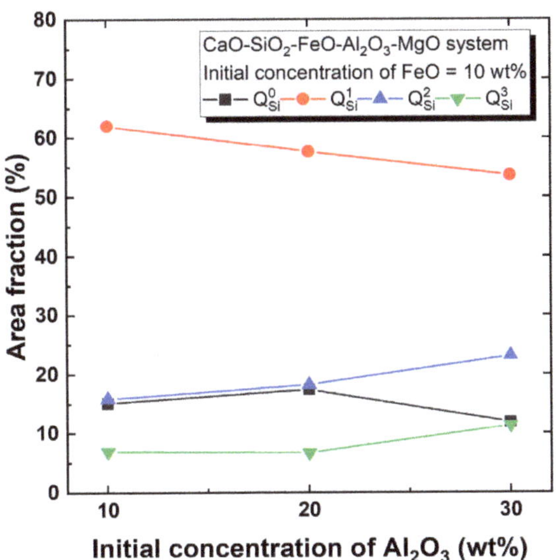

Figure 7. Relationship between relative area fractions of silicate tetrahedral structure (Q_{Si}^n) and initial concentration of Al_2O_3 in the CaO–SiO_2–FeO–Al_2O_3–MgO slag system at a fixed initial concentration of 10 wt% FeO.

4. Conclusions

Understanding the thermophysical properties of molten CaO–SiO_2–FeO–Al_2O_3–MgO systems is significant for FeO reduction by Al dross addition in the EAF process. In the present study, the viscosity of a CaO–SiO_2–FeO–Al_2O_3 system with a high concentration of MgO, which reached saturation at 1823 K, was measured by varying the FeO and Al_2O_3 concentrations at a fixed CaO/SiO_2 ratio. Structural changes in the molten slag system with composition variations were investigated using FT-IR and Raman spectroscopy. The following conclusions were drawn from the present study.

1. Decreases in viscosity at higher temperatures were commonly observed in the CaO–SiO_2–FeO–Al_2O_3–MgO slag system within the temperature range of 1823–1873 K. Based on the Arrhenius equation, the activation energy of viscous shear for the present slag system was evaluated. The highest activation energy (837.9 kJ/mol) was observed for the CaO–SiO_2–MgO ternary slag system. The change in the thermodynamically equilibrated phase of the slag system would be dominant in determining the activation energy.

2. The effect of FeO and Al_2O_3 on the slag viscosity was evaluated based on the silicate and aluminate network structures in the molten slag. An increase in the slag viscosity was observed with increasing Al_2O_3 concentration at 1873 K from 1.03 dPa·s to 1.9 dPa·s, from 0.6 dPa·s to 1.2 dPa·s, and from 0.4 dPa·s to 1.1 dPa·s when FeO was 0, 10, and 20 wt%, respectively. According to FT-IR spectroscopy, [AlO_4] tetrahedral units were formed with increasing Al_2O_3 concentration. In contrast, a decrease in viscosity was observed with increasing FeO concentration at 1873 K. Higher FeO concentrations at a fixed Al_2O_3 content resulted in an increase in Q_{Al}^2 and a decrease in Q_{Al}^3, indicating the depolymerization of the aluminate network structure.

3. According to the intermediate-range order structural investigation by Raman spectroscopy, the silicate network structure was polymerized with increasing Al_2O_3 concentration. Quantitative evaluation of the Q^n_{Si} structural units revealed an increase in Q^2_{Si} and Q^3_{Si} units with a decrease in Q^1_{Si} units with increasing Al_2O_3 concentration, indicating the polymerization of the silicate structure. The association of the $[AlO_4]$ tetrahedral units with the $[SiO_4]$ tetrahedral silicate network induced the polymerization of the slag structure and an increase in the viscosity of the molten slag.

Author Contributions: Conceptualization, Y.K. and D.-J.M.; methodology, Y.K.; formal analysis, Y.K.; investigation, Y.K.; data curation, Y.K.; writing—original draft preparation, Y.K.; writing—review and editing, D.-J.M. All authors have read and agreed to the published version of the manuscript.

Funding: This research was funded by Ministry of Science, ICT, and Future Planning of Korea, grant number GP2020-013.

Acknowledgments: Youngjae Kim acknowledges financial support from the Basic Research Project no. GP2020-013 of the Korea Institute of Geoscience and Mineral Resources (KIGAM), funded by the Ministry of Science, ICT, and Future Planning of Korea.

Conflicts of Interest: The authors declare no competing financial interest.

References

1. Lee, J.; An, S.B.; Shin, M.; Sim, K.J. Valorization of electrical arc furnace oxidizing slag. In *Celebrating the Megascale*; Mackey, P.J., Grimsey, E.J., Jones, R.T., Brooks, G.A., Eds.; Springer: Cham, Switzerland, 2014; pp. 347–355, ISBN 9781118889619.
2. Kim, H.S.; Kim, K.S.; Jung, S.S.; Hwang, J.I.; Choi, J.S.; Sohn, I. Valorization of electric arc furnace primary steelmaking slags for cement applications. *Waste Manag.* **2015**, *41*, 85–93. [CrossRef]
3. Sohn, I.; Hwang, J.I.; Choi, J.S.; Jeong, Y.S.; Lee, H.C. Development of ECO Slag Processing Technology for Iron Recovery and Value-Added Products in Steelmaking. In Proceedings of the 7th European Slag Conference (EUROSLAG 2013), Ijmuiden, Netherlands, 9–11 October 2013; EUROSLAG Publication: IJmuiden, Netherlands, 2013; pp. 292–305.
4. Ahmedzade, P.; Sengoz, B. Evaluation of steel slag coarse aggregate in hot mix asphalt concrete. *J. Hazard. Mater.* **2009**, *165*, 300–305. [CrossRef]
5. Muhmood, L.; Vitta, S.; Venkateswaran, D. Cementitious and pozzolanic behavior of electric arc furnace steel slags. *Cem. Concr. Res.* **2009**, *39*, 102–109. [CrossRef]
6. Manso, J.M.; Polanco, J.A.; Losañez, M.; González, J.J. Durability of concrete made with EAF slag as aggregate. *Cem. Concr. Compos.* **2006**, *28*, 528–534. [CrossRef]
7. Motz, H.; Geiseler, J. Products of steel slags an opportunity to save natural resources. *Waste Manag.* **2001**, *21*, 285–293. [CrossRef]
8. Manso, J.M.; Gonzalez, J.J.; Polanco, J.A. Electric arc furnace slag concrete. *J. Mater. Civ. Eng.* **2004**, *16*, 639–645. [CrossRef]
9. Pellegrino, C.; Faleschini, F. Electric arc furnace slag concrete. In *Sustainability Improvements in the Concrete Industry*; Springer: Berlin/Heidelberg, Germany, 2016; pp. 77–106, ISBN 9783319285405.
10. Jung, S.S.; Sohn, I. Crystallization control for remediation of an Fe_tO-rich CaO-SiO_2-Al_2O_3-MgO EAF waste slag. *Environ. Sci. Technol.* **2014**, *48*, 1886–1892. [CrossRef] [PubMed]
11. Kim, Y.; Min, D.J. Effect of FeO and Al_2O_3 on the MgO Solubility in CaO-SiO_2-FeO-Al_2O_3-MgO Slag System at 1823 K. *Steel Res. Int.* **2012**, *83*, 852–860. [CrossRef]
12. Yoon, C.M.; Park, Y.; Min, D.J. Effects of SiO_2 and B_2O_3 on MgO solubility and ionic structure in MgO-and-spinel doubly saturated aluminate slag. *Ceram. Int.* **2020**, *46*, 17062–17075. [CrossRef]
13. Yoon, C.M.; Park, Y.; Min, D.J. Thermodynamic Study on MgO Solubility in High-Alumina-Content Slag System. *Metall. Mater. Trans. B Process Metall. Mater. Process. Sci.* **2018**, *49*, 2322–2331. [CrossRef]
14. Lee, S.; Min, D.J. Viscous Behavior of FeO-Bearing Slag Melts Considering Structure of Slag. *Steel Res. Int.* **2018**, *89*, 1–6. [CrossRef]
15. Park, Y.; Min, D.J. Effect of Iron Redox Equilibrium on the Foaming Behavior of MgO-Saturated Slags. *Metall. Mater. Trans. B Process Metall. Mater. Process. Sci.* **2018**, *49*, 1709–1718. [CrossRef]
16. Kondratiev, A.; Jak, E.; Hayes, P.C. Predicting slag viscosities in metallurgical systems. *JOM* **2002**, *54*, 41–45. [CrossRef]
17. Lee, Y.S.; Kim, J.R.; Yi, S.H.; Min, D.J. Viscous behaviour of CaO-SiO_2-Al_2O_3-MgO-FeO slag. In Proceedings of the 7th Int. Conference on Molten Slags Fluxes and Salts, Cape Town, South Africa, 25–28 January 2004; The South African Institute of Mining and Metallurgy: Cape Town, South Africa, 2004; pp. 225–230.
18. Seok, S.H.; Min, D.J. Study on the Viscous Behavior of the Smelting Reduced Steelmaking Slag. *Korean J. Met. Mater.* **2007**, *45*, 360–367.
19. Zhang, G.; Wang, N.; Chen, M.; Wang, Y. Viscosity and Structure of CaO-SiO_2-"FeO"-Al_2O_3-MgO System during Iron-Extracting Process from Nickel Slag by Aluminum Dross. Part 2: Influence of Al_2O_3/SiO_2 Ratio. *Steel Res. Int.* **2018**, *89*, 1800273. [CrossRef]

20. Zhang, G.; Wang, N.; Chen, M.; Li, H. Viscosity and Structure of CaO-SiO$_2$-"FeO"-Al$_2$O$_3$-MgO System during Iron-Extracting Process from Nickel Slag by Aluminum Dross. Part 1: Coupling Effect of "FeO" and Al$_2$O$_3$. *Steel Res. Int.* **2018**, *89*, 1800272. [CrossRef]
21. Shen, X.; Chen, M.; Wang, N.; Wang, D. Viscosity property and melt structure of CaO-MgO-SiO$_2$-Al$_2$O$_3$-FeO slag system. *ISIJ Int.* **2019**, *59*, 9–15. [CrossRef]
22. Kim, G.H.; Sohn, I. Role of B$_2$O$_3$ on the Viscosity and Structure in the CaO-Al$_2$O$_3$-Na$_2$O-Based System. *Metall. Mater. Trans. B* **2013**, *45*, 86–95. [CrossRef]
23. Kim, Y.; Morita, K. Relationship between Molten Oxide Structure and Thermal Conductivity in the CaO–SiO$_2$–B$_2$O$_3$ System. *ISIJ Int.* **2014**, *54*, 2077–2083. [CrossRef]
24. Kim, Y.; Yanaba, Y.; Morita, K. Influence of structure and temperature on the thermal conductivity of molten CaO-B$_2$O$_3$. *J. Am. Ceram. Soc.* **2017**, *100*, 5746–5754. [CrossRef]
25. Avramov, I. Viscosity activation energy. *Phys. Chem. Glas. Eur. J. Glas. Sci. Technol. Part B* **2007**, *48*, 61–63.
26. Turkdogan, E.T.; Bills, P.M. A critical review of viscosity of CaO-MgO-Al$_2$O$_3$-SiO$_2$ melts. *Br. Ceram. Soc. Bull.* **1960**, *39*, 682–687.
27. Machin, J.S.; Yee, T.B.; Hanna, D.L. Viscosity Studies of System CaO-MgO-Al$_2$O$_3$-SiO$_2$: III, 35, 45, and 50% SiO$_2$. *J. Am. Ceram. Soc.* **1952**, *35*, 322–325. [CrossRef]
28. Park, J.H.; Min, D.J.; Song, H.S. Amphoteric behavior of alumina in viscous flow and structure of CaO-SiO$_2$($-$MgO)- Al$_2$O$_3$ slags. *Metall. Mater. Trans. B Process Metall. Mater. Process. Sci.* **2004**, *35*, 269–275. [CrossRef]
29. Wang, Z.; Sun, Y.; Sridhar, S.; Zhang, M.; Guo, M.; Zhang, Z. Effect of Al$_2$O$_3$ on the Viscosity and Structure of CaO-SiO$_2$-MgO-Al$_2$O$_3$-FetO Slags. *Metall. Mater. Trans. B Process Metall. Mater. Process. Sci.* **2015**, *46*, 537–541. [CrossRef]
30. Mysen, B.O.; Virgo, D.; Kushiro, I. The structural role of aluminum in silicate melts; a Raman spectroscopic study at 1 atmosphere. *Am. Mineral.* **1981**, *66*, 678–701.
31. Mysen, B. Relationships between silicate melt structure and petrologic processes. *Earth Sci. Rev.* **1990**, *27*, 281–365. [CrossRef]
32. Mysen, B.O.; Richet, P. *Silicate Glasses and Melts*, 1st ed.; Elsevier: Amsterdam, The Netherlands, 2005; ISBN 0444520112.
33. Sohn, I.; Min, D.J. A review of the relationship between viscosity and the structure of calcium-silicate-based slags in ironmaking. *Steel Res. Int.* **2012**, *83*, 611–630. [CrossRef]
34. Kim, H.; Kim, W.H.; Sohn, I.; Min, D.J. The effect of MgO on the viscosity of the CaO-SiO$_2$-20 wt% Al$_2$O$_3$-MgO slag system. *Steel Res. Int.* **2010**, *81*, 261–264. [CrossRef]
35. Talapaneni, T.; Yedla, N.; Pal, S.; Sarkar, S. Experimental and Theoretical Studies on the Viscosity–Structure Correlation for High Alumina-Silicate Melts. *Metall. Mater. Trans. B Process Metall. Mater. Process. Sci.* **2017**, *48*, 1450–1462. [CrossRef]
36. Kim, T.S.; Park, J.H. Structure-viscosity relationship of low-silica calcium aluminosilicate melts. *ISIJ Int.* **2014**, *54*, 2031–2038. [CrossRef]
37. McMillan, P.; Piriou, B. Raman spectroscopic of calcium aluminate glases and crystals. *J. Non. Cryst. Solids* **1983**, *55*, 221–242. [CrossRef]
38. McMillan, P.; Piriou, B.; Navrotsky, A. A Raman spectroscopic study of glasses along the joins silica-calcium aluminate, silica-sodium aluminate, and silica-potassium aluminate. *Geochim. Cosmochim. Acta* **1982**, *46*, 2021–2037. [CrossRef]
39. Higby, P.L.; Ginther, R.J.; Aggarwal, I.D.; Friebele, E.J. Glass formation and thermal properties of low-silica calcium aluminosilicate glasses. *J. Non. Cryst. Solids* **1990**, *126*, 209–215. [CrossRef]
40. Huang, C.; Behrman, E.C. Structure and properties of calcium aluminosilicate glasses. *J. Non. Cryst. Solids* **1991**, *128*, 310–321. [CrossRef]
41. Bykov, V.N.; Osipov, A.A.; Anfilogov, V.N. Structure of High-Alkali Aluminosilicate Melts from the High-Temperature Raman Spectroscopic Data. *Glas. Phys. Chem.* **2003**, *29*, 105–107. [CrossRef]
42. Huang, W.J.; Zhao, Y.H.; Yu, S.; Zhang, L.X.; Ye, Z.C.; Wang, N.; Chen, M. Viscosity property and structure analysis of FeO-SiO$_2$-V$_2$O$_3$- TiO$_2$-Cr$_2$O$_3$ slags. *ISIJ Int.* **2016**, *56*, 594–601. [CrossRef]
43. Yao, L.; Ren, S.; Wang, X.; Liu, Q.; Dong, L.; Yang, J.; Liu, J. Effect of Al$_2$O$_3$, MgO, and CaO/SiO$_2$ on Viscosity of High Alumina Blast Furnace Slag. *Steel Res. Int.* **2016**, *87*, 241–249. [CrossRef]

Article

Chelation-Assisted Ion-Exchange Leaching of Rare Earths from Clay Minerals

Georgiana Moldoveanu and Vladimiros Papangelakis *

Department of Chemical Engineering and Applied Chemistry, University of Toronto, 200 College Street, Toronto, ON M5S 3E5, Canada; georgiana.moldoveanu@utoronto.ca
* Correspondence: vladimiros.papangelakis@utoronto.ca; Tel.: +1-416-978-1093

Abstract: The effect of biodegradable chelating agents on the recovery of rare earth elements (REE) from clay minerals via ion-exchange leaching was investigated, with the aim of proposing a cost-effective, enhanced procedure that is environmentally benign and allows high REE recovery while reducing/eliminating ammonium sulfate usage. A processing route employing a lixiviant system consisting of simulated sea water (equivalent to about 0.5 mol/L NaCl) in conjunction with chelating agents was also explored, in order to offer a process alternative for situations with restricted access to fresh water (either due to remote location or to lower the operating costs). Screening criteria for the selection of chelating agents were established and experiments were conducted to assess the efficiency of selected reagents in terms of REE recovery. The results were compared to extraction levels obtained during conventional ion-exchange leaching procedures with ammonium sulfate and simulated sea water only. It was found that stoichiometric addition of N,N'-ethylenediaminedisuccinic acid (EDDS) and nitrilotriacetic acid-trisodium form (NTA-Na$_3$) resulted in 10–20% increased REE extraction when compared to lixiviant only, while achieving moderate Al co-desorption and maintaining neutral pH values in the final solution.

Keywords: weathered crust elution-deposited rare earth ore; rare earth recovery; ion-exchange leaching; chelation; chelating agents; polydentate ligands

Citation: Moldoveanu, G.; Papangelakis, V. Chelation-Assisted Ion-Exchange Leaching of Rare Earths from Clay Minerals. *Metals* **2021**, *11*, 1265. https://doi.org/10.3390/met11081265

Academic Editors: Jean François Blais, Srecko Stopic and Dariush Azizi

Received: 29 June 2021
Accepted: 6 August 2021
Published: 11 August 2021

Publisher's Note: MDPI stays neutral with regard to jurisdictional claims in published maps and institutional affiliations.

1. Introduction

1.1. Background

Rare earth elements (REEs) are a collection of fourteen of the fifteen naturally-occurring lanthanides (excluding promethium), further grouped, depending on the atomic number, into "light" rare earth elements (LREEs)—La, Ce, Pr, and Nd, and "middle & heavy" (HREEs)—Sm, Eu, Gd, Tb, Dy, Ho, Er, Tm, Yb, and Lu. Yttrium (Y) and scandium (Sc) are also considered "rare earths", as they occur alongside lanthanides in the same ore deposits and have similar properties [1]. Due to their unique physical and chemical properties, REEs became progressively more indispensable to the modern industry, with increasing demand in specific fields such clean energy, aerospace, and sustainable technology sectors. It is estimated [2] that the demand for REEs from clean technologies will reach 51.9 thousand metric tons (kt) rare earth oxide REO in 2030, with Nd and Dy, respectively, comprising 75% and 9% of the demand. Adamas Intelligence [3] forecasted that magnet rare earth oxide demand (Nd, Pr, Dy, and Tb) will increase at a compound annual growth rate of 9.7%, and the value of global magnet rare earth oxide consumption will rise fivefold by 2030, from $2.98 billion in 2019 to $15.65 billion at the end of the decade.

REEs occur as accessory minerals in various rocks, but the most commercially significant sources, as reviewed by Kanazawa and Kamitani [4], are fluorocarbonates (bastnaesite), phosphates (monazite and xenotime) and weathered crust elution-deposited rare earth ores (ion-adsorption clays). Carbonate and phosphate sources, despite being high grade, are associated with elevated recovery costs due to mining, beneficiation, and the need of aggressive conditions to dissolve the REEs [5]. Ion-adsorption type deposits are substantially lower grade than other lanthanide sources; however, this disadvantage is largely offset by the easier mining and processing costs, the relatively low content of radioactive elements (Th, U), and high HREE content [6]. The ion-adsorption ores contain 0.03 to 0.3 wt.% REEs, out of which generally 60–80% occur as physically adsorbed species on clays, recoverable by ion-exchange leaching [7]. Despite the low grades, ion-adsorption ores account for ~35% of China's total REE output and ~80% of the world's HREE production [8,9]. While at the present China is the only country to commercially produce REE from ion-adsorption ores, recent geological surveys have led to the discovery and investigation of similar deposits in South America [10], Thailand [11], and Africa [12].

The conventional method of processing the ion-adsorption ores is by ion-exchange leaching using monovalent sulfate or chloride salt solutions at ambient temperature ([7,13–16]). During leaching, the physisorbed REE are substituted on the substrate by the exchange ions and transferred into solution as soluble sulfates or chlorides.

Ammonium sulfate is the established lixiviant for the recovery of lanthanides from ion-adsorption ores by either heap or in-situ leaching, due to its high extraction efficiency and low product contamination [16]. However, recent trends in ion-adsorption ore research are focused on minimizing the usage of ammonium sulfate in an effort to reduce ammonia pollution of surface and ground waters, either by adding certain leaching-enhancing additives to the conventional $(NH_4)_2SO_4$ lixiviant or by evaluating alternative leaching reagents. Tian et al. [17] investigated small additions of natural organic reagents such as Sesbania gum (plant-derived), Luo et al. [18] assessed humic and fulvic acid additions, while Zhang et al. [19] proposed a novel "targeted solution injection" method for in-situ leaching, that would optimize the use of ammonium sulfate. Regarding alternative lixiviant use, Rocha et al. [10] evaluated NaCl and NH_4Cl, whereas Xiao et al. [20,21] assessed the use of an $MgSO_4$-$CaSO_4$ combination; however, the use of non-ammonium-based reagents such as sodium or magnesium salts generally leads to decrease in REE production and poor product purity due to their lower extraction efficiency [15] and possibility of co-precipitation/entrainment during subsequent processing stages. Nevertheless, as the grade of ion-adsorption ores is generally low and the recovery of REEs needs to be maximized in order to economically justify the process, the use of the most efficient extraction lixiviant (i.e., ammonium sulfate) is advisable, but employed in conjunction with operating practices designed to ensure minimum environmental impact.

The application of coordination chemistry (i.e., the capacity of certain ligands to form stable complexes with metals) to leaching is a well-known and implemented technique in mining and metallurgy, especially for extraction of PGMs, Ag, Cu, and U from ores [22]. For example, as an alternative to the well-established routes of gold leaching with cyanate [23], thiosulphate [24] or thiourea [25], Senanayake [26] comprehensively describes gold leaching by copper(II) in ammoniacal thiosulphate solutions in the presence of various additives. Similarly, while ammonia-based reagents are the traditional abiotic leaching media for copper [27], Oraby and Ecksteen reported on selective leaching of Cu from a Cu–Ag concentrate in the presence of glycine [28] and leaching of Au in a H_2O_2–glycine medium [29,30]. These processes are, however, dissolution-based leaching, where the ligands are assisting the main lixiviant (either acid or base), by extending the solubility window of metal species in solution via complex formation and thereby enhancing the extraction efficiency.

The use of chelating agents (mostly aminopolycarboxylic and polycarboxylic acids and their salts) for mobilization and removal of toxic metals from metal-contaminated soils is a widely studied and applied low-cost, efficient, soil remediation technique conducted either in-situ (soil flushing) or ex-situ (heap/column leaching [31,32]; chelating agents are able to desorb (mobilize) metals from soil solid phases by forming strong water-soluble compounds stable over a wide range of pH, which are subsequently removed by enhanced phytoextraction or soil washing techniques [33–35]. Synthetic amino-polycarboxylic acids (amino-PCAs) such as ethylenediaminetetraacetic acid (EDTA) and diethylenetrinitrilopentaacetic acid (DTPA) and their analogues are the most widely used industrial chelating agents, with applications in pulp and paper, cleaning, chemical processing, agriculture, and water treatment [35,36]. In their comprehensive review, Eivazihollagh et al. [31] describe the application of chelating agents to various fields such as wastewater treatment and soil remediation, mineral flotation, organometallic catalysis, and metal recovery, to name a few, and also present the main routes employed for ligand recovery/reuse. Despite obvious advantages such as low cost and good chelating efficiency across the spectrum, these reagents are toxic and exhibit resistance to conventional biological or physico-chemical water treatment destruction methods and show extended persistence in the environment (i.e., low biodegradability).

Lanthanide recovery from ion-adsorption ores is generally performed either as batch, in-situ, or heap leaching, following similar concepts of toxic metals removal during soil remediation procedures. Although the preferential chelation of lanthanides with various specific ligands is a well-known chemistry fact, the applications were initially limited to laboratory-scale techniques ([37–41]). Various applications of chelating agents to REE extraction, especially from ion-adsorption clays, have been developed lately. Li et al. [42] evaluated ammonium citrate to leach the weathered crust elution-deposited rare earth ores. Wang et al. [43] investigated various carboxylic acids as additives to 0.3 mol/L NH$_4$Cl for the leaching of REEs from ion-adsorption ores, while Zhang et al. [44] explored the leaching of rare earth from ion-adsorption ores by ammonium acetate. More recent studies involving the use of chelating/complexing reagents during ion-exchange leaching of rare earths were conducted by Chai et al. [45], who explored ammonium carboxylate–ammonium citrate mixture as lixiviant, and Chen et al. [46], who evaluated formate salts. Similarly, studies conducted by Cristiani et al. made use of the good complex-forming capacity of polyamines to evaluate the efficiency of functionalized clays as sorbents capable of the uptake/removal of heavy metals from polluted aqueous effluents [47] and lanthanides from leachates of electronic wastes [48].

The aim of the present study was to investigate a cost-effective, enhanced ion-exchange leaching procedure that is environmentally benign and allows high REE recovery while reducing ammonium sulfate usage, by employing biodegradable chelating agents in conjunction with the main lixiviant. Additionally, in an effort to further reduce/eliminate ammonia-based leaching, a processing route employing a lixiviant system consisting of simulated sea water (equivalent to about 0.5 mol/L NaCl) in conjunction with chelating agents was explored. Although ion-exchange leaching with NaCl (usually 1 mol/L, according [16]) generally leads to lower total rare earth (TREE) extraction than during leaching with ammonium sulphate, and application of chloride-based reagents has its own challenges (e.g., higher reagent costs and equipment corrosion risks), the use of a naturally occurring, inexpensive, and readily available lixiviant such as seawater is worth evaluating. It is expected that chelating agents will improve TREE extraction with seawater and offer a cost-effective process alternative for situations where access to fresh water and large quantities of chemical reagents is restricted (either due to remote location or to lower the operating costs).

The chelating agents selected are known for good chelating abilities and biodegradability; while some of them have been evaluated for REE ion-exchange leaching before (e.g., citric acid, EDTA, acetate-based), the others are newly applied. More specifically, the authors established screening criteria for the selection of optimal chelating agents, conducted experiments in order to evaluate the efficiency of selected reagents to maintain high REE extraction in the presence of lower lixiviant concentration or simulated seawater, and compared the results with REE extraction levels obtained during conventional ion-exchange leaching procedures.

1.2. Selection of Chelating Agents

Lanthanides are hard acids with strong preference for electronegative atoms, consequently they bond very well to hard bases, i.e., ligands containing oxygen ([22,49]). In aqueous solutions, complexation always involves substitution of the metal–oxygen bond from solvation water with another metal–oxygen bond from a ligand and the bonds are predominantly electrostatic [50]. At a molecular level, Kettle [51] explains the affinity between lanthanides and oxygen-containing ligands via the Ligand Field Theory: the 4f orbitals of REEs are well shielded by the 5d and 6s orbitals and do not participate in bonding, undergoing only minimal crystal field splitting; unlike the case for transitional metals (d-block elements), the interactions of lanthanides with ligands are rather dominated by steric and electrostatic effects. Because of this, lanthanides are considered weak field and have affinity towards weak field ligands (e.g., O-containing chelating agents), forming weak field, high spin complexes.

According to Choppin [52], rare earths have high coordination numbers of 8–12 and are thus capable of forming stronger complexes with organic poly-functional ligands than with inorganic ligands due to the possibility of forming multiple metal–oxygen bonds. Furthermore, Smith and Martell [53] indicated that the stability constant values for polydentate ligands are greater than for monodentate ones (i.e., if a bond to one of the donor atoms is broken, the others will hold) and increase with the number of coordinating groups, explaining thus why EDTA is such an efficient (albeit not selective) chelating agent.

In accordance with the challenges described in Section 1.1, the main factors governing the selection of ligands are delineated as following:

1. Extraction strength: capable of forming stable, strong complexes with the target metals
2. Selectivity towards target metals (i.e., low impurity co-extraction)
3. Low toxicity and high biodegradability
4. Cost-effective

Based on these considerations, the most efficient chelating agents (highest binding power) should therefore contain more than one oxygen-containing functional group (carboxyl and/or hydroxyl). The reagents selected for this study are commonly available polydentate compounds known to exhibit good chelating power for heavy metals, low toxicity and superior biodegradability:

(a) Poly carboxylic acids (PCA) and amino-poly carboxylic acids recently under study in Europe as alternative chelating agents for heavy metals removal from wastewaters and contaminated soils [34]: citric acid, nitrilotriacetic acid (NTA), aspartic acid, *N*,*N*'-ethylenediaminedisuccinic acid (EDDS) and ethylenediaminetetraacetic acid (EDTA)
(b) Natural amino acids investigated for hydrometallurgical applications due to their complexing action towards transitional metals ([28–30]): glycine and asparagine.

In order to avoid introducing additional impurity cations in the system (and thus possibly contaminate the final rare earth product), the acid form (HL) of the chelating agents was selected for this study; however, as it is reported that the amount of available free ligand increases with increasing pH due to improved dissociation [32,35], the tri-sodium form of NTA (NTA-Na$_3$) was also evaluated for comparison purposes. NTA is reported to undergo fast degradation in natural conditions due to action of various bacteria strains from the Proteobacteria subclass, with a half-life of degradation for 100 µg/L NTA of ~31 h (WHO, 1996), while ~80% of EDDS converts to CO_2 in 20 days [54,55]. The citric and L-aspartic acids, as well as the small molecule amino acids selected are naturally occurring compounds with known applications in nutritional supplements and food industries, hence of no toxicity.

The efficacy of a chelating agent is usually rated with the overall stability constant of formation of the ligand–metal complexes (given the symbol β). Smith and Martell [53] and Cotton [49] reported that, despite their high coordination numbers (8–12), lanthanides form monodentate and bidentate complexes with ligands having four or fewer coordinating sites, but only monodentate complexes with higher coordinating ligands.

The logβ values can be used to rank different ligands towards a specific metal (the higher the logβ, the stronger, more stable the complex). Table 1 presents the available data on constants of formation (logβ) for complexes of rare earths with the selected chelating agents, as reviewed by various authors; for the multidentate ligands that form only 1:1 complexes, $\log K_1 = \log\beta$.

Table 1. Constants of formation ($\log\beta$) for complexes of trivalent rare earth ions and aluminum with the chelating agents employed in the present study (25 °C, 1 atm, 0.1 mol/L KNO$_3$ ionic strength).

	$\log\beta$						
	Hexadentate ($\log K_1 = \log\beta$)		Tetradentate ($\log K_1 = \log\beta$)		Tridentate ($\log\beta$)		Bidentate ($\log\beta$)
M^{3+}	EDTA [1]	EDDS [1]	NTA [1]	Citric [2]	Aspartic [2]	Asparagine [3]	Glycine [4]
La	15.5	11.8	10.3	9.5	8.3	7.1	6.1
Ce	16	12.4	10.7	9.6	8.8	7.2	6.4
Pr	16.4	13.1	11.0	9.7	9.1	7.6	6.9
Nd	16.6	13.7	11.2	9.8	9.5	7.8	7.1
Sm	17.1	14.5	11.5			8.0	
Eu	17.3	14.8	11.5	9.8			
Gd	17.4	14.9	11.5	9.9		8.2	
Tb	17.9	15.0	11.6				
Dy	18.3	15.1	11.7			8.6	
Ho	18.6	15.4	11.8				
Er	18.9	16.1	12.0				
Tm	19.3	16.4	12.2				
Yb	19.6	17.0	12.3			8.9	
Lu	19.9	17.6	12.4				
Al	16.3	13.4	11.4	11.7		9.3	6.4

[1] [53]; [2] [56]; [3] [57]; [4] [58].

Data for aluminum is included due to the fact that Al is the main impurity to interfere with the ion-exchange leaching process. Impurities associated with the ion-adsorption ores are usually Na, K, Mg, Ca, Mn, Zn, Al, and Fe [7]; while most of these cations occur as part of the mineral matrix and do not leach out during the mild REE leaching conditions, a significant amount of Al, due to its trivalent state, is physically adsorbed and liable to be desorbed along with the lanthanides during the process [10].

2. Materials and Methods

2.1. Materials and Analytical Procedures

For the preparation of all solutions used in the present work, de-ionized water and ACS-grade reagents were used. Ion-adsorption clay ores of African origin (courtesy of Tantalus Rare Earths AG) were tested in this study. X-ray Fluorescence (XRF, Brucker AXS S2 Ranger, Brucker, Madison, WI, USA) was employed to determine the overall bulk chemical composition of the solids, whereas the REE content was determined by aqua regia digestion (HCl:HNO$_3$ 3:1 v/v) at 220 °C for 1 h (Ethos EZ microwave system, Milestone, Sorisole, Italy) followed by inductively coupled pslasma optical emission spectrometry on the filtered solution diluted with 5% HNO$_3$ (ICP-OES, Agilent 720 series, Agilent Technologies, Santa Clara, CA, USA). The composition of all liquid phases following leaching was analyzed by ICP-OES (Agilent 720 series).

2.2. Batch Leaching Tests

The baseline experiments involved 50 g clays, 0.125 mol/L (NH$_4$)$_2$SO$_4$ (i.e., 0.25 mol/L NH$_4^+$ exchange ions) as the main lixiviant, ambient conditions, liquid to solid (L:S) ratio of 2:1 (v/w), moderate stirring to ensure slurry suspension (300–500 rpm), and 30 min total time, following experimental procedures developed earlier (Moldoveanu and Papangelakis 2012; 2013). At the beginning of the experiment, each of the chelating agents listed in Table 1 were added to the slurry in 1:1 and 2:1 stoichiometric excess, respectively, with respect to the total content of adsorbed cations (i.e., REEs and impurities). Previous work determined the desorption kinetics to be very fast (<15 min) and independent of leaching conditions such as temperature, pH, and agitation, which influence only terminal extraction levels (Moldoveanu and Papangelakis, 2013). At the end of each experiment, the pregnant leach solution (PLS) was separated by vacuum filtration. The solid residue was washed twice with deionized water of adjusted pH 5 (L:S = 2:1), dried in the oven at 50 °C (overnight), weighed, and stored, while the mother liquor and wash water collected after filtration were diluted with 5% (v/v) HNO$_3$ and analyzed for REE content.

The following formula was employed to quantify the total REE (TREE) extraction (as %):

$$\% \text{ TREE extracted} = \frac{\text{TREE}_{\text{in, clay}} - \text{TREE}_{\text{aq,final}}}{\text{TREE}_{\text{in,clay}}} \times 100$$

where: TREE$_{\text{in,clay}}$ = mass of TREE contained in the initial amount of leached ionic clays (mg) and TREE$_{\text{aq,final}}$ = mass of TREE contained in the final leaching solution + wash solution (mg), (determined as concentration, units of mg/L, and converted to mass by multiplying with the volume of respective solution).

3. Results and Discussion

3.1. Ore Composition

The bulk chemical composition of the ion-adsorption ore presented in Table 2 is characteristic of the typical weathered ores containing mixed alumino-silicates, mainly kaolinite/halloysite (Al$_2$(Si$_2$O$_5$)(OH)$_4$), quartz (SiO$_2$) and mica (KAl$_3$Si$_3$O$_{10}$(OH)$_2$), consistent with the overall compositions described in literature ([16,59,60]); the total REE (TREE) content, was determined to be 0.13% (w/w).

Table 2. Overall chemical composition of clays (XRF, major elements only, >0.1 wt%).

Oxide	SiO$_2$	Al$_2$O$_3$	Fe$_2$O$_3$	TiO$_2$	K$_2$O	MnO	ZrO$_2$	CaO
Composition (wt%)	40.3	32.4	19.6	2.5	2.3	0.3	0.2	0.2

The individual and relative REE content, respectively, as shown in Table 3, indicates that the ore contains about 78.5% LREE and 21.5% medium and heavy REE.

Table 3. Individual REE content and relative REE distribution in ore (ICP-OES).

REE	Y	La	Ce	Pr	Nd	Sm	Eu	Gd	Tb	Dy	Ho	Er	Tm	Yb	Lu	TREE
ppm	222	321	168	64	250	44	19	59	6	36	4	24	9	73	5	1306
%	17	24.6	13	4.9	19	3.4	1.4	4.6	0.5	2.8	0.3	1.8	0.7	5.6	0.4	100

3.2. Batch Ion-Exchange Leaching Tests with Ammonium Sulfate and Chelating Agents

The ion-exchange leaching performance with 50% less ammonium sulfate (i.e., 0.25 mol/L NH$_4^+$) in the presence of the chelating reagents will be compared to the baseline case of 0.5 mol/L NH$_4^+$ alone, as well as with the TREE extraction levels achieved with EDTA (method adapted from [61,62]).

Figure 1 shows the comparative TREE extraction levels obtained for addition of chelating agents in 1:1 and 2:1 stoichiometric ratios with respect to the total content of adsorbed cations (i.e., REE and impurities), following the procedure described in Section 2.2. It can be observed that 1:1 addition of chelating agents to 0.25 mol/L NH$_4^+$ (as sulfate) resulted in 6–10% increased extraction when compared to lixiviant alone, except for glycine and asparagine, which showed no/very little improvement. This finding is in accordance with the values of logβ shown in Table 1 and the fact that the stability of a chelate increases with the number of functional groups on the ligand ([49,50]). Moreover, EDDS, NTA and aspartic acid reached levels of extraction close to that achieved by 0.5 mol/L NH$_4^+$, denoting sustained high TREE recovery with 50% less lixiviant. The use of NTA-Na$_3$ showed only marginal improvement when compared to the acidic form (NTA), but it may prove a better environmental choice due to lower final acidity levels. EDTA led to the highest TREE extraction, but this reagent was employed only as a measure of maximum extraction achievable and is not being considered as an option due to the reasons explained in Section 1.

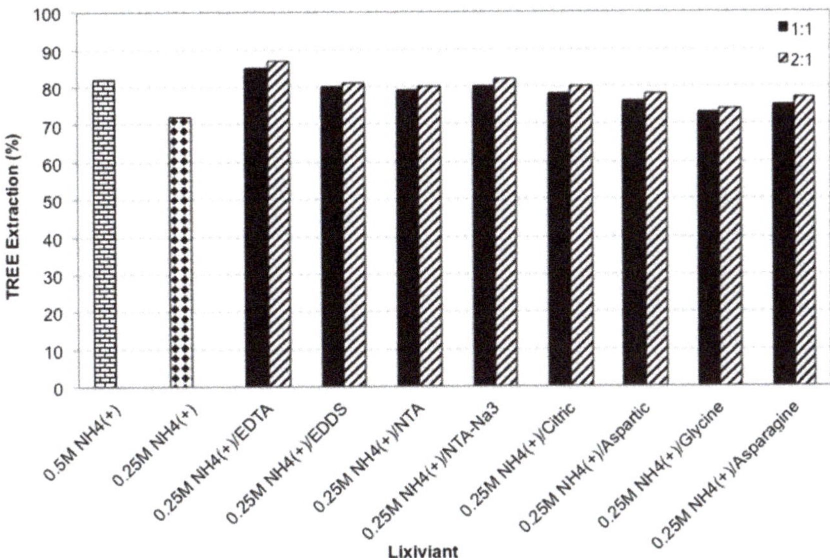

Figure 1. Influence of chelating agent addition on the TREE extraction with ammonium sulfate (ambient conditions, 30 min, L/S = 2/1).

Figure 1 indicates that the 2:1 stoichiometric excess did not lead to appreciable improvement in TREE extraction; this finding is in conformity with data reported that lanthanides form ML and ML_2 complexes with ligands having four or fewer coordinating sites but only ML complexes with higher dentate ligands ([49,53]). As higher excess of chelating agent does not seem necessary/useful, the 1:1 ratio brings the additional benefit of minimum environmental impact.

In terms of individual REE behavior, a certain selectivity was noticed towards Y and the light REEs (La to Gd, with the exception of Pr and Ce), as these elements exhibited up to 30% higher extraction than the heavy REEs. This trend was explained by the higher charge density associated with the HREEs, which leads to stronger adsorption on clays. Individual REE extraction levels with 0.25 mol/L NH_4^+ and 1:1 chelating agents are shown in Table S1 of the Supplementary Materials.

3.3. Process Implications—Seawater as Lixiviant

The immediate implication of the chelation-assisted ion-exchange leaching process pertains to the possibility of using seawater for leaching solution preparation, instead of fresh water and ammonium sulfate. The average natural seawater composition, in terms of major elements, is given in Table 4, and shows Na and Mg chlorides and sulfates as the main components; although the authors showed in previous studies that Na and Mg are less efficient than NH_4^+ as exchange cations for REEs ([13,15]), it is hypothesized that the presence of ligands has the potential to improve seawater's performance and increase the REE recovery. This would reduce the overall hydrometallurgical plant freshwater consumption, eliminate the ammonia pollution, lessen recycling requirements, and open up the possibility of returning the final purified streams to the sea without risk of contaminating the soil.

Table 4. Average natural seawater composition—major elements [63].

Element	Na	Mg	Ca	K	Cl	S (as SO_4^-)	pH
Concentration (mol/L)	0.45	0.05	0.01	0.01	0.53	0.04	7–8

The authors prepared synthetic seawater (SSW) of similar composition and performed comparative leaching experiments of ion-adsorption clays with SSW only (corresponding roughly to 0.45 mol/L exchange cations as NaCl), and SSW + selected chelating agents. These cases were compared to the TREE extraction levels achieved when using the very efficient ammonium sulfate lixiviant (both 0.5 mol/L and 0.25 mol/L NH_4^+, respectively); the results are depicted in Figure 2.

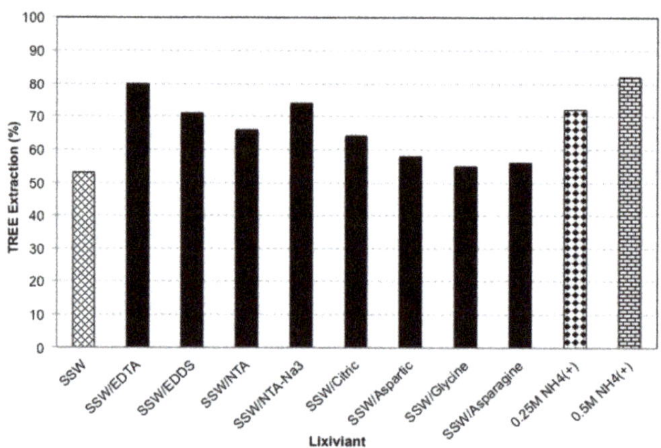

Figure 2. Influence of chelating agent addition on the TREE extraction with simulated seawater (ambient conditions, 30 min, L/S = 2/1, 1:1 chelating agents).

It can be observed that 1:1 addition of chelating agents to SSW resulted in noticeably increased extraction when compared to leaching with SSW alone (again, with the exception of glycine and asparagine) from ~5% for aspartic acid to ~20% for EDDS and NTA-Na3 (notwithstanding 30% for EDTA), reaching levels close to ones achieved with 0.25 mol/L NH_4^+. Individual REE extraction levels with SSW and 1:1 chelating agents are shown in Table S2 of the Supplementary Materials.

3.4. Behaviour of Aluminum

Previous studies conducted by our group [15], as well as other researchers ([7,64,65]) revealed that the main impurity associated with the ion-adsorption ores is Al, which follows identical desorption kinetics to REEs during leaching with $(NH_4)_2SO_4$, with no selectivity window. The potentially high aluminum concentration in the leachate has a negative impact on the whole downstream REE recovery process, as it leads to excessive consumption of the precipitation reagent (generally oxalic acid, [66]).

The authors evaluated the Al concentration in the leachate (as mg/L) produced during ion-extraction leaching of REEs with ammonium sulfate and SSW alone, and with 1:1 ratio chelating agents under the conditions selected; the results are summarized in Figure 3. The case for 2:1 excess is not shown here as it held no added benefit for REE extraction, as shown in Figure 1, but it extracted slightly more Al. Aluminum extraction levels with 0.25 mol/L NH_4^+ and 2:1 chelating agents is shown in Figure S1 of the Supplementary Materials.

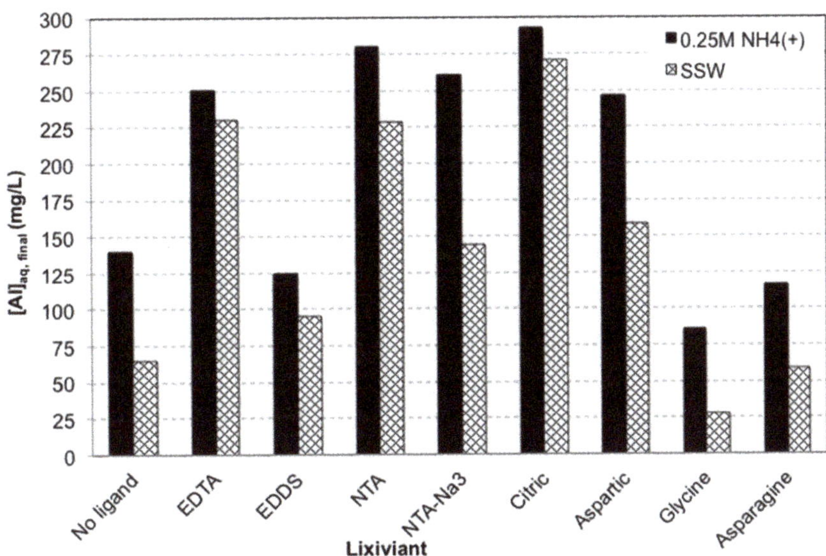

Figure 3. Aluminum concentration in the leachate (ambient conditions, 30 min, L/S = 2/1, 1:1 chelating agents).

The comparative chart shows that EDTA, NTA (in both forms), citric and aspartic acids prove very good chelating agents for aluminum, achieving much higher Al concentration in the leachate than in the presence of ammonium sulfate alone (0.25 mol/L NH_4^+) or SSW, and therefore offering no selectivity towards REE. The lower Al concentration levels obtained with glycine and asparagine (in both cases) must be due to the inferior chelating power of these compounds, as also indicated in the case of REEs. EDDS appears to suppress Al desorption, which, combined with the good REE extraction levels shown in Figures 2 and 3, make it the recommended choice for the chelation-assisted ion-exchange leaching. NTA-trisodium, although extracting more aluminum than EDDS, performed better than the other chelating agents in this aspect, and could be considered a viable alternative to the more expensive EDDS, especially when used in conjunction with seawater.

3.5. Influence of Solution pH on Rare Earths and Aluminum Extraction

The final (equilibrium) pH in the PLS is an important factor, as it impacts the metal recovery levels and the environment (both during the in-situ leaching procedure and also upon discharge—especially if seawater is to be employed). Table 5 lists the initial pH of the lixiviant solutions (containing the chelating agent but prior to clay addition) and the final (equilibrium) pH of the filtrate; the pH$_{final}$ is a combination of the pH$_{initial}$ and the buffering effect of the clays (due to the existence of H^+ and OH^- groups also adsorbed on the surface). It was decided not to adjust the pH for a specific value, as the final/equilibrium values were not considered high enough to initiate the hydrolysis process in the presence of the chelating agents (known to expand the pH solubility window for lanthanide species due to the formation of stable aqueous complexes). Additionally, the operating costs will be further lowered in the absence of an unnecessary pH adjusting step.

Table 5. Initial and final (equilibrium) solution pH values.

Chelating Agent	$(NH_4)_2SO_4$ (0.25 M NH_4^+)		SSW	
	$pH_{initial}$	pH_{final}	$pH_{initial}$	pH_{final}
No ligand	5.7	4.1	8.8	3.8
EDTA	2.8	3	2.6	2.1
EDDS	8.5	7.1	6.4	6
NTA	2.3	3.3	2	1.9
NTA-Na$_3$	8.5	6.1	9.2	5.5
Citric	2.5	2.6	1.8	2
Aspartic	3.1	3.7	4.1	3.9
Glycine	5.8	4.3	6.8	4.1
Asparagine	4.8	4.1	6.7	3.8

It can be observed that, with the exception of EDDS and NTA-Na3, the pH_{final} was generally below four for both ammonium sulfate and SSW, which could explain the high levels of aluminum in the PLS (as these pH levels are below the Al hydrolysis threshold pH of ~4.5). EDTA, citric acid and NTA led to even more acidic levels of around pH 2; these high levels of final acidity would demand thorough in-situ washing in order to bring the soil towards more neutral pH values as well as pre-treatment prior to discharge (in the case of seawater use).

The use of EDDs and Na-Na3 resulted in more neutral values of pH_{final} in both lixiviant systems, which, combined with good TREE recovery and relatively suppressed Al co-extraction, may make these the recommended ligands to be employed, especially with sweater. There is no risk of potential TREE loss at these pH values, due to the fact that chelating agents are known to extend the solubility window beyond the hydrolysis values, which is in the range 6.5–7 for lanthanides (depending on the specific REE). Moreover, the higher pH values obtained in the case of EDDS and NTA-trisodium have a beneficial effect on the dissociation extent of the chelating agent and subsequent availability of the ligand for REE.

3.6. Process Considerations Involving Different Types of Clay Ores

The major components of the ion-adsorption clays usually employed for research (most of them of Chinese origin) are mainly kaolinite/halloysite, with probable fractions of chlorite and illite, which explains the generally low total rare earth element content (usually 0.03–0.3 wt%) due to low cation exchange capacity (CEC) of these clays. Kaolinite and chlorite have a CEC of 5–15 meq/100 g, illite 25–40 meq/100 g, while the CEC of montmorillonite is 80–120 meq/100 g [67]. Based on these figures, it can be inferred that ion-adsorption ores containing clay fractions with higher CEC, such as montmorillonite, smectite, vermiculite, etc., would have a higher initial overall TREE content.

Alshameri et al. [68] evaluated kaolinite (Kao), montmorillonite (Mt), muscovite (Ms) and illite (Ilt) for their adsorptive/and regeneration behaviors towards La^{3+} and Yb^{3+} (as proxies for light and heavy REEs, respectively). They concluded that montmorillonite exhibited the highest adsorption and regeneration efficiencies for both La^{3+} and Yb^{3+} and noticed a decrease in the order of Mt > Ms > Ilt > Kao. Less intuitively, however, it was reported that Kao had highest extraction efficiencies for both REEs, in the order of Kao > Ilt > Mt > Ms. This behavior was linked to the structure and surface properties of the clays: while the overall high CEC of "pristine" clays allowed for elevated REE adsorption, the lower desorption from 2:1 clays (such as Mt and Ms) was probably due to the difficulty of the desorbing agent in accessing non-surface adsorption sites. Based on these studies, we can conclude that the consumption of more chelating agent during hypothetical leaching of high CEC clays is not a predictable assumption, despite the elevated initial REE content in clays. Only proper experimental assessment can determine the actual REE extraction and chelating reagent consumption from clays other than the ones employed in the present study.

The model for the lanthanide desorption mechanism from clay materials was proposed and described by the authors in a previous paper [13]. The process is a simple ion-exchange reaction between the rare earths physically adsorbed on clays and exchange cations from solution (such as NH_4^+, Na^+, Mg^{2+}), and the main driving force is the difference in hydration enthalpy between REEs and the exchange cation (i.e., cations with a more negative hydration enthalpy, such as lanthanides, have more affinity towards the aqueous phase). The chelating reagents are expected to preferentially coordinate the rare earths once in solution; due to their large molecular structure it is not expected that the ligands will adsorb on the clay surface.

4. Conclusions

The present study investigated the effect of chelating agents on the recovery of rare earth elements from clay minerals via ion-exchange leaching, in order to propose an enhanced procedure that is environmentally benign and allows high REE recovery while reducing or eliminating ammonium sulfate usage.

The authors established screening criteria for the selection of optimal chelating agents, conducted experiments in order to evaluate the efficiency of the selected reagents and compared the results with REE extraction levels obtained during conventional ion-exchange leaching procedures with ammonium sulfate. The main reasons for ligand selection were rapid biodegradability, non-toxicity, and high values of stability constants of formation for complexes.

It was found that 1:1 addition of EDDS, NTA (both the acid and tri-sodium form), aspartic and citric acid to 0.25 M NH_4^+ (as sulfate) resulted in 6–10% increased extraction when compared to lixiviant alone, while 2:1 stoichiometric excess did not lead to appreciable improvement in TREE extraction. Although seawater alone did not perform well as lixiviant, 1:1 addition of chelating agents to SSW resulted in noticeably increased TREE extraction (e.g., 20% for EDDS and NTA-Na3), reaching levels close to the ones achieved with 0.25 mol/L NH_4^+. Glycine and asparagine did not enhance TREE recovery in either lixiviant system, due to the inferior chelating power of these compounds.

All chelating agents investigated (again, with the exception of glycine and asparagine) achieved considerably higher Al concentration in the leachate than in the presence of ammonium sulfate or SSW alone, and therefore offered no selectivity towards REE, although EDDS and NTA-Na3 appear to slightly suppress Al desorption.

The main implication of this study is the possibility to use simple seawater with added chelating agents as an extracting agent. From a process perspective, the use of EDDS or NTA-Na3 in conjunction with lower NH_4^+ concentrations and especially seawater appears to be the recommended option, as these systems led to high TREE extraction, moderate Al co-desorption and neutral pH values in the PLS. This has the potential to reduce the overall hydrometallurgical plant freshwater consumption, limit/eliminate the ammonia pollution, and open up the possibility of returning the final purified streams to the sea without risk of contaminating the soil—offering an environmentally benign ion-exchange leaching process.

Supplementary Materials: The following are available online at https://www.mdpi.com/article/10.3390/met11081265/s1, Figure S1: Comparison of Aluminum extraction at various ratios of lixiviant (ammonium sulphate, Am, containing 0.25 mol/L NH4+ ions) and chelating agents. Table S1: Individual rare earth extraction levels with 0.25 mol/L (NH4)2SO4 (AMS) in the presence of various chelating agents (1:1 ratio). Table S2: Individual rare earth extraction levels with simulated sea water (SSW) containing ~0.48 mol/L NaCl in the presence of various chelating agents (1:1 ratio).

Author Contributions: Conceptualization, G.M and V.P.; methodology, G.M.; validation, V.P.; formal analysis, V.P.; investigation, G.M.; data curation, G.M.; resources, G.M.; writing-original draft preparation, G.M.; writing, review and editing, V.P.; visualization, G.M.; supervision, V.P.; project administration, V.P. Both authors have read and agreed to the published version of the manuscript.

Funding: This research received no external funding.

Institutional Review Board Statement: Not Applicable.

Informed Consent Statement: Not applicable.

Data Availability Statement: The data presented in this study are available upon request from the corresponding author.

Conflicts of Interest: The authors declare no conflict of interest.

References

1. Balaram, V. Rare earth elements: A review of applications, occurrence, exploration, analysis, recycling, and environmental impact. *Geosci. Front.* **2019**, *10*, 1285–1303. [CrossRef]
2. Zhou, B.; Li, Z.; Chen, C. Global Potential of Rare Earth Resources and Rare Earth Demand from Clean Technologies. *Minerals* **2017**, *7*, 203. [CrossRef]
3. Adamas Intelligence. Rare Earth Elements: Market Issues and Outlook. 2019. Available online: www.adamasintel.com/wp-content/uploads/2019/07/Adamas-Intelligence-Rare-Earths-Market-Issues-and-Outlook-Q2-2019.pdf (accessed on 8 August 2021).
4. Kanazawa, Y.; Kamitani, M. Rare Earth Minerals and Resources in the World. *J. Alloys Compd.* **2006**, *408*, 1339–1343. [CrossRef]
5. Gupta, C.K.; Krishnamurthy, N. Extractive metallurgy of rare earths. *Int. Mater. Rev.* **1999**, *37*, 197–248. [CrossRef]
6. Murakami, H.; Ishihara, S. REE Mineralization of Weathered Crust and Clay Sediment on Granitic Rocks in the Sanyo Belt, SW Japan and the Southern Jiangxi Province, China. *Resour. Geol.* **2008**, *58*, 373–401. [CrossRef]
7. Chi, R.; Tian, J. *Weathered Crust Elution-deposited Rare Earth Ores*; Nova Science Publishers: New York, NY, USA, 2008.
8. Yang, J.J.; Lin, A.; Li, X.L.; Wu, Y.; Zhou, W.; Chen, Z. China's Ion-adsorption Rare Earth Resources, Mining Consequences and Preservation. *Environ. Dev.* **2013**, *8*, 131–136. [CrossRef]
9. Ganguli, R.; Cook, D.R. Rare Earths: A review of the Landscape. *MRS Energy Sustain. A Rev. J.* **2018**, *5*. [CrossRef]
10. Rocha, A.; Schissel, D.; Sprecher, A.; de Tarso, P.; Goode, J. Process Development for the Serra Verde Weathered Crust Elution-deposited Rare Earth Deposit in Brazil. In Proceedings of the 52nd Conference of Metallurgists (COM 2013), Metallurgical Society of the Canadian Institute of Mining, Metallurgy and Petroleum (MetSoc-CIM), Montreal, QC, Canada, 27–31 October 2013.
11. Sanematsu, K.; Kon, Y.; Imai, A.; Watanabe, K.; Watanabe, Y. Geochemical and Mineralogical Characteristics of Ion-adsorption Type REE Mineralization in Phuket, Thailand. *Miner. Depos.* **2013**, *48*, 437–451. [CrossRef]
12. Tantalus Rare Earths, AG. Technical Report and Updated Resource Estimate. 2014. Available online: https://intel.rscmme.com/report/Tantalus_Rare_Earths_AG_Tantalus_17-12-2014 (accessed on 8 August 2021).
13. Moldoveanu, G.; Papangelakis, V. Recovery of rare earth elements adsorbed on clay minerals: I. Desorption mechanism. *Hydrometallurgy* **2012**, *117–118*, 71–78. [CrossRef]
14. Moldoveanu, G.; Papangelakis, V. Recovery of Rare Earth Elements Adsorbed on Clay Minerals: II. Leaching with Ammonium Sulfate. *Hydrometallurgy* **2013**, *131–132*, 158–166. [CrossRef]
15. Moldoveanu, G.; Papangelakis, V. The Influence of Both Cation and Anion Type on Leaching Efficiency of Lanthanides from Ion-adsorption Ores. IMPC 2016: XXVIII International Mineral Processing Congress Proceedings. In Proceedings of the Metallurgical Society of the Canadian Institute of Mining, Metallurgy and Petroleum (MetSoc-CIM), Montreal, QC, Canada, 11–15 September 2016; ISBN 978-1-926872-29-2.
16. Chi, R.; Tian, J.; Luo, X.; Xu, Z.; He, Z. Basic Research on the Weathered Crust Elution-deposited Rare Earth Ores. In Proceedings of the 52nd Conference of Metallurgists (COM 2013), Metallurgical Society of the Canadian Institute of Mining, Metallurgy and Petroleum (MetSoc-CIM), Montreal, QC, Canada, 27–31 October 2013; pp. 189–199.
17. Tian, J.; Tang, X.; Yin, J.; Chen, J.; Luo, X.; Rao, J. Enhanced Leachability of a Lean Weathered Crust Elution-Deposited Rare-Earth Ore: Effects of Sesbania Gum Filter-Aid Reagent. *Metall. Mater. Trans. B* **2013**, *44B*, 1070–1077. [CrossRef]
18. Luo, X.; Ma, P.; Luo, C.; Chen, X.; Feng, B.; Yan, Q. The Effect of LPF on the Leaching Process of a Weathered Crust elution-Deposited Rare Earth Ore. In Proceedings of the 53nd Conference of Metallurgists (COM 2014), Metallurgical Society of the Canadian Institute of Mining, Metallurgy and Petroleum (MetSoc-CIM), Vancouver, BC, Canada, 28 September–1 October 2014.
19. Zhang, Z.; He, Z.; Yu, J.; Xu, Z.; Chi, R. Novel Solution Injection Technology for In-Situ Leaching of Weathered Crust Elution-Deposited Rare Earth Ores. *Hydrometallurgy* **2016**, *164*, 248–256. [CrossRef]
20. Xiao, Y.; Feng, Z.; Huang, X.; Huang, L.; Chen, Y.; Wang, L.; Long, Z. Recovery of rare earths from weathered crust elution-deposited rare earth ore without ammonia-nitrogen pollution: I. Leaching with Magnesium Sulfate. *Hydrometallurgy* **2015**, *153*, 58–65.
21. Xiao, Y.; Feng, Z.; Huang, X.; Huang, L.; Chen, Y.; Wang, L.; Long, Z. Recovery of rare earth from the ion-adsorption type rare earths ore: II. Compound leaching. *Hydrometallurgy* **2016**, *163*, 83–90.
22. Ogwuebu, C.; Chileshe, V. Coordination chemistry in mineral processing. *Miner. Process. Extr. Metall. Rev.* **2000**, *21*, 497–525. [CrossRef]
23. La Brooy, S.R.; Linge, H.G.; Walker, G.S. Review of Gold extraction from Ores. *Miner. Eng.* **1994**, *7*, 1213–1241. [CrossRef]
24. Xu, B.; Kong, W.; Li, Q.; Yang, Y.; Liu, X. A Review of Thiosulfate Leaching of Gold: Focus on Thiosulfate Consumption and Gold Recovery from Pregnant Solution. *Metals* **2017**, *7*, 222. [CrossRef]
25. Li, J.; Miller, J.D. A Review of Gold Leaching in Acid Thiourea Solutions. *Miner. Process. Extr. Metall. Rev.* **2006**, *27*, 177–214. [CrossRef]

26. Senanayake, G. Gold leaching by copper(II) in ammoniacal thiosulphate solutions in the presence of additives. Part I: A review of the effect of hard–soft and Lewis acid-base properties and interactions of ions. *Hydrometallurgy* **2012**, *115–116*, 1–20. [CrossRef]

27. Tavakoli Mohammadi, M.R. Ammonia leaching in the copper industry: A review. In Proceedings of the XXVI International Mineral Processing Congress—IMPC 2012, New Delhi, India, 24–28 September 2012.

28. Oraby, E.A.; Ecksteen, J.J. The selective leaching of copper from a gold–copper concentrate in glycine solutions. *Hydrometallurgy* **2014**, *150*, 14–19. [CrossRef]

29. Oraby, E.A.; Ecksteen, J.J. The leaching of gold, silver and their alloys in alkaline glycine–peroxide solutions and their adsorption on carbon. *Hydrometallurgy* **2015**, *152*, 199–203. [CrossRef]

30. Oraby, E.A.; Ecksteen, J.J. The leaching and adsorption of gold using low concentration amino acids and hydrogen peroxide: Effect of catalytic ions, sulphide minerals and amino acid type. *Miner. Eng.* **2015**, *70*, 36–42.

31. Eivazihollagh, A.; Svanedal, I.; Edlund, H.; Norgren, M. On chelating surfactants: Molecular perspectives and application prospects. *J. Mol. Liq.* **2019**, *278*, 688–705. [CrossRef]

32. Leštan, D.; Luo, C.; Li, X. The use of chelating agents in the remediation of metal-contaminated soils: A review. *Environ. Pollut.* **2008**, *153*, 3–13. [CrossRef]

33. Kociałkowski, W.Z.; Diatta, J.B.; Grzebisz, W. Evaluation of Chelating Agents as Heavy Metals Extractants in Agricultural Soils under Threat of Contamination. *Pol. J. Environ. Stud.* **1999**, *8*, 159–184.

34. Kołodyńska, D. Chelating Agents of a New Generation as an Alternative to Conventional Chelators for Heavy Metal Ions Removal from Different Waste Waters. In *Expanding Issues in Desalination*; Ning, R.Y., Ed.; InTech: London, UK, 2011. [CrossRef]

35. Kołodyńska, D. Application of a new generation of complexing agents in removal of heavy metal ions from different wastes. *Environ. Sci. Pollut. Res.* **2013**, *20*, 5939–5949. [CrossRef] [PubMed]

36. Haynes, R.J.; Swift, R.S. An Evaluation of the use of DTPA and EDTA as extractants for micronutrients in moderately acidic soils. *Plant Soil* **1983**, *74*, 111–122. [CrossRef]

37. Fukuda, Y.; Nakao, A.; Hayashi, K. Syntheses and specific structures of higher-order mixed chelate lanthanide complexes containing terpyridine, acetylacetonate, and nitrate ligands. *J. Chem. Soc. Dalton Trans.* **2002**, *4*, 527–533. [CrossRef]

38. Wheelwright, E.J.; Spedding, F.H. *The Use of Chelating Agents in the Separation of the Rare Earth Elements by Ion-Exchange Methods*; Ames Laboratory ISC Technical Reports; 1955; Paper 101; Available online: http://lib.dr.iastate.edu/ameslab_iscreports/101 (accessed on 8 August 2021).

39. Mackey, J.L. A Study of the Rare-Earth Chelate Stability Constants of Some Aminopolyacetic Acids. Retrospective Theses and Dissertations, Iowa State University of Science and Technology, Ames, IA, USA, 1960; p. 2797. Available online: http://lib.dr.iastate.edu/rtd (accessed on 8 August 2021).

40. Rowlands, D.L.G. Observations on the Separation Factors of the Rare Earths with Various Ligands. *J. Inorg. Nucl. Chem.* **1967**, *29*, 809–814. [CrossRef]

41. Karraker, R.H. Stability Constants of Some Rare-Earth-Metal Chelates. Retrospective Theses and Dissertations, Iowa State University of Science and Technology, Ames, IA, USA, 1969; p. 1969. Available online: http://lib.dr.iastate.edu/rtd (accessed on 8 August 2021).

42. Li, Q.; He, Z.; Zhang, Z.; Zhang, T.; Zhong, C.; Chi, R. Study on the recovery of rare earth from weathered crust elution-deposited rare earth ore by citrate coordination leaching. *Chin. Rare Earths* **2015**, *36*, 18–22.

43. Wang, L.; Liao, C.; Yang, Y.; Xu, Y.; Xiao, Y.; Yan, C. Effects of organic acids on the leaching process of ion-adsorption type rare earth ore. *J. Rare Earths* **2017**, *35*, 1233–1238. [CrossRef]

44. Zhang, H.; Zhang, Z.; Liu, D.; Chai, X.; Chi, R. Permeability Characteristics of Weathered Crust Elution- Deposited Rare Earth Ore In-Situ Leaching with Ammonium Acetate. *Min. Metall. Eng.* **2019**, *39*, 110–114.

45. Chai, X.; Li, G.; Zhang, Z.; Chi, R.; Chen, Z. Leaching Kinetics of Weathered-crust Elution-deposited Rare Earth Ore with Compound Ammonium Carboxylate. *Minerals* **2020**, *10*, 516. [CrossRef]

46. Chen, Z.; Zhang, Z.; Chi, R. Leaching Process of Weathered-crust Elution-deposited Rare Earth Ore with Formate Salts. *Front. Chem.* **2020**, *8*, 1017. [CrossRef] [PubMed]

47. Cristiani, C.; Iannicelli Zubiani, E.M.; Dotelli, G.; Finocchio, E.; Gallo Stampino, P.; Lichelli, M. Polyamine-Based Organo-Clays for Polluted Water Treatment: Effect of Polyamine Structure and Content. *Polymers* **2019**, *11*, 897. [CrossRef]

48. Cristiani, C.; Bellotto, M.; Dotelli, G.; Latorrata, S.; Ramis, G.; Gallo Stampino, P.; Iannicelli Zubiani, E.M.; Finocchio, E. Rare Earths (La, Y, and Nd) Adsorption Behaviour towards Mineral Clays and Organoclays: Monoionic and Trionic Solutions. *Minerals* **2021**, *11*, 30. [CrossRef]

49. Cotton, S. *Lanthanide and Actinide Chemistry*; John Wiley & Sons Ltd.: Chichester, UK, 2006.

50. Choppin, G.R. Comparison of the solution chemistry of the lanthanides and actinides. *J. Less-Common Met.* **1983**, *93*, 232–330. [CrossRef]

51. Kettle, S.F.A. *Physical Inorganic Chemistry-a Coordination Chemistry Approach*; Springer: Berlin/Heidelberg, Germany, 1996; pp. 238–268.

52. Choppin, G.R. Structure and thermodynamics of lanthanide and actinide complexes in solution. *Pure Appl. Chem.* **1971**, *27*, 23–41. [CrossRef]

53. Smith, R.M.; Martell, A.E. Critical Stability Constants, Enthalpies and Entropies for the Formation of Metal Complexes of Aminopolycarboxylic Acids and Carboxylic Acids. *Sci. Total Environ.* **1987**, *64*, 125–147. [CrossRef]

54. Jones, P.W.; Williams, D.R. Chemical Speciation Used to Assess [S,S']-ethylenediaminedisuccinic acid (EDDS) as a readily-biodegradable replacement for EDTA in radiochemical decontamination formulations. *Appl. Radiat. Isot.* **2001**, *54*, 587–593. [CrossRef]

55. Tandy, S.; Ammann, A.; Schulin, R.; Nowack, B. Biodegradation and speciation of residual SS-ethylenediaminedisuccinic acid (EDDS) in soil solution left after soil washing. *Environ. Pollut.* **2006**, *142*, 191–199. [CrossRef]

56. Moeller, T.; Martin, D.F.; Thompson, L.C.; Ferrus, R.; Feistel, G.R.; Randall, W.J. The Coordination Chemistry of Yttrium and the Rare Earth Metal Ions. *Chem. Rev.* **1965**, *65*, 1–50. [CrossRef]

57. Berthon, G. The Stability Constants of Complexes of Amino Acids with Polar Side Chains. *Pure Appl. Chem.* **1995**, *67*, 1117–1240. [CrossRef]

58. Kiss, T.; Sovago, I.; Gergely, A. Critical Survey of Stability Constants of Complexes of Glycine. *Pure Appl. Chem.* **1991**, *63*, 597–638. [CrossRef]

59. Weaver, C.; Pollard, L. *The Chemistry of Clay Minerals*, 1st ed.; Elsevier Science Ltd.: Amsterdam, NY, USA, 1973.

60. Peng, S. *Geological Characteristics and the Prospecting Criteria of the Granite-Weathering Crust Ion-Adsorption Type REE Deposits in Nanling Area, South China*; Materials Science Forum, Trans Tech Publications: Zurich, Switzerland, 1991; pp. 33–42.

61. Ding, Z.; Wang, Q.; Hu, J. Extraction of heavy metals from water-stable soil aggregates using EDTA. *Procedia Environ. Sci.* **2013**, *18*, 679–685. [CrossRef]

62. Minkina, T.M.; Mandzhieva, S.S.; Burachevskaya, M.V.; Bauer, T.V.; Sushkova, S.N. Method of determining loosely bound compounds of heavy metals in the soil. *MethodsX* **2018**, *5*, 217–226. [CrossRef] [PubMed]

63. Atkinson, M.J.; Bingman, C. Elemental composition of commercial sea salts. *J. Aquaric. Aquat. Sci.* **1997**, *VIII*, 39–43.

64. He, Z.; Zhang, Z.; Yu, J.; Xu, Z.; Chi, R. Process optimization of rare earth and aluminum leaching from weathered crust elution-deposited rare earth ore with compound ammonium salts. *J. Rare Earths* **2016**, *34*, 413–419. [CrossRef]

65. Yu, B.; Hu, Z.; Zhou, F.; Feng, J.; Chi, R. Lanthanum (III) and Yttrium (III) Adsorption on Montmorillonite: The Role of Aluminum Ion in Solution and Minerals. *Miner. Process. Extr. Metall. Rev.* **2020**, *41*, 107–116. [CrossRef]

66. Chi, R.; Xu, Z. A solution chemistry approach to the study of rare earth element precipitation by oxalic acid. *Metall. Mater. Trans. B* **1999**, *30B*, 189–195. [CrossRef]

67. Meunier, A. *Clays*; Springer: Berlin/Heidelberg, Germany, 2005.

68. Alshameri, A.; He, H.; Xin, C.; Zhu, J.; Xinghu, W.; Zhu, R.; Wang, H. Understanding the role of natural clay minerals as effective adsorbents and alternative source of rare earth elements: Adsorption operative parameters. *Hydrometallurgy* **2019**, *185*, 149–161. [CrossRef]

Article

Influence of Dysprosium Compounds on the Extraction Behavior of Dy from Nd-Dy-Fe-B Magnet Using Liquid Magnesium

Sun-Woo Nam [1,2], Sang-Min Park [1], Mohammad Zarar Rasheed [1], Myung-Suk Song [1], Do-Hyang Kim [2] and Taek-Soo Kim [1,*]

[1] Research Institute of Advanced Manufacturing Technology, Korea Institute of Industrial Technology, 12, Gaetbeol-ro, Incheon 21999, Korea; sunwoo@kitech.re.kr (S.-W.N.); jhsm8920@kitech.re.kr (S.-M.P.); zarar@kitech.re.kr (M.Z.R.); mssong@kitech.re.kr (M.-S.S.)

[2] Department of Materials Science and Engineering, Yonsei University, Seoul 03722, Korea; dohkim@yonsei.ac.kr

* Correspondence: tskim@kitech.re.kr

Abstract: During the liquid metal extraction reaction between a Nd-Dy-Fe-B magnet and liquid Mg, Nd rapidly diffuses out of the magnet, whereas Dy is not extracted due to the reaction with the matrix and the formation of Dy_2Fe_{17} phase. In addition, the Dy_2O_3 phase exists at the grain boundaries. Until now, only the effect of the Dy_2O_3 phase on the extraction of Dy has been reported. In this study, the effect of the Dy_2Fe_{17} phase on the extraction of Dy from the Nd-Dy-Fe-B magnet was investigated in liquid Mg. The formation of the Dy_2Fe_{17} phase during the reaction between Mg and matrix ($RE_2Fe_{14}B$) was first examined using a thermodynamical approach and confirmed by microstructural analysis. It was observed that Dy extraction was dominated by Dy_2Fe_{17} phase decomposition from 3 h to 24 h, followed by Dy_2O_3 phase dominant reaction with Mg. Comparing the activities of the Dy_2Fe_{17} phase and the Dy_2O_3 phase, the reaction of Dy_2Fe_{17} is dominant, as compared to the Dy_2O_3 phase. Finally, at 48 h, the high Dy extraction percentage of 93% was achieved. As a result, in was concluded that the Dy_2Fe_{17} phase acts as an obstacle in the extraction of Dy. In the future, if research to control the Dy_2Fe_{17} phase proceeds, it will be of great importance to advance the recycling of Dy.

Keywords: recycling; pyrometallurgy; dysprosium; liquid metal extraction; phase transformation

Citation: Nam, S.-W.; Park, S.-M.; Rasheed, M.Z.; Song, M.-S.; Kim, D.-H.; Kim, T.-S. Influence of Dysprosium Compounds on the Extraction Behavior of Dy from Nd-Dy-Fe-B Magnet Using Liquid Magnesium. *Metals* **2021**, *11*, 1345. https://doi.org/10.3390/met11091345

Academic Editor: Dariush Azizi

Received: 30 June 2021
Accepted: 13 August 2021
Published: 26 August 2021

Publisher's Note: MDPI stays neutral with regard to jurisdictional claims in published maps and institutional affiliations.

1. Introduction

Rare earth (RE) permanent magnets have recently been employed in high-tech industrial applications, such as electric vehicles, renewable energy, robotics, and their utilization is increasing as the world shifts towards a green economy [1,2]. Even though there are various types of magnets, such as ferrite, AlNiCo, etc., the reason for the domination of the permanent market by the RE magnets is due to their superior magnetic properties, such as high remnant magnetization and coercivity [3]. The rare-earth sublattice (4f electrons) comprises essential components to stabilize the magnetization direction for the crystal axes, i.e., high magnetic anisotropy.

Recently, the focus on carbon neutrality has led to a sharp increase in the consumption of permanent magnets to achieve net-zero carbon dioxide emission. Among the RE permanent magnets, the Nd-Dy-Fe-B magnets are widely used because of their relatively low cost and high productivity, compared to the SmCo magnets. The Nd-Dy-Fe-B magnets are generally known to be composed of about 30 wt.% REEs contents. Moreover, Dy is integrated into the magnets to enhance their thermal stability and corrosion resistance [4]. Thus, the demand for Dy has been also gradually increasing with the development of green technology. However, the production of heavy REEs (HREEs) has undergone serious balance problems related to political, geological, and technical issues as the HREEs like

239

Dy are only produced from the ion-adsorbed ores in southern China [5]. In the case of ion-absorbed ores, the REEs are absorbed on the clay, which is formed with the hydrated cations. As opposed to other rare earth ore, such as the bastnaesite, monazite, xenotime, and so on, a large number of HREEs can be effectively obtained without the emission of radioactive substances by using the chemicals. However, the usage of acids and base solutions is a cause of environmental concerns.

Therefore, diversifying the supply of HREEs is significant to improve the supply chain for sustainability. For recycling, the ways to recover end of life (EOL) magnets or magnet scraps are suggested. During the magnet fabrication process, the amount of magnet scrap can be generated to about 20% to 30% [6]. Besides, the EOL magnets can be accumulated, depending on the applications, such as small electronics, electric vehicles, and wind turbines with various shapes and sizes.

One of the recycling processes for recovering HREEs from EOL magnets or scraps, the pyrometallurgy method, is a potential alternative to collect REEs from Nd-Dy-Fe-B magnet [7–9]. Among these processes, liquid metal extraction (LME) is based on a selective reaction with target metals by using solvent metals, such as Mg, Ag, Bi, and Cu [10–13]. It has the advantage of being an environmentally friendly chemical-free process without the emission of wastes and no requirement of additional reduction processes due to the direct recovery of REEs in metal form.

The Mg is a strong candidate for being an extraction agent in the LME process. It can be selectively reacted with REEs (Nd, Dy) without Fe and B intermetallic compounds due to higher chemical affinity with REEs (Nd, Dy)) compared to with Fe [13]. Previous studies have shown that Nd is easily extracted from the Nd-Dy-Fe-B magnet and the reaction mechanism is successfully demonstrated [10,14–16]. On the other hand, the low extraction efficiency of Dy was reported because of the small amount of Dy present in the Nd-Dy-Fe-B magnet and quite a different reaction behavior with Mg. Akahori et al. thermodynamically demonstrated that the oxidation of Dy can be affected by the decreasing extraction by forming Dy_2O_3 and not $DyNdO_3$. It is shown that preventing the oxidized phases is decisive in improving extraction efficiency [16]. To understand the Dy extraction mechanism, Kim et al. investigated the development of (Nd and Dy)-oxide phases in the microstructures as a result of formation during the process. They offer experimental evidence that the limited extraction of Dy is caused by the formation of Dy oxide [17]. Park et al. conducted a comparison of the extraction efficiency with increasing oxide composition. It is shown that Dy was not easily extracted in the form of Dy-oxide and Dy_2Fe_{17}, while Nd was completely extracted [18]. Even though the scrap was heavily oxidized due to the small scrap size, the Dy_2Fe_{17} phase remained as the result of the decomposition of REFeB grain. It is supported by the experimental results of Nam et al., that the phase transformation of direct reaction between DyFeB and Mg phenomenologically shows that Dy_2Fe_{17} is first formed in the pure DyFeB phase as a byproduct, while liquid Mg is infiltrated into the grain [19]. However, the reasoning behind the influence of the Dy_2Fe_{17} phase in the extraction process is still unclear.

In this work, the entire extraction behavior of Dy is systematically investigated, depending on the time, activity with Mg, intermediate phase, and oxides, infiltrating Mg into the Nd-Dy-Fe-B specimen. The generation and decomposition of intermediate phases are observed in detail and the Dy_2Fe_{17} phase is clearly identified. This demonstrates that the interplay between Dy_2Fe_{17} and oxides affects the extraction behavior of Dy.

2. Experimental

The permanent magnets for the liquid metal extraction (LME) process were supplied by Jahwa Electronics Co. Ltd., Cheongju, Korea. Pure Mg was purchased from JC Magnesium Co., Burnaby, BC, Canada. The chemical composition of the magnet is shown in Table 1 and were determined by X-ray fluorescence (XRF; Thermo Fisher Science ARL PERFORM'X, Middlesex County, MA, USA). The preparation size of the magnet sample and pure Mg is $10 \times 10 \times 3$ mm and $10 \times 10 \times 10$ mm, respectively. To confirm the reaction behavior, the magnets were placed at the bottom of a mild steel crucible with Mg on top. The crucible was then placed inside a high-frequency induction furnace and heated in an atmosphere-controlled chamber for a reaction time between 30 min and 48 h. The Mg to magnet mass ratio was 15 to 1 and LME reactions were observed at 900 °C. This has been known, as an ideal condition, to maximize the extraction ratio [20]. For analyzing the characteristics of the magnet obtained by furnace cooling following the change in periods at 900 °C, the reaction sample was cut into properly sized samples using a diamond wheel cutter with a thickness of 0.3 mm. The specimens were ground using abrasive papers, which were scaled from 200 to 4000 grit SiC. Subsequently, the samples were polished using a 0.1 μm Al_2O_3 suspension. After polishing, the specimens were cleaned in an ultrasonic cleaner for 5 min by steeping them in ethanol. Then, the specimens were dried using a high-pressure air spray gun. The microstructure of the samples was characterized using an FE-SEM (JSM-5310, JEOL, Tokyo, Japan) and a transmission electron microscope (TEM; JEM-F200, JEOL, Tokyo, Japan). The thickness of the diffusion layer in the magnet was measured using the BSE mode in FE-SEM and the concentrations of the REE (Nd, Dy) and Mg were investigated using XRF and EDS analyses in both the diffusion layer and Mg-zone.

Table 1. Chemical composition of the magnet by XRF.

Elements	Fe (wt.%)	Nd (wt.%)	Dy (wt.%)	Minor (wt.%)
Magnet	69.0	25.8	3.50	1.72

3. Results and Discussion

Table 1 shows the chemical composition of the magnet which is the well-controlled type of oxygen content at 800 ppm. The minor components are Cu, Al, Co, Nb, etc.

Considering the magnet composition, a 100% extracted ratio of REE into Mg can be calculated to be 1.730% in Nd and 0.233% in Dy, respectively. Table 2 shows the characterization of the extracted concentration of Nd and Dy with increasing reaction time by XRF measurement. Figure 1 reveals converted values with the extraction efficiency by using Equations (1) and (2). While all of Nd is completely reacted at 6 h, the extraction of Dy is 72% at the same reaction time.

$$\text{Percentage of extraction Nd } (\%) = \left(\frac{C_{\text{Nd in Mg}}}{1.730} \times 100 \right) \tag{1}$$

$$\text{Percentage of extraction of Dy } (\%) = \left(\frac{C_{\text{Dy in Mg}}}{0.233} \times 100 \right) \tag{2}$$

Table 2. The concentration of elements in the Mg region by XRF.

Time (Hours)	Concentration(wt.%)	C_{Nd}	C_{Dy}
0.5		0.902	0.041
1		1.08	0.063
3		1.48	0.117
6		1.73	0.155
12		1.73	0.158
24		1.73	0.193
48		1.73	0.219

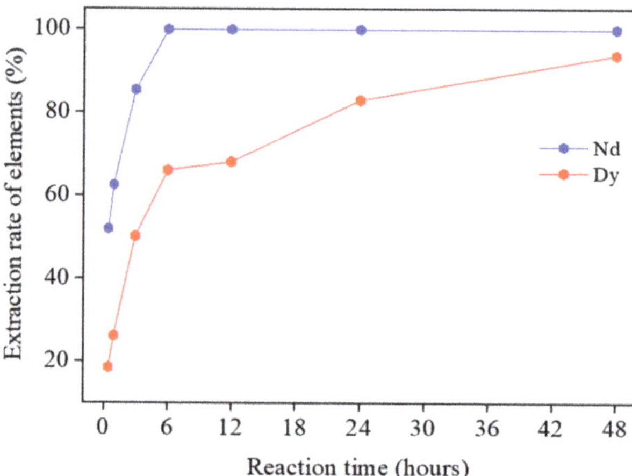

Figure 1. Variation of Nd and Dy extraction efficiency with increasing time.

To understand the extraction behavior of Dy in detail, the microstructures are observed with increasing reaction time on the magnet side. Figure 2 shows that liquid Mg diffused inside the magnet along the grain boundaries and formed ligaments (dark regions) around the magnet (grey region), forming a reaction zone inside the magnet. In morphology, the pattern of Mg infiltration is developed as the changing size and the number of Mg zone with increasing reaction time. Because the Mg can be infiltrated in the entire magnet area in just 30 min, the REEs in the magnet can be totally extracted by the strengthened Mg ligament networking. The liquid Mg diffuses into a magnet to form a diffusion layer on the surface which, in turn, alters the chemical composition of the surface layer. This change in the composition of the surface layer decreases the surface energy and melting point of the solid, allowing it to transform into a liquid state by melting.

The liquid Mg first reacts with RE-rich and RE-oxide phases through the grain boundary in magnets as the diffusion path. Due to the low melting point in the RE-rich phase, which is only 600 °C, the initial extraction curves of Nd and Dy are rapidly increased in Figure 1 [21]. The high reaction temperature of 900 °C promotes the liquid-liquid reaction between RE-rich and Mg. On the other hand, the liquid–solid reaction arises from the high melting points, which are 2230 °C in Nd_2O_3 and 2408 °C in Dy_2O_3. Thus, the oxides remained with small particles in the microstructures due to the relatively slow reactivity. Approximately 1 μm-sized white particles, which are identified as the RE oxide phase, were observed, as shown in Figure 2a–g.

With increasing reaction time, it is observed that the Mg infiltration is gradually increased and the distribution of oxides is decreased. This means that the oxides can be decomposed despite the slow reactivity with Mg. Considering the total extraction time of Nd at 6 h, the remained oxides indicate Dy_2O_3 after 6 h. It was thermodynamically demonstrated that the Nd_2O_3 reacts faster with Mg when comparing Gibbs free energy of Nd_2O_3 and Dy_2O_3 [16].

Figure 3 shows the change in the matrix ($RE_2Fe_{14}B$) with increasing reaction time. At 30 min, the co-existence between needle-shaped particles and the oxide phase is observed in the matrix. The reaction with the particles in the matrix complies with the tendency of oxides in the grain boundary. To identify the phase of needle-shaped particles, the XRD and the STEM-EDS experiments are conducted. The XRD analysis is conducted on the reacted specimen for 3 h. The phases are defined with Fe, Fe_2B, Mg, RE_2O_3, $Mg_{12}RE$, and RE_2Fe_{17} in Figure 4. Because of the gathering of all of the phase information about both grain boundary and the matrix, the Dy oxide phases and RE_2O_3 are discovered in the grain boundary, while the remaining phases in the matrix are defined to be Fe, Fe_2B, Mg, and $Mg_{12}RE$. Interestingly, the needle-shaped particles are characterized with the

RE-Fe intermetallic phases using phase elimination. Because of the chemical similarity, the RE_2Fe_{17} phase cannot be clearly determined to be Nd_2Fe_{17} or Dy_2Fe_{17}, even though the Nd is known to generate only the Nd-Mg-based intermetallic phase, and the formation of Nd_2Fe_{17} is not reported [14].

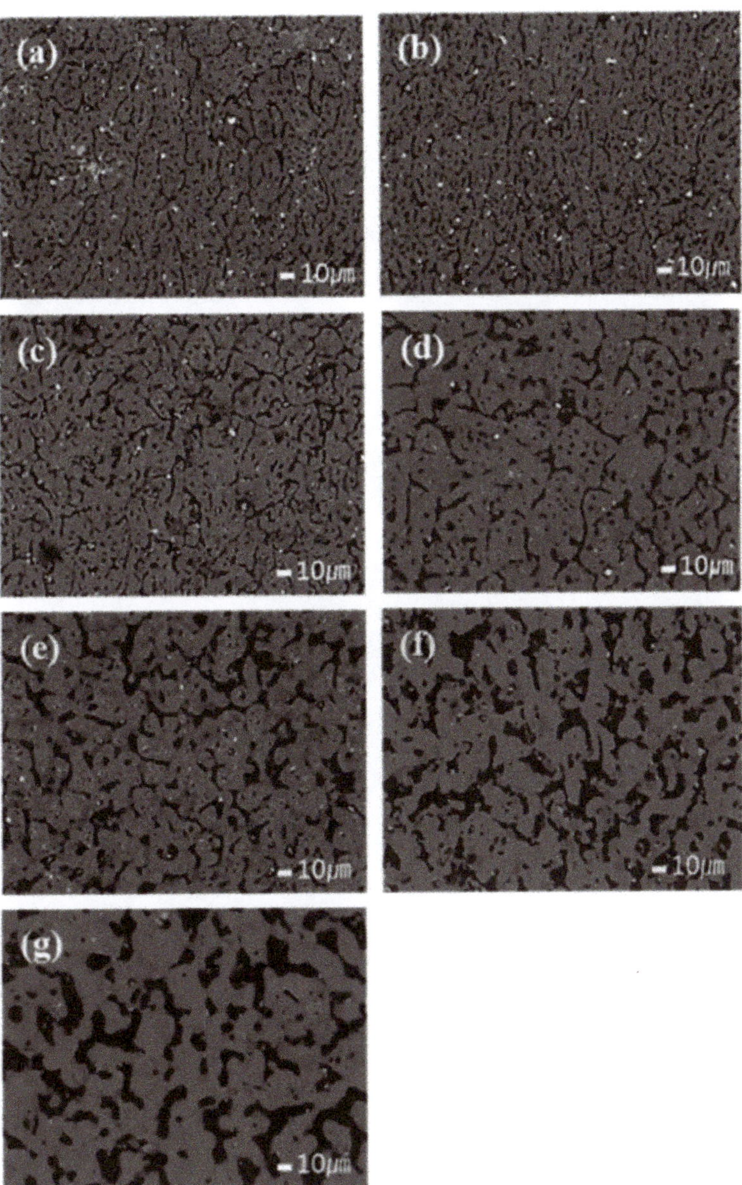

Figure 2. The microstructure of the reaction region in the magnet with Mg for different reaction times: (**a**) 30 min, (**b**) 1 h, (**c**) 3 h, (**d**) 6 h, (**e**) 12 h, (**f**) 24 h, and (**g**) 48 h.

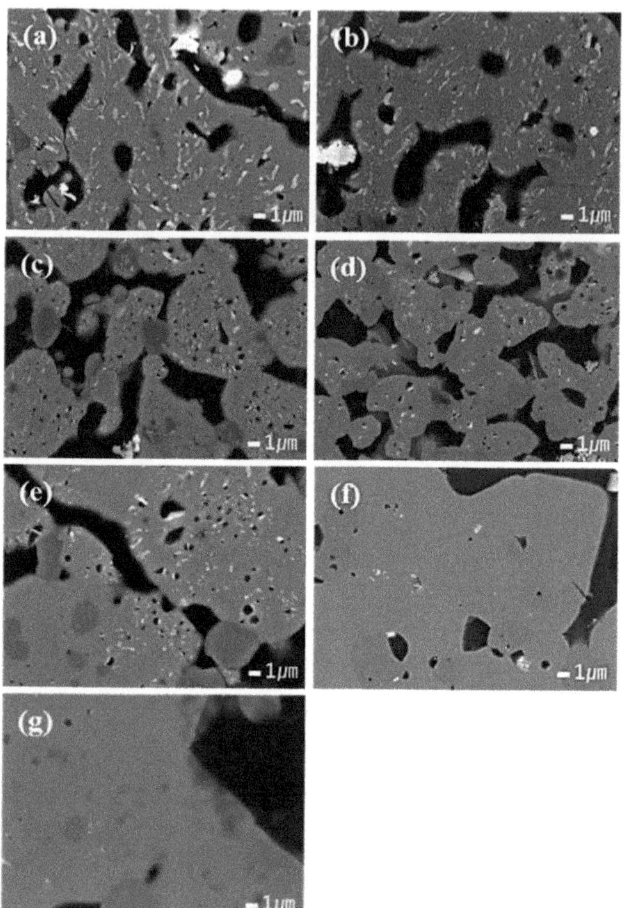

Figure 3. Microstructure of the matrix in reaction region with respect to the reaction Table (**a**) 30 min, (**b**) 1 h, (**c**) 3 h, (**d**) 6 h, (**e**) 12 h, (**f**) 24 h, and (**g**) 48 h.

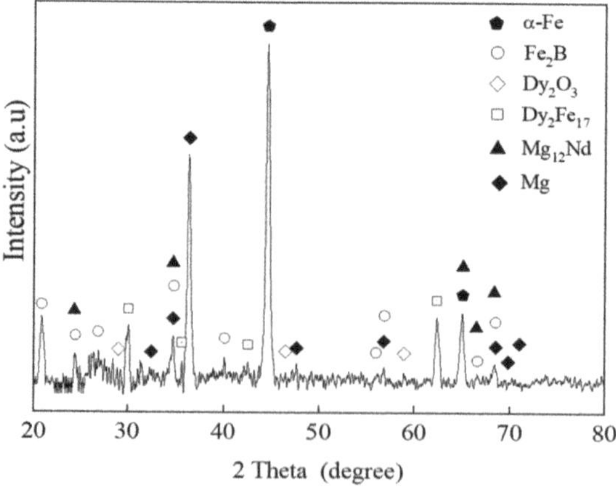

Figure 4. XRD pattern of the matrix at a reaction time at 3 h.

To precisely define the intermetallic phase, Figure 5 shows the results of characterized phases in the matrix by STEM-EDS. To observe the morphologies, the areas in the reacted specimen at 3 h are divided into three zones, marked as 1, 2, and 3. The chemical compositions for each zone are collected in Table 3 by EDS in STEM. While there is little Nd and Dy inside zone 3, as most of the REEs were swept to the Mg side, it is observed that REE-Fe intermetallic compounds in zones 1 and 2 are indeed Dy_2Fe_{17}. These results match well with the XRD results.

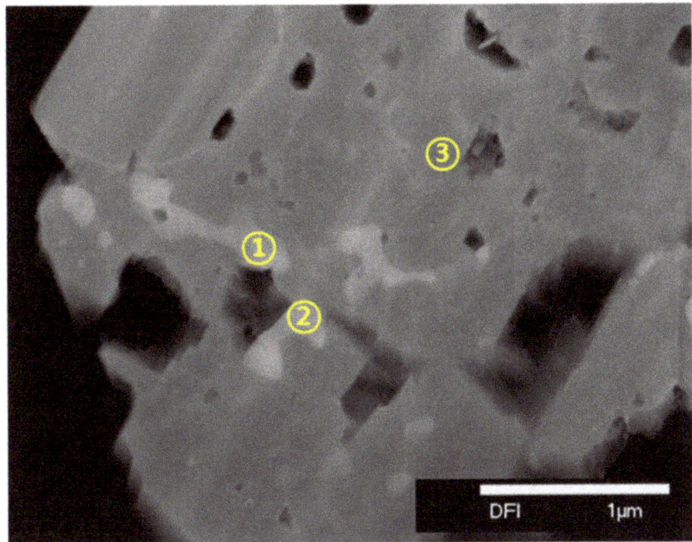

Figure 5. TEM image of the matrix at a reaction time at 3 h.

Table 3. TEM-EDS results of reaction zone after 3 h.

No.	Fe (wt.%)	Nd (wt.%)	Dy (wt.%)
1	90.2	0.660	9.00
2	91.1	0.550	8.16
3	99.2	0.160	0.660

The formation of Dy_2Fe_{17} can be inferred by the heat of the mixing values between Dy-Mg and Nd-Mg. Even though the values of heat of the mixing between Mg and REEs (Nd and Dy) are the same with -6 cal/mol during the reaction, the values between Dy-Fe and Nd-Fe is different to be $\Delta H_{DyFe}^{mix} = -3$ cal/mol, and $\Delta H_{NdFe}^{mix} = +1$ cal/mol, respectively [22]. Thus, the diffusion of Dy to Mg can be interfered with due to its reaction with Fe as Fe exhibits a higher affinity towards Dy as compared to Nd. Similar to the results of the analysis, the Dy-Fe intermetallic compounds, Dy_2Fe_{17}, are observed in the results of the phase analysis on the matrix.

To investigate the reactivity between Mg and Dy_2Fe_{17}, the thermodynamic calculations were carried out using Equation (3):

$$2Mg_{(l)} + Dy_2Fe_{17(s)} \rightarrow 2MgDy_{(s)} + 17Fe_{(s)} \tag{3}$$

The standard Gibbs free energy change, $\Delta G°$, of the reaction between Mg and Dy_2Fe_{17}, shown in (3), is 85,778.2 J/mol (at 900 °C) [23], and the Gibbs free energy change for reaction (3), ΔG, is expressed as follows:

$$\Delta G = \Delta G^\circ + RT \ln K \left(\frac{\alpha_{Fe}^{17} \, \alpha_{MgDy}^{2}}{\alpha_{Mg}^{2} \, \alpha_{Dy2Fe17}} \right) \tag{4}$$

The activity of the chemical species i at temperature T/K is α_i and the condition for reaction (3) and (4) to proceed are:

$$\Delta G = 85,778.2 + 2RT \ln \alpha MgDy \; < 0 \tag{5}$$

The activity of α_{Mg}, $\alpha_{Dy2Fe17}$, α_{Dy2O3}, and α_{Fe} are defined as 1. The condition for reaction (5) to progress is as follows at 900 °C:

$$\alpha MgDy \; < 0.01218$$

It is indicated that the reaction priority can be derived, considering the grain boundary as well as the matrix, during Mg infiltration. According to the Ellingham diagram, the reaction of REEs with Mg is inevitably hindered because of the high-affinity properties with oxygen in REEs give rise to RE_2O_3 formation around the grain boundary in advance. Despite the same group, there is a large difference in reactivity of Nd_2O_3 and Dy_2O_3 with Mg which is $\alpha < 0.0433$ and $\alpha < 0.00535$, respectively [16]. Moreover, the Dy-Fe reactivity is considered with $\alpha < 0.01218$ in the matrix, which is five times higher than Dy_2O_3 reactivity. It is suggested that the decomposition of Dy_2Fe_{17} in the matrix leads to the Dy extraction process after Dy in the RE-rich phase on the grain boundary can be initially swept to Mg.

To check the tendency of reaction priority, based on the thermodynamics, the reaction time dependence of volume fractions in Dy_2O_3 and Dy_2Fe_{17} is experimentally revealed in Figure 6a,b, respectively. The phase fractions are estimated by the SEM image analysis. It is observed that the decomposition rate of Dy_2Fe_{17} is gradually decreasing with increasing reaction time while Dy_2O_3 is relatively stable except for 48 h. Considering the curves of extraction efficiency in Figure 1, the slope of extraction efficiency is drastically increased before 3 h because the RE-rich reaction with initial infiltrated Mg into the grain boundary is almost complete. Steadily expanding Mg reaction zone into the matrix, while the Nd and Nd_2O_3 are totally reacted, the slope of Dy extraction is simultaneously found to be gradual. The decomposition of Dy_2Fe_{17} is mainly contributed to improving Dy extraction because the volume of Dy_2Fe_{17} is started to be dramatically decreased from 3 h to 6 h in Figure 6b. Nevertheless, the reason for the gradual slope is that the relatively small amount of Dy compared to in the RE-rich phase. Finally, Dy_2O_3 phases, which are the most stable phase, started to decompose with increasing time once the decomposition of Dy_2Fe_{17} was almost complete. Infiltrating Mg into the magnet with increasing time, the oxide and intermetallic phases are generated depending on the reaction zones, which are the grain boundary and the matrix. Before decomposing Dy_2O_3, the Dy_2Fe_{17} phases are induced by a reaction between magnets and liquid Mg in the matrix via RE-rich reaction. It is suggested that the Dy_2Fe_{17} phases are attributed to the first main hurdles, considering thermodynamic activity and the analysis of extraction behavior, prior to the decomposing Dy_2O_3 phases.

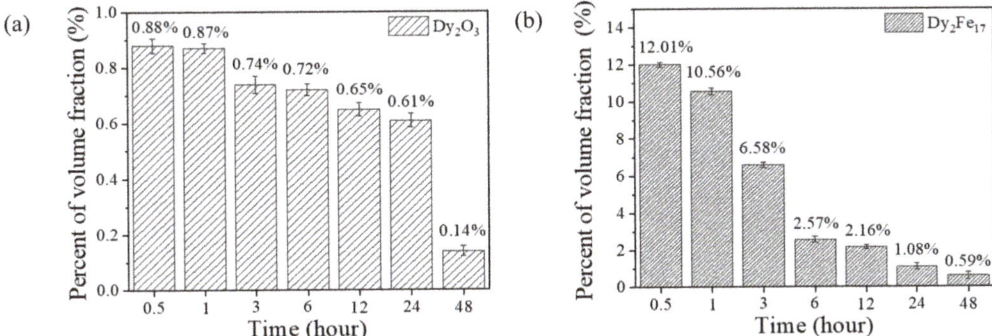

Figure 6. The volume fraction of remained phase with reaction time (**a**) Dy_2O_3 (**b**) Dy_2Fe_{17}.

4. Conclusions

The phase transformation in the microstructures of magnets during the LME process is systematically investigated with increasing reaction time. The liquid Mg is diffused into the magnet area and it gradually expanded its reaction zones inside the grain boundary and matrix. The REEs in the RE-rich phases rapidly swept to the Mg side. Since then, in spite of the magnet with well-controlled oxygen contents, the RE-oxide phases with a foam of RE_2O_3 are distributed in the grain boundary and the RE-Fe intermetallic compounds are simultaneously discovered with needle-shaped particles in the matrix. The RE-Fe phases are precisely defined with Dy_2Fe_{17} by quantitative analysis and comparing thermodynamic reactivity. While the extraction of Nd is complete in the RE-rich reaction without the formation of intermetallic compounds, the extraction curves of Dy behave quite differently due to Mg reactivity. In terms of thermodynamics, the formation of Dy-Fe intermetallic compounds is inferred, and their reactivity is compared with oxides. The extraction behavior of Dy shows that the de-oxidized Dy in the RE-rich phase of the grain boundary is quickly reacted with Mg in 3 h. Even though there is a Dy_2O_3 phase in the same area, the Mg reaction preferentially arises with Dy_2Fe_{17} in the matrix due to differences in stability and reactivity. Afterwards, the Dy extraction is finalized to the reaction of Dy_2O_3 with Mg. It is suggested that the appearance of Dy_2Fe_{17} phases is intrinsic in the decomposing behavior of magnet, unlike Nd, and is attributed to the first hurdles prior to the decomposing Dy_2O_3 phases.

Author Contributions: Conceptualization, T.-S.K. and S.-W.N.; methodology, S.-W.N., S.-M.P. and M.Z.R.; validation, T.-S.K. and S.-W.N.; analysis, S.-W.N. and S.-M.P.; writing—original draft preparation, S.-W.N.; writing—review and editing T.-S.K. and M.-S.S. supervision T.-S.K. and D.-H.K.; project administration, T.-S.K.; All authors have read and agreed to the published version of the manuscript.

Funding: This research was supported by a grant from project of development of environment friendly pyrometallurgy process for high purity HREE and materialization (Project number: 20000970) by Korea evaluation Institute of Industrial Technology (KEIT) in Republic of Korea.

Institutional Review Board Statement: Not applicable.

Informed Consent Statement: Not applicable.

Data Availability Statement: Not applicable.

Conflicts of Interest: The authors declare no confliction of interest.

References

1. Balaram, V. Rare earth elements: A review of applications, occurrence, exploration, analysis, recycling, and environmental impact. *Geosci. Front.* **2019**, *10*, 1285–1303. [CrossRef]
2. John, S. *Rare Earths and China: A Review of Changing Criticality in the New Economy*; IFRI: Paris, France, 2019.

3.   Roskill. *Rare Earth Magnet Applications to Account for ~40% of Total RE Demand by 2030, up from 29% in 2020*; Green Car Congress: London, UK, 2021. Available online: https://www.greencarcongress.com/2021/02/20210203-roskill.html (accessed on 29 June 2021).

4.   Cui, X.G.; Cui, C.Y.; Cheng, X.N.; Xu, X.J. Effect of Dy_2O_3 intergranular addition on thermal stability and corrosion resistance of Nd–Fe–B magnets. *Intermetallics* **2014**, *55*, 118–122. [CrossRef]

5.   Sugimoto, S. Current status and recent topics of rare earth permanent magnet. *J. Phys. D Appl. Phys.* **2011**, *44*, 064001. [CrossRef]

6.   Zakotnik, M.; Harris, I.R.; Williams, A.J. Multiple recycling of NdFeB-type sintered magnets. *J. Alloys Compd.* **2009**, *469*, 314–321. [CrossRef]

7.   Saito, T.; Sato, H.; Ozawa, S.; Yu, J.; Motegi, T. The extraction of Nd from waste Nd–Fe–B alloys by the glass slag method. *J. Alloys Compd.* **2003**, *353*, 189–193. [CrossRef]

8.   Asabe, K.; Saguchi, A.; Takahashi, W.; Suzuki, R.O.; Ono, K. Recycling of Rare Earth Magnet Scraps: Part I Carbon Removal by High Temperature Oxidation. *Mater. Trans.* **2001**, *421*, 2487–2491. [CrossRef]

9.   Suzuki, R.O.; Saguchi, A.; Takahashi, W.; Yagura, T.; Ono, K. Recycling of Rare Earth Magnet Scraps: Part II Oxygen Removal by Calcium. *Mater. Trans.* **2001**, *42*, 2492–2498. [CrossRef]

10.   Xu, Y.; Chumbley, L.S.; Laabs, F.C. Liquid metal extraction of Nd from NdFeB magnet scrap. *J. Mater. Res.* **2000**, *15*, 2296–2304. [CrossRef]

11.   Takeda, O.; Okabe, T.H.; Umetsu, Y. Recovery of neodymium from a mixture of magnet scrap and other scrap. *J. Alloys Compd.* **2006**, *408*, 387–390. [CrossRef]

12.   Nam, S.W.; Kim, D.K.; Kim, B.S.; Kim, D.H.; Kim, T.S. Extraction Mechanism of Rare Earth Elements Contain in Permanent Magnets Using Molten Bismuth. *Sci. Adv. Mater.* **2017**, *9*, 1987–1992. [CrossRef]

13.   Moore, M.; Gebert, A.; Stoica, M.; Uhlemann, M.; Wolfgang, L. A route for recycling Nd from Nd-Fe-B magnets using Cu melts. *J. Alloys Compd.* **2015**, *647*, 997–1006. [CrossRef]

14.   Chae, H.J.; Kim, Y.D.; Kim, B.S.; Kim, J.G.; Kim, T.S. Experimental investigation of diffusion behavior between molten Mg and Nd–Fe–B magnets. *J. Alloys Compd.* **2014**, *586*, s143–s149. [CrossRef]

15.   Kim, Y.S.; Chae, H.J.; Seo, S.J.; Park, K.T.; Kim, B.S.; Kim, T.S. Prediction of Diffusion Behaviors Between Liquid Magnesium and Neodymium-Iron-Boron Magnets. *Sci. Adv. Mater.* **2017**, *8*, 134–137. [CrossRef]

16.   Akahori, T.; Miyamoto, Y.; Saeki, T.; Okamoto, M.; Okabe, T.H. Optimum conditions for extracting rare earth metals from waste magnets by using molten magnesium. *J. Alloys Compd.* **2017**, *703*, 337–343. [CrossRef]

17.   Kim, Y.S.; Nam, S.W.; Park, K.T.; Kim, T.S. Extraction Behavior of Rare Earth Element (Nd, Dy) from Rare Earth Magnet by Using Molten Magnesium. *Sci. Adv. Mater.* **2017**, *9*, 2166–2172. [CrossRef]

18.   Park, S.M.; Nam, S.W.; Lee, S.H.; Song, M.S.; Kim, T.S. Effect of Oxidation Behavior of (Nd, Dy)-Fe-B Magnet on Heavy Rare Earth Extraction Process. *J. Korea Powder Metall. Inst.* **2021**, *28*, 91–96. [CrossRef]

19.   Nam, S.W.; Park, S.M.; Kim, D.H.; Kim, T.S. Thermodynamic Calculations and Parameter Variations for Improving the Extraction Efficiency of Dy in Ternary Alloy System. *Met. Mater. Int.* **2021**, *27*, 538–544. [CrossRef]

20.   Park, S.M.; Nam, S.W.; Cho, H.Y.; Lee, S.H.; Hyun, S.K.; Kim, T.S. Effect of Mg Ratio on the Extraction of Dy from (Nd,Dy)-Fe-B permanent magnet using liquid Mg. *Arch. Metall. Mater.* **2020**, *65*, 1281–1285.

21.   Luo, Y.; Zhang, N.; Gragam, C.D., Jr. Variation of hardness with temperature in sintered NdFeB magnets. *J. Appl. Phys.* **1987**, *61*, 3442. [CrossRef]

22.   Takeuchi, A.; Inoue, A. Akira Classification of Bulk Metallic Glasses by Atomic Size Difference, Heat of Mixing and Period of Constituent Elements and Its Application to Characterization of the Main Alloying Element. *Mater. Trans.* **2005**, *46*, 2817–2829. [CrossRef]

23.   Nagai, T.; Shirai, S.; Maeda, M. Thermodynamic measurement of Dy + Fe binary system by double Knudsen cell mass spectrometry. *J. Chem. Thermodyn.* **2013**, *65*, 78–82. [CrossRef]

MDPI
St. Alban-Anlage 66
4052 Basel
Switzerland
Tel. +41 61 683 77 34
Fax +41 61 302 89 18
www.mdpi.com

Metals Editorial Office
E-mail: metals@mdpi.com
www.mdpi.com/journal/metals